DIGITAL SIGNAL PROCESSING
LABORATORY

DIGITAL SIGNAL PROCESSING LABORATORY:

LabVIEW-Based FPGA Implementation

NASSER KEHTARNAVAZ AND
SIDHARTH MAHOTRA

UNIVERSITY OF TEXAS AT DALLAS

BrownWalker Press
Boca Raton

Digital Signal Processing Laboratory:
LabVIEW-Based FPGA Implementation

BrownWalker Press
Boca Raton, Florida
USA • 2010

ISBN-10: 1-59942-550-5 *(paper)*
ISBN-13: 978-1-59942-550-4 *(paper)*

ISBN-10: 1-59942-551-3 *(ebook)*
ISBN-13: 978-1-59942-551-1 *(ebook)*

www.brownwalker.com

Cover photo @Cutcaster.com/Kheng Guan Toh

Publisher's Cataloging-in-Publication Data

Kehtarnavaz, Nasser.
Digital signal processing laboratory : LabVIEW-based FPGA implementation /
Nasser Kehtarnavaz & Sidharth Mahotra.
p. cm.
ISBN: 978-1-59942-550-4
1. Signal processing—Digital techniques. 2. Field programmable gate arrays. 3. LabVIEW.
I. Title.
TK5102.9 .K443 2010
612.382`2—dc22

2010938309

CONTENTS

Contents

Contents

Contents

Chapter 11: FPGA Hardware Implementation   319

Lab 11 - Part 1: Real-Time FPGA Hardware Implementation of FIR Filters   321

Lab 11 - Part 2: Real-Time FPGA Hardware Implementation of IIR Filters   335

Lab 11 - Part 3: Real-Time FPGA Hardware Implementation of Short Time Fourier Transform   355

Contents

Lab 14: FPGA Implementation of MP3 Player   435

Preface

I have been teaching DSP Lab or Applied DSP courses for many years. I started with DSP assembly programming of TI DSP processors. Then, as DSP compilers became more efficient, I put more emphasis on the C programming of floating-point DSP processors and later on the C programming of fixed-point DSP processors. More recently, I introduced the hybrid programming approach in DSP lab courses. Hybrid programming involves the combination of graphical programming and textual programming. I used the LabVIEW graphical programming environment due to its interactivity features and its ability to build DSP systems in a time-efficient manner within a one-semester course or a short design cycle time. Again, this year I changed the content of my DSP lab course in order to teach students how to implement DSP algorithms on FPGA processors. The reason for this change has been the ever increasing demand by industry for FPGA implementation skills of DSP algorithms. This laboratory textbook is the outcome of our newly designed labs at the University of Texas at Dallas based on the LabVIEW FPGA Module which provides the necessary tools to carry out simulation, synthesis, and real-time implementation in a hybrid manner and in one easy-to-use environment.

I would like to express my gratitude to National Instruments for their financial support and feedback enabling us to write this book for the benefit of DSP students as well as DSP engineers. I would also like to thank the UTD graduate students Chandrasekhar Patlolla and Jeremy Brodt for their FPGA implementation codes as part of the last three application chapters. Last but not least, I would like to greatly appreciate my graduate student Sidharth Mahotra for co-authoring this book with me. Without his participation, I would not have been able to put together this book within a short period of one year.

Nasser Kehtarnavaz
August 2010

CHAPTER 1
Introduction

Nowadays, Digital Signal Processing (DSP) algorithms are used in a wide variety of embedded systems such as cell-phones, digital cameras, TVs, digital audio and video players, and biomedical devices. As far as the hardware implementation of DSP algorithms is concerned, various processors are used within embedded systems. These processors include GPP (General Purpose Processor), DSP (Digital Signal Processor, also called DSP), FPGA (Field Programmable Gate Array), GPU (Graphics Processing Unit), and ASIC (Application-Specific Integrated Circuit). Among these processors, the first four offer programmability, which is the ability or flexibility to use the same hardware engine to implement different DSP algorithms.

There are many textbooks that are written for digital signal processing laboratory courses to provide signal processing students with hands-on implementation experience using GPP and DSP processors, e.g. [1-3]. However, despite the growing utilization of FPGAs in embedded systems, there exists no laboratory textbook for signal processing students to gain hands-on experience in implementing digital signal processing algorithms on FPGAs in such courses. This laboratory textbook is thus written in order to address this need by providing the teaching material to achieve FPGA implementation in digital signal processing laboratory courses.

This book covers FPGA implementation as an alternative or complement to DSP implementation that is often performed in such laboratory courses. The intention here is not to give preference to one platform over another for implementing signal processing algorithms as this choice is very much application dependent. It is well known that FPGAs offer high data rates over DSPs. So, in high data throughput applications such as video processing, FPGAs provide throughput advantages over DSPs due to the massive parallelism capability they offer. However, it should be realized that digital signal processing algorithms vary greatly in their scope, some requiring mostly mathematical operations, e.g. FFT, while others use many conditional operations, e.g. communication protocols. Thus, depending on the algorithm, DSPs may provide efficiency advantages over FPGAs. FPGAs are fixed-point machines. Hence, for applications requiring floating-point operations, floating-point DSPs may offer area advantages over FPGAs. Of course, cost effectiveness in terms of the performance/dollar metric and power consumption are also other important factors for choosing one platform over the other.

In general, it is easier or it involves less coding effort for signal processing students to write C codes and thus to achieve DSP implementation than writing VHDL (Very-high-speed integrated circuit Hardware Description Language)/Verilog codes to achieve FPGA implementation as they are often familiar with C programming but not with VHDL program-

1

ming. In fact, despite the availability of FPGA implementation books, e.g. [4], the programming aspect is seen to be the major reason why FPGA implementation has not been widely adopted in digital signal processing laboratory courses. Although several high level tools have been introduced for performing FPGA implementation, the efficiency achieved by them is limited and not all aspects of FPGA implementation such as I/O or buffering are addressed by these tools. In this book, we have taken the above issue into consideration by presenting a hybrid mode of programming to perform FPGA implementation, that is, by combining VHDL textual programming with the more intuitive graphical programming of LabVIEW. This hybrid mode of programming allows signal processing students to utilize a wealth of existing VHDL codes while at the same time to maintain a high level or easier graphical programming approach towards implementing signal processing algorithms on FPGAs.

The audience of this book is meant to be those students who have already taken DSP theory courses and are already familiar with digital signal processing concepts. The book is not written to include an exhaustive set of signal processing topics; rather the emphasis is placed on covering the steps that one needs to take in order to reach FPGA implementation of DSP algorithms in a short amount of time as part of a semester-long DSP lab course or within a short design cycle time. The book is also beneficial to practicing engineers who wish to implement signal processing algorithms in FPGA-based embedded systems.

Considering that there are various FPGAs manufactured by different companies, in our approach, we have separated the behavioral FPGA implementation from the actual hardware FPGA implementation. This decoupling approach allows students to still learn about FPGA implementation through simulation despite the fact that they may not have access to any FPGA hardware or access to a different FPGA hardware than the one used in the book. More specifically, up to Chapter 11, FPGA implementation issues are addressed via simulation or PC emulation regardless of any specific FPGA hardware in mind. Then, only in Chapter 11, we have used the NI FlexRIO FPGA board to perform actual hardware implementation which naturally differs when using other FPGA boards.

1.1 FPGA Overview

FPGAs are processors that allow implementation of logic functions or truth tables via circuits that are reprogrammable. The reprogrammabilty is provided by processing blocks that are spread throughout the FPGA fabric. In many FPGA processors, the lines or the interconnects connecting logic blocks are also programmable. The architecture and name of the basic logic block differ from one vendor to another, for example Xilinx uses the name Configurable Logic Block (CLB) and Altera uses Logic Array Block (LAB). Modern FPGAs also have on-chip I/O blocks to allow easy interfacing with external devices. Figure 1.1 shows a diagram depicting an overview of a modern FPGA architecture. The FPGA architecture that is used in the book for actual hardware implementation in Chapter 11 is the Xilinx Virtex-5 FPGA which appears on the NI FlexRIO FPGA board. The details of this architecture and other FPGA architectures are fully covered in [5, 6].

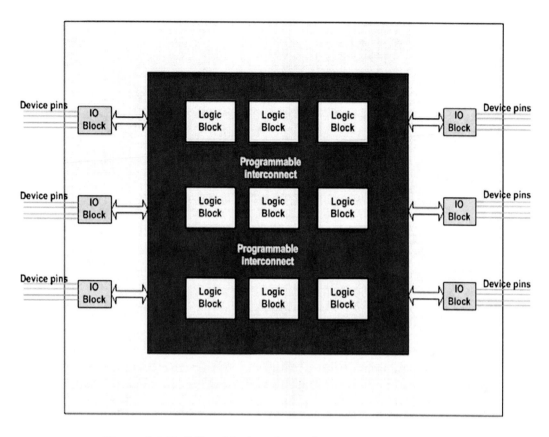

Figure 1.1: Building blocks of a modern FPGA architecture

The salient features of the Vertix-5 FPGA architecture are:

- **I/O Blocks:** These blocks provide the interface between internal configurable logic blocks and pins that come out of the processor.
- **Configurable Logic Blocks**: These programmable logic blocks provide combinatorial and synchronous logic. They are also used to provide distributed memory. A CLB on Virtex-5 contains a pair of slices. Each slice uses look-up tables or LUTs (four), storage elements (four), multiplexers, and carry logic to provide logic, arithmetic, and ROM functions. Additionally, some slices support the additional functions of storing data using distributed RAM and shifting data using 32-bit registers. Since the basic element in a slice is LUT, for a better understanding, it helps to see how a logic function gets implemented via LUT.

To implement a particular function using LUT, the output columns of a truth table are loaded into memory as a LUT. The input columns in the truth table are then used as select inputs to multiplexers. For example, consider the truth table of a half adder circuit displayed in Table 1.1. In this table, A and B denote the inputs, and S and C the sum and carry outputs, respectively.

A	B	S	C
0	0	0	0
0	1	1	0
1	0	1	0
1	1	0	1

Table 1.1: Truth table of a half adder circuit

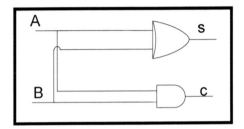

Figure 1.2: Half adder circuit

Alternatively, the same truth table can be realized using a combination of multiplexers and LUT as shown in Figure 1.3. The multiplexers shown in this figure use n (two here) data select lines to select and pass one of the 2^2 or 4 data input lines to the output. For example, for the first entry A=B=0, the input line 1 for both of the multiplexers is passed to the output, which means a 0 for both S and C.

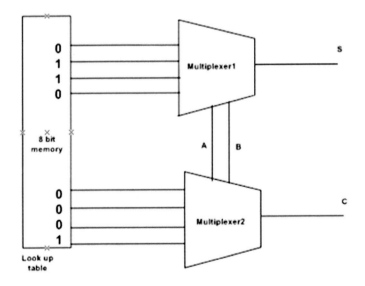

Figure 1.3: LUT implementation of half adder

- **Block RAM:** These are 36Kbit dual port random access memory modules. In most conventional memories, a read would follow a write, thus taking a minimum of two clock cycles. However, with Block RAM, writing and reading is achieved in one clock cycle. On Virtex-5, these modules can also be cascaded to form larger memories. Another advantage of these modules is their incorporation of dedicated logic, which allows creation of synchronous and asynchronous FIFO (First In First Out) buffers.
- **Clock Management Tile (CMT):** These blocks provide high performance clocking.
- **DSP48E:** These are slices that include 25x18 two's complement multipliers and 48-bit adder/accumulator allowing parallel implementation support for DSP algorithms, in particular for Multiply + Accumulate (MAC) operations. Virtex-5 has 640 DSP48E slices that are spread throughput the processor.

1.2 Organization of Chapters

The book includes 14 chapters and 14 labs. After this introduction, the LabVIEW programming environment is presented in Chapter 2. Lab 1 and Lab 2 in Chapter 2 provide a tutorial for students to get familiar with LabVIEW programming. Lab 1 covers the basic features of LabVIEW, while Lab 2 covers more advanced features of LabVIEW for signal processing. These labs may be skipped if students are already familiar with LabVIEW. Considering that signal processing students are not normally familiar with VHDL programming, Chapter 3 gives an introduction to VHDL programming followed by Lab 3 covering VHDL example codes for signal processing. This lab may be skipped if students are already familiar with VHDL programming. The hybrid programming approach is then presented in Chapter 4 using the LabVIEW FPGA Module and NI IP Integration Node. Lab 4 in this chapter includes examples of hybrid codes. In Chapter 5, fixed-point and floating-point number representations are discussed followed by Lab 5 showing fixed-point effects. The topic of analog to digital signal conversion is presented in Chapter 6 followed by Lab 6 covering signal sampling experiments. Chapter 7 includes the convolution operation and Lab 7 covers the FPGA implementation of convolution. Chapter 8 presents digital filtering followed by Lab 8 showing how to achieve the FPGA implementation of digital filters. In Chapter 9, adaptive filtering is presented and in Lab 9 the FPGA implementation of adaptive filters. Chapter 10 presents frequency domain processing including Fast Fourier Transform (FFT) and Discrete Cosine Transform (DCT). Lab 10 in this chapter covers the FPGA implementation of these transforms. In Chapter 11, it is shown how to carry out the actual real-time hardware implementation of FIR and IIR filters as well as Short Time Fourier Transform (STFT) on the NI FlexRIO FPGA board. This chapter can be skipped if such a hardware board is not available. Chapters 12 through 14 and Labs 12 through 14, respectively, include three DSP systems or application projects that are designed using the LabVIEW FPGA Module together with their hardware implementations on the NI FlexRIO FPGA board. These applications include Discrete Wavelet Transform (DWT), software-defined radio, and MP3 decoder.

1.3 Software Installation

For the labs, LabVIEW and its following tools are needed. First, install LabVIEW 2009, the latest version at the time of this writing, by running *setup.exe* on the LabVIEW Core DVD 1 and 2. Also, install the toolkits **Digital Filter Design** and **System Identification** from the

Core DVDs by selecting appropriate options. Next, install **LabVIEW FPGA Module** and the toolkits of **Advanced Signal Processing, Sound and Vibration Measurement Suite, Spectral Measurements** and **Adaptive Filter** from the LabVIEW DVD titled **RF and Wireless Communication Software** and **Signal and Image Processing Software**. The LabVIEW **IP Intergation Node** is also needed for the labs which can be downloaded from the site http://decibel.ni.com/content/docs/DOC-5907. If using the NI FlexRIO FPGA board for actual hardware implementation covered in Chapter 11, run *setup.exe* of the DVD shipped with the board titled **NI PXI Platform Services** and **NI FlexRIO Adapter Module Development Kit**. Then, install the NI Device Driver DVD selecting Virtex 7965 FPGA. To compile FPGA designs even in the absence of this target, download and run *setup.exe* for NI RIO 3.4 drivers from http://joule.ni.com/nidu/cds/view/p/id/1607/lang/en.

All the codes appearing in the book can be downloaded from the book website at www.utdallas.edu/~kehtar/FPGA (username=FPGA, password=LabVIEW).

It should be noted that although some of the appearances of figures may change as a result of future releases of LabVIEW, the same steps in the labs can still be carried out as future LabVIEW versions are expected to be compatible with the 2009 version used in the book.

1.4 Bibliography

[1] N. Kehtarnavaz, *Real-Time Digital Signal Processing Based on the TMS320C6000*, Elsevier/Newnes, 2004.

[2] T. Welch, C. Wright, and M. Morrow, *Real-Time Digital Signal Processing from Matlab to C with the TMS320C6x DSK*, CRC Press, 2006.

[3] S. Mitra, *Digital Signal Processing Laboratory Using Matlab*, McGraw Hill, 1999.

[4] U. Meyer-Baese, *Digital Signal Processing with Field Programmable Gate Arrays*, Springer, 2007.

[5] *Xilinx Virtex-5 SXT Data Sheet*, http://www.xilinx.com/products/virtex5/sct.htm

[6] R. Woods, J. McAllister, Y. Yi, and G. Lightbody, *FPGA-Based Implementation of Signal Processing Systems*, Wiley, 2008.

CHAPTER 2

LabVIEW Programming Environment

LabVIEW constitutes a graphical programming environment that allows one to design and analyze a DSP system in a shorter time as compared to text-based programming environments. LabVIEW graphical programs are called Virtual Instruments (VIs). VIs run, based on the concept of data flow programming, that is the execution of a block or a graphical component is dependent on the flow of data, or more specifically a block executes when data are made available at all of its inputs. Output data of the block are then sent to all other connected blocks. Data flow programming allows multiple operations to be performed in parallel since its execution is determined by the flow of data and not by sequential lines of code.

2.1 Virtual Instruments (VIs)

A VI consists of two components which include a Front Panel (FP) and a Block Diagram (BD). A FP provides the graphical user-interface while a BD incorporates the graphical code. When a VI is located within the block diagram of another VI, it is called a subVI, similar to the concept of a function in C. LabVIEW VIs are modular meaning that any VI or subVI can be run by itself.

2.1.1 Front Panel and Block Diagram

A FP contains the user interfaces of a VI shown in a BD. Inputs to a VI are represented by so called controls. Knobs, pushbuttons and dials are a few examples of controls. Outputs from a VI are represented by so called indicators. Graphs, LEDs (light indicators) and meters are a few examples of indicators. As a VI runs, its FP provides a display or user-interface of controls (inputs) and indicators (outputs).

A BD contains terminal icons, nodes, wires, and structures. Terminal icons are interfaces through which data are exchanged between a FP and a BD. They correspond to controls or indicators that appear on a FP. Whenever a control or indicator is placed on a FP, a terminal icon gets added to the corresponding BD. A node represents an object which has input and/or output connectors and performs a certain function. SubVIs and functions are examples of nodes. Wires establish the flow of data in a BD. Structures are used to control the flow of a program such as repetitions or conditional executions. Figure 2.1 shows how a FP and a BD window look like.

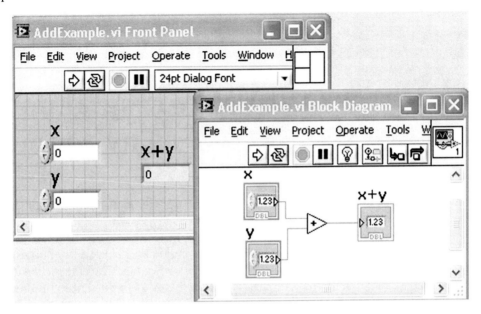

Figure 2.1: Front Panel and Block Diagram windows

2.1.2 Icon and Connector Pane

A VI icon is a graphical representation of a VI. It appears in the top right corner of a BD or a FP window. When a VI is inserted in a BD as a subVI, its icon gets displayed. A connector pane defines inputs (controls) and outputs (indicators) of a VI. The number of inputs and outputs can be changed by using different connector pane patterns. In Figure 2.1, a VI icon is shown at the top right corner of the BD and its corresponding connector pane having two inputs and one output is shown at the top right corner of the FP.

2.2 Graphical Environment

2.2.1 Functions Palette

The Functions palette (Figure 2.2) provides various function VIs or blocks for building a system. This palette can be displayed by right clicking on an open area of a BD. Note that this palette can only be displayed in a BD.

2.2.2 Controls Palette

The Controls palette, see Figure 2.3, provides controls and indicators of a FP. This palette can be displayed by right clicking on an open area of a FP. Note that this palette can only be displayed in a FP.

2.2.3 Tools Palette

The Tools palette (**View>>Tools palette**) provides various operation modes of the mouse cursor for building or debugging a VI. The Tools palette and the frequently used tools are shown in Figure 2.4.

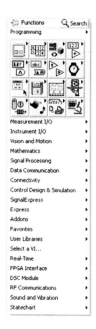

Figure 2.2: Functions palette

Figure 2.3: Controls palette

Icon	Tool
✕ ▭	Automatic Tool Selection
🖑	Operating tool
▸	Positioning tool
A	Labeling tool
◆	Wiring tool
⊕	Probe tool

Figure 2.4: Tools palette

Each tool is utilized for a specific task. For example, the Wiring tool is used to wire objects in a BD. If the automatic tool selection mode is enabled by clicking the **Automatic Tool Selection** button (highlighted on selection), the best matching tool gets selected based on a current cursor position.

2.3 Building a Front Panel

In general, a VI is put together by going back and forth between a FP and a BD, placing inputs/outputs on the FP and building blocks on the BD.

2.3.1 Controls

Controls make up the inputs to a VI. Controls grouped in the Numeric Controls palette (**Controls>>Express>>Numeric control**) are used for numerical inputs, grouped in the Buttons & Switches palette (**Controls>>Express>>Buttons**) for Boolean inputs, and grouped in the Text Controls palette (**Controls>>Express>> Text Ctrls**) for text and enumeration inputs. These control options are displayed in Figure 2.5.

2.3.2 Indicators

Indicators make up the outputs of a VI. Indicators grouped in the Numeric Indicators palette (**Controls>>Express>>Num Inds**) are used for numerical outputs, grouped in the LEDs palette (**Controls>>Express>>LEDs**) for Boolean outputs, grouped in the Text Indicators palette (**Controls>>Express>>Text Inds**) for text outputs, and grouped in the Graph Indicators palette (**Controls>>Express>>Graph Indicators**) for graphical outputs. These indicator options are displayed in Figure 2.6.

2.3.3 Align, Distribute and Resize Objects

The menu items on the toolbar of a FP (see Figure 2.7) provide options to align and orderly distribute objects on the FP. Normally, when controls and indicators are placed on a FP, one uses these options to tidy up their appearance.

Figure 2.5: Control palettes

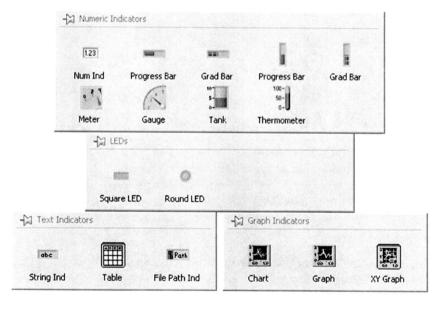

Figure 2.6: Indicator palettes

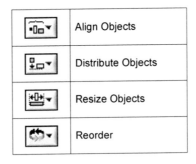

	Align Objects
	Distribute Objects
	Resize Objects
	Reorder

Figure 2.7: Menu for align, distribute, resize, and reorder objects

2.4 Building a Block Diagram

2.4.1 Express VI and Function

Express VIs denote higher-level VIs that are configured to incorporate lower-level VIs or functions. These VIs are displayed as expandable nodes with a blue background. Placing an Express VI in a BD brings up a configuration dialog window allowing adjustment of its parameters. As a result, Express VIs demand less wiring, but often they do not allow a complete configuration of their parameters manually. The configuration window of an Express VI can be brought up by double clicking on the VI.

Basic operations such as addition or subtraction are represented by functions. Figure 2.8 shows three examples corresponding to three types of a BD object (VI, Express VI, and function).

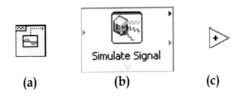

Simulate Signal

(a)   (b)   (c)

Figure 2.8: Block Diagram objects (a) VI, (b) Express VI, and (c) function

A subVI or an Express VI can be displayed as icons or expandable nodes. Icons are used to save space in a BD, while expandable nodes are used to provide easier wiring or better readability. Expandable nodes can be resized to show their connection nodes more clearly. Three appearances of a VI/Express VI are shown in Figure 2.9.

2.4.2 Terminal Icons

FP objects get displayed as terminal icons in a BD. A terminal icon exhibits an input or output as well as its data type. Figure 2.10 shows two terminal icon examples consisting of a double precision numerical control and indicator. As shown in this figure, terminal icons can be displayed as data type terminal icons to conserve space in a BD (**right click>>View as icon**).

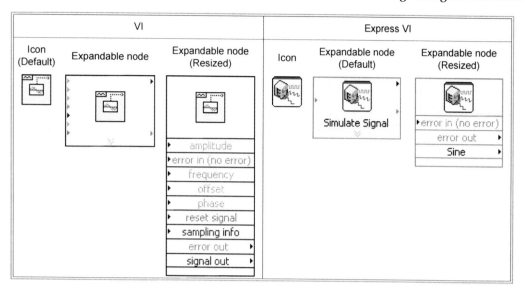

Figure 2.9: Icon vs. expandable node

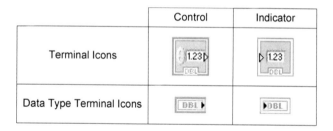

Figure 2.10: Terminal icon examples displayed in a BD

2.4.3 Wires

Wires transfer data from one node to another in a BD. Based on the data type of a data source, the color and thickness of its connecting wires change. Wires for the basic data types used in LabVIEW are shown in Figure 2.11. Other than the data types shown in this figure, there are some other specific data types. For example, the dynamic data type is always used for Express VIs, and the waveform data type, which corresponds to the output from a waveform generation VI, is a special cluster of components incorporating trigger time, time interval, and data value.

2.4.4 Structures

A structure is represented by a graphical enclosure. The graphical code enclosed by a structure gets repeated or executed conditionally. A loop structure is equivalent to a 'for loop' or a 'while loop' statement encountered in text-based programming languages, while a case structure is equivalent to an 'if-else' statement.

Wire Type	Scalar	1D Array	2D Array	Color
Numeric				Orange (Floating point) Blue (Integer)
Boolean				Green
String				Pink

Figure 2.11: Basic types of wires [2]

2.4.4.1 For Loop

A For Loop structure is used to perform repetitions. As illustrated in Figure 2.12, the displayed border indicates a For Loop structure, where the count terminal 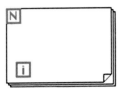 represents the number of times the loop is to be repeated. It is set by wiring a value from outside of the loop to it. The iteration terminal ▣ denotes the number of completed iterations, which always starts at zero. Thus, the iteration terminal goes from 0 to N-1 (similar to C).

Figure 2.12: For Loop

LabVIEW 2009 allows stopping a For Loop before N number of iterations by using the Conditional Terminal feature, which is activated by right clicking the For Loop, and selecting Conditional Terminal. The conditional For Loop is shown in Figure 2.13.

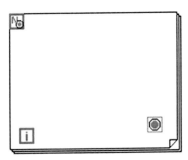

Figure 2.13: For Loop with Conditional Terminal

For the conditional loop, the count terminal gets a red dot and a conditional terminal ◉ appears within the For Loop border. If a stop control is wired to this terminal, the loop is stopped before N iterations are completed. This is similar to the break statement in C.

2.4.4.2 While Loop

A While Loop structure allows repetitions depending on a condition (Figure 2.14). The conditional terminal  initiates a stop if the condition is true. Similar to a For Loop, the iteration terminal ⊡ provides the number of completed iterations, always starting at zero.

Figure 2.14: While Loop

2.4.4.3 Case Structure

A Case structure (Figure 2.15) allows running different sets of operations depending on the value it receives through its selector terminal, which is indicated by ⍰. In addition to Boolean type, the input to a selector terminal can be of integer, string, or enumerated type. This input determines which case to execute. The case selector ◄ True ▼► shows the status being executed. Cases can be added or deleted as needed.

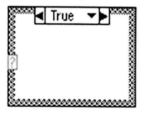

Figure 2.15: Case structure

2.5 MathScript

The MathScript feature of LabVIEW allows performing textual programming in conjunction with graphical programming. It includes various built-in functions and uses matrices and arrays as fundamental data types with built-in operations for data manipulation. User-defined functions can also be added to it. It is compatible with the M-file script syntax or MATLAB codes. MathScript possesses an interactive and a programming interface named MathScript Interactive Window and MathScript Node, respectively.

A MathScript Interactive Window is shown in Figure 2.16. Three interfaces of **Command Window**, **Output Window**, and **MathScript Window** are shown in this figure. The Command Window interface is used to enter commands, perform script debugging or to view help statements for built-in functions. The Output Window interface is used for viewing output values. The MathScript Window interface is used to display variables, edit scripts, and display command history. Script editing allows the execution of a group of commands. A MathScript Node **(Mathematics>>Script and formula>>Mathscript)** represents a textual code via a blue rectangle as shown in Figure 2.16. Its inputs and outputs are defined on the border of this rectangle for transferring data between the graphical environment and a textual MathScript code. For example, as indicated in Figure 2.17, the input variables on the left side, namely **lowcutoff, uppcutoff** and **order**, transfer values to the M-file script and the output variables on the right side, namely **F** and **sH**, transfer values to the graphical environment. This way M-file script variables get represented within the graphical programming environment.

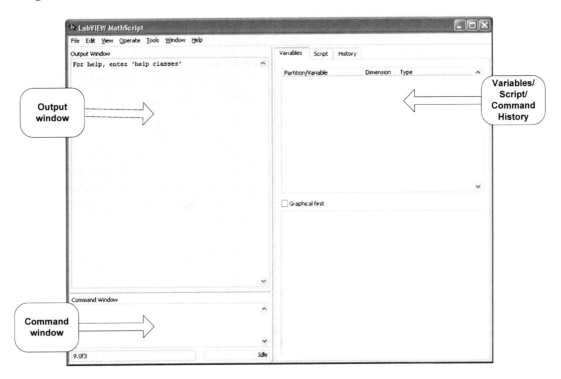

Figure 2.16: MathScript Interactive Window

2.6 Grouping Data: Array & Cluster

An array represents a group of elements having the same data type. An array consists of data elements having a dimension up to $2^{31}-1$ (corresponding to type double). For example, if a random number is generated in a loop, it makes sense to build the output as an array since the length of the data element is fixed at 1 and the data type is not changed during iterations.

A cluster consists of a collection of different data type elements, similar to the structure data type in text-based programming languages. Clusters allow one to reduce the number of wires on a BD by bundling different data type elements together and passing them to only one terminal. An individual element can be added to or extracted from a cluster by using the cluster functions such as **Bundle by Name** and **Unbundle by Name**.

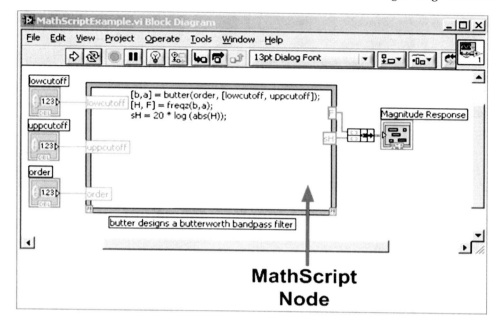

Figure 2.17: MathScript Node Interface

2.7 Debugging and Profiling VIs

2.7.1 Probe Tool

The Probe tool is used for debugging VIs as they run. The value on a wire can be checked while a VI is running. Note that the Probe tool can only be accessed in a BD window.

The Probe tool can be used together with breakpoints (right clicking on a structure/function and selecting **Set breakpoint**) and execution highlighting (identified by ![icon] in the BD of a VI) to identify the source of an incorrect or an unexpected outcome. A breakpoint is used to pause the execution of a VI at a specific location while execution highlighting helps one to visualize the flow of data during program execution.

2.7.2 Profile Tool

The Profile tool can be used to gather timing and memory usage information, i.e. how long a VI takes to run and how much memory it consumes. It is necessary to make sure a VI is stopped before setting up a Profile window.

2.7.3 New Features in LabVIEW 2009 [6]

Some of the new features of LabVIEW 2009 are highlighted below:

- More options for 2D and 3D plots including feather plot, scatter plot, mesh plot, etc.
- **Parallel For Loop:** This loop allows executing For Loops in parallel assuming that the loops do not have dependency among each other, leading to higher through-puts.

- **Probe watch window:** It provides a one step solution to multiple probing by putting all the floating probes into one window, thus easing the debugging process.

An effective way to become familiar with LabVIEW programming is by going through hands-on examples. Thus, in the labs covered next in this chapter, most of the key programming features of LabVIEW are covered via building some simple VIs. More detailed information on LabVIEW programming can be found in [1-6].

2.8 Bibliography

[1] National Instruments, *LabVIEW Getting Started with LabVIEW*, Part Number 323427A-01, 2003.

[2] National Instruments, *LabVIEW User Manual*, Part Number 320999E-01, 2003.

[3] National Instruments, *LabVIEW Performance and Memory Management*, Part Number 342078A-01, 2003.

[4] National Instruments, *Introduction to LabVIEW Six-Hour Course*, Part Number 323669B-01, 2003.

[5] R. Bishop, *Learning With Labview 7 Express*, Prentice Hall, 2003.

[6] National instruments, http://www.ni.com/labview/whatsnew/features.htm

[7] National instruments, Inside LabVIEW Mathscript, http://zone.ni.com/devzone/ cda/tut/p/id/3502

Lab 1:

Getting Familiar with LabVIEW

The objective of this first lab is to provide an initial hands-on experience in building a VI. For detailed explanations of the LabVIEW features mentioned here, the reader is referred to [1]. LabVIEW 2009 can be launched by double clicking on the LabVIEW 2009 icon. The dialog window shown in Figure L1.1 should appear.

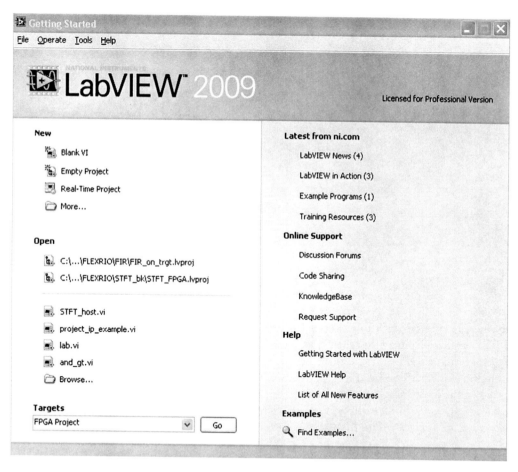

Figure L1.1: Starting LabVIEW

L1.1 Building a Simple VI

To become familiar with the LabVIEW programming environment, it is found to be more effective if one starts by going through a simple example. The example presented here consists of calculating the sum and average of two input values. This example is described in a step-by-step fashion below.

L1.1.1 VI Creation

To create a new VI, click on the **Blank VI** under **New**. This step can also be done by choosing **File >> New VI** from the menu. As a result, a blank FP and a blank BD window appear, as shown in Figure L1.2. It should be remembered that a FP and a BD coexist when building a VI.

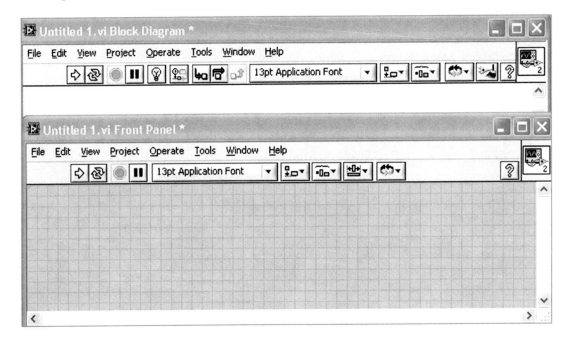

Figure L1.2: Blank VI

Clearly, the number of inputs and outputs to a VI is dependent on its function. In this example, two inputs and two outputs are needed, one output generating the sum and the other the average of two input values. The inputs are created by locating two Numeric Controls on the FP. This is done by right-clicking on an open area of the FP to bring up the Controls palette, followed by choosing **Controls>>Modern>>Numeric>>Numeric Control.** Each numeric control automatically places a corresponding terminal icon on the BD. Double clicking on a numeric control highlights its counterpart on the BD, and vice versa.

Next, let us label the two inputs as **x** and **y**. This is achieved by using the Labeling tool from the **Tools** palette, which can be displayed by choosing **View >> Tools Palette** from the menu bar. Choose the Labeling tool and click on the default labels, **Numeric** and **Numeric 2**, in order to edit them. Alternatively, if the automatic tool selection mode is enabled by clicking **Automatic Tool Selection** in the **Tools** palette, the labels can be edited by simply double clicking on the default labels. Editing a label on the FP changes its corresponding terminal icon label on the BD, and vice versa.

Similarly, the outputs are created by locating two Numeric Indicators (**Controls>> Modern>> Numeric>> Numeric Indicator**) on the FP. Each numeric indicator automatically places a corresponding terminal icon on the BD. Edit the labels of the indicators to read **Sum** and **Average**.

For a better visual appearance, objects on a FP window can be aligned, distributed, and resized using the buttons appearing on the FP toolbar. To do this, select the objects to be aligned or distributed and apply the appropriate option from the toolbar menu. Figure L1.3 shows the configuration of the FP just created.

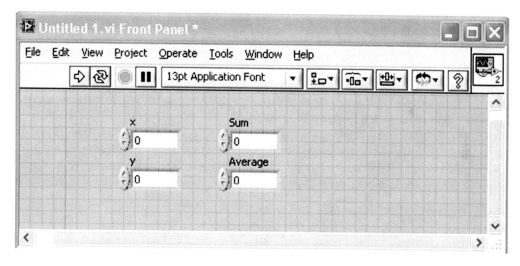

Figure L1.3: FP configuration

Now, let us build a graphical code on the BD to perform the summation and averaging operations. Note that <Ctrl-E> toggles between a FP and a BD window. If one finds the objects on a BD are too close to insert other functions or VIs in-between, a horizontal or vertical space can be inserted by holding down the <Ctrl> key to create space horizontally and/or vertically. As an example, Figure L1.4 (b) illustrates a horizontal space inserted between the objects shown in Figure L1.4 (a).

Next, place an **Add** function (**Programming>>Numeric>>Add**) and a **Divide** function (**Programming>>Numeric>> Divide**) on the BD. The divisor, in our case 2, needs to be entered as a **Numeric Constant (Programming>>Numeric>>Numeric Constant)** and connected to the y terminal of the Divide function using the Wiring tool.

To have a proper data flow, functions, structures and terminal icons on a BD need to be wired. The Wiring tool is used for this purpose. To wire these objects, point the Wiring tool at a terminal of a function or a subVI to be wired, left click on the terminal, drag the mouse to a destination terminal and left click once again.

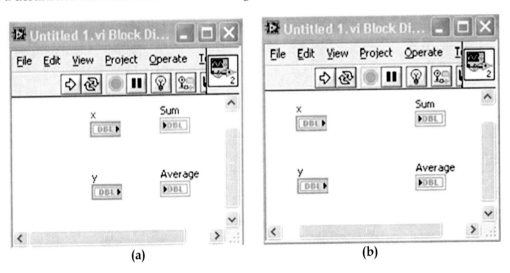

(a)   (b)

Figure L1.4: Inserting horizontal/vertical space: (a) creating space while holding down the <Ctrl> key, and (b) inserted horizontal space

Figure L1.5 illustrates the wires placed between the terminals of the numeric controls and the input terminals of the add function. Notice that the label of a terminal is displayed whenever the cursor is moved over it if the automatic tool selection mode is enabled. Also, note that the Run button ⇨ on the toolbar remains broken until the wiring process is completed.

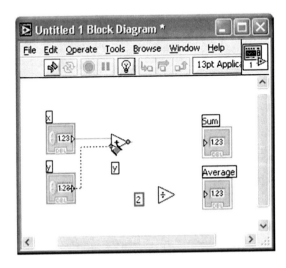

Figure L1.5: Wiring BD objects

For better readability of a BD, wires which are hidden behind objects or crossed over other wires can be cleaned up by right-clicking on them and choosing **Clean Up Wire** from the shortcut menu. Any broken wires can be cleared by pressing <Ctrl-B> or **Edit>>Remove Broken Wires**.

The label of a BD object, such as a function, can be shown (or hidden) by right-clicking on the object and checking (or unchecking) **Visible Items>>Label** from the shortcut menu. Also, a terminal icon corresponding to a numeric control or indicator can be shown as a data type terminal icon. This is done by right-clicking on the terminal icon and unchecking **View As Icon** from the shortcut menu. Figure L1.6 shows an example where the numeric controls and indicators are shown as data type terminal icons. The notation DBL represents double precision data type.

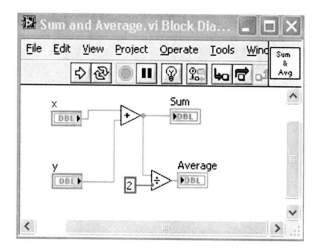

Figure L1.6: Completed BD

It is worth pointing out that there exists a shortcut to build the above VI. Instead of choosing the numeric controls, indicators, or constants from the Controls or Functions palette, one can right-click on the terminals of a BD object such as a function or a subVI. As an example of this approach, create a blank VI and locate an Add function. Right-click on its x terminal and choose **Create>>Control** from the shortcut menu to create and wire a numeric control or input. This locates a numeric control on the FP as well as a corresponding terminal icon on the BD. The label is automatically set to x. Create a second numeric control by right-clicking on the y terminal of the Add function. Next, right-click on the output terminal of the Add function and choose **Create>> Indicator** from the shortcut menu. A data type terminal icon, labeled as x+y, is created on the BD as well as a corresponding numeric indicator on the FP.

Next, right-click on the y terminal of the Divide function to choose **Create>> Constant** from the shortcut menu. This creates a Numeric Constant as the divisor and wires its y terminal. Type the value 2 in the numeric constant. Right-click on the output terminal of the Divide function, labeled as x/y, and choose **Create>>Indicator** from the shortcut menu. In case a wrong option is chosen, the terminal does not get wired. A wrong terminal option can be easily changed by right-clicking on the terminal and choosing **Change to Control** or **Change to Constant** from the shortcut menu.

To save the created VI for later use, choose **File >>Save** from the menu or press <Ctrl-S> to bring up a dialog window to enter a name. Type Sum and Average as the VI name and click **Save**.

To test the functionality of the VI, enter some sample values in the numeric controls on the FP and run the VI by choosing **Operate >> Run**, by pressing <Ctrl-R>, or by clicking the Run button on the toolbar. From the displayed output values in the numeric indicators, the functionality of the VI can be verified. Figure L1.7 illustrates the outcome after running the VI with two inputs 10 and 20.The sum is 10+20=30, and average 30/2=15, which can get verified from the FP.

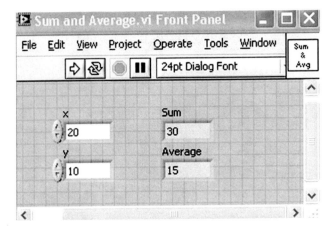

Figure L1.7: VI verification

L1.1.2 SubVI Creation

If a VI is to be used as part of a higher level VI, its connector pane needs to be configured. A connector pane assigns inputs and outputs of a subVI to its terminals through which data are exchanged. A connector pane can be displayed by right-clicking on the top right corner icon of a FP and selecting **Show Connector** from the shortcut menu.

The default pattern of a connector pane is determined based on the number of controls and indicators. In general, the terminals on the left side of a connector pane pattern are used for inputs, and the ones on the right side for outputs. Terminals can be added to or removed from a connector pane by right-clicking and choosing **Add Terminal** or **Remove Terminal** from the shortcut menu. If a change is to be made to the number of inputs/outputs or to the distribution of terminals, a connector pane pattern can be replaced with a new one by right-clicking and choosing **Patterns** from the shortcut menu. Once a pattern is selected, each terminal needs to be reassigned to a control or an indicator by using the Wiring tool, or by enabling the automatic tool selection mode.

Figure L1.8 (a) illustrates assigning a terminal of the **Sum and Average.vi** to a numeric control. The completed connector pane is shown in Figure L1.8(b). Notice that the output terminals have thicker borders. The color of a terminal reflects its data type.

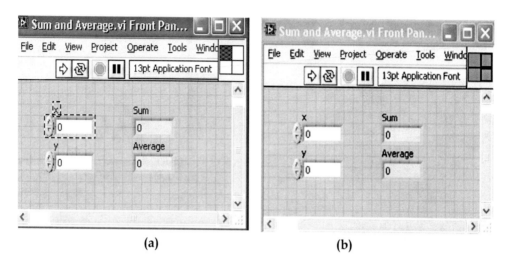

(a)   (b)

Figure L1.8: Connector pane: (a) assigning a terminal to a control, and (b) terminal assignment completed

Considering that a subVI icon is displayed on the BD of a higher level VI, it is important to edit the subVI icon for it to be explicitly identified. Double clicking on the top right corner icon of a BD brings up the Icon Editor. The tools provided in the Icon Editor are very similar to those encountered in other graphical editors, such as Microsoft Paint. An editing of the icon for the Sum and Average VI is illustrated in Figure L1.9.

A subVI can also be created from a section of a VI. To do so, select the nodes on the BD to be included in the subVI, as shown in Figure L1.10 (a). Then, choose **Edit >> Create SubVI**. This inserts a new subVI icon.Figure L1.10 (b) illustrates the BD with an inserted subVI. This subVI can be opened and edited by double clicking on its icon on the BD. Save this subVI as Sum and Average.vi. This subVI performs the same function as the original Sum and Average VI.

L1.2 Using Structures and SubVIs

Let us now consider another example to demonstrate the use of structures and subVIs. In this example, a VI is used to show the sum and average of two input values in a continuous fashion. The two inputs can be altered by the user. If the average of the two inputs becomes greater than a preset threshold value, a LED warning light is lit.

As the first step to build such a VI, build a FP as shown in Figure L1.11 (a). For the inputs, consider two Knobs **(Controls >> Modern >> Numeric >> Knob)**.

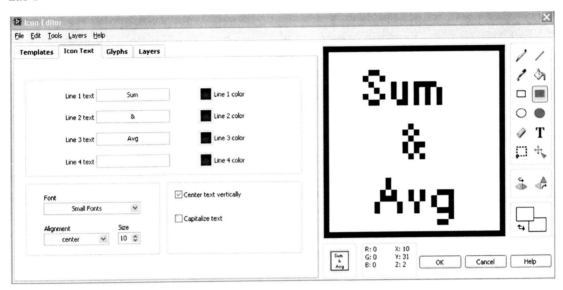

Figure L1.9: Editing subVI icon

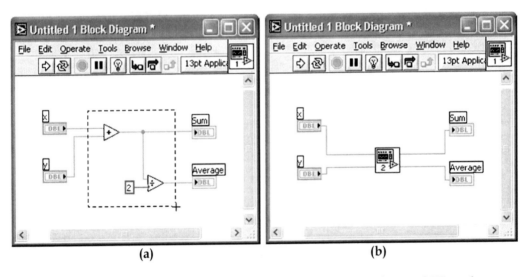

Figure L1.10: Creating a subVI: (a) selecting nodes to make a subVI, and (b) inserted subVI icon

Adjust the size of the knobs by using the Positioning tool. Properties of knobs such as precision and data type can be modified by right-clicking and choosing **Properties** from the shortcut menu. A Knob Properties dialog box is brought up and an **Appearance** tab gets shown by default. Edit the label of one of the knobs to read **Input 1**. By right clicking on the knob and selecting **Properties>>Display format>>Decimal**, Input 1 gets expressed in decimal number values.

Label the second knob as **Input 2** and repeat all the adjustments as done for the first knob. As the final step of configuring the FP, align and distribute the objects using the appropriate buttons on the FP toolbar.

To set the outputs, locate and place a Numeric Indicator, a Round LED **(Controls>>Modern >>Boolean>>Round LED)**, and a Gauge **(Controls>>Modern>>Numeric>>Gauge)**. Edit the labels of the indicators as shown in Figure L1.11(a).

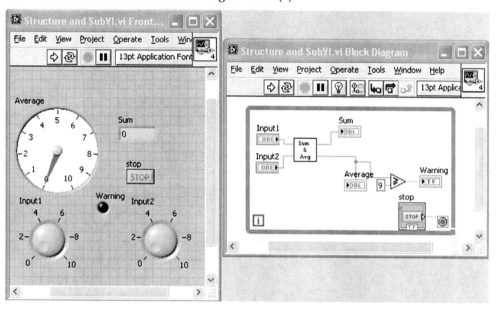

(a)   (b)
Figure L1.11: Example of structure and subVI (a) FP, and (b) BD

Now let us build the BD as shown in Figure L1.11(b). There are five control and indicator icons already appearing on the BD. Right-click on an open area of the BD to bring up the **Functions** palette and then choose **Select a VI**. This brings up a file dialog box. Navigate to the Sum and Average.vi in order to place it on the BD. This subVI is displayed as an icon on the BD. Wire the numeric controls, **Input 1** and **Input 2**, to the **x** and **y** terminals, respectively. Also, wire the **Sum** terminal of the subVI to the numeric indicator labeled **Sum**, and the **Average** terminal to the gauge indicator labeled **Average**.

A **Greater or Equal?** function can be used **(Functions>>Programming>>Comparison>> Greater or Equal?)** in order to compare the average output of the subVI with a threshold value. Create a wire branch on the wire between the Average terminal of the subVI and its indicator via the Wiring tool. Then, extend this wire to the x terminal of the Greater or Equal? function. Right-click on the y terminal of the Greater or Equal? function and choose **Create>> Constant** in order to place a Numeric Constant. Enter 9 in the numeric constant. Then, wire the Round LED, labeled as Warning, to the x>=y? terminal of this function to provide a Boolean value.

In order to run the VI continuously, a **While Loop** structure is used. Choose **Functions>> Programming >>Structures>> While Loop** to create a While Loop. Change the size by dragging the mouse to enclose the objects in the While Loop as illustrated in Figure L1.12. If the While Loop is created by using **Functions >> Programming>> Structures >>While Loop**, then the Stop Button is not included as part of the structure. This button can be created by right-clicking on the conditional terminal and choosing **Create >>Control** from the shortcut menu. A Boolean condition can be wired to a conditional terminal, instead of a stop button, in order to stop the loop programmatically.

As the final step, tidy up the wires, nodes and terminals on the BD using the **Align object** and **Distribute object** options on the BD toolbar. Then, save the VI in a file named **Structure and SubVI.vi**.

Now run the VI to verify its functionality. After clicking the Run button on the toolbar, adjust the knobs to alter the inputs. Verify whether the average and sum are displayed correctly in the gauge and numeric indicators.

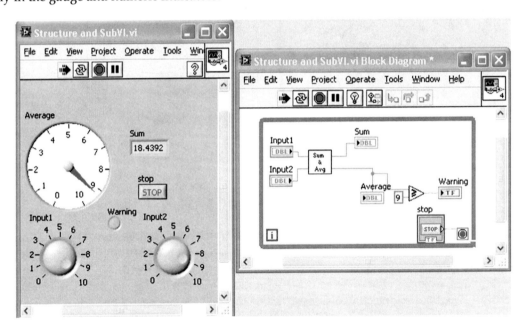

Figure L1.12: FP and BD as VI runs

When the average value of the two inputs becomes greater than the preset threshold value of 9, the warning LED will light up, see Figure L1.12. Click the stop button on the FP to stop the VI. Otherwise, the VI keeps running until the conditional terminal of the While Loop becomes true.

L1.3 Create an Array with Indexing

Auto-indexing enables one to read/write each element from/to a data array in a loop structure. In this section, this feature is covered.

Let us first create a **For Loop (Functions>>Programming>>Structures>>>For Loop)**. Right-click on its count terminal and choose **Create >>Constant** from the shortcut menu to set the number of iterations. Enter 10 so that the code inside it gets repeated 10 times. Note that the current loop iteration count, which is read from the iteration terminal, starts at index 0 and ends at index 9.

Place a **Random Number (0-1) function (Functions>>Programming>>Numeric>>Random Number (0-1))** inside the For Loop and wire the output terminal of this function, number (0 to 1), to the border of the For Loop to create an output tunnel. The tunnel appears as a box with the array symbol [ ] inside it. For a For Loop, auto-indexing is enabled by default whereas for a While Loop, it is disabled by default. Create an indicator on the tunnel by right-clicking and choosing **Create>>Indicator** from the shortcut menu. This creates an array indicator icon outside the loop structure on the BD. Its wire appears thicker due to its array data type. Also, another indicator representing the array index gets displayed on the FP. This indicator is of array data type and can be resized as desired. In this example, the size of the array is specified as 10 to display all the values, considering that the number of iterations of the For Loop is set to be 10.

Create a second output tunnel by wiring the output of the Random Number (0-1) function to the border of the loop structure, then right-click on the tunnel and choose **Disable indexing** from the shortcut menu to disable auto-indexing. By doing this, the tunnel becomes a filled box representing a scalar value. Create an indicator on the tunnel by right-clicking and choosing **Create>>Indicator** from the shortcut menu. This sets up an indicator of scalar data type outside the loop structure on the BD.Next, create a third indicator on the **Number (0 to 1)** terminal of the Random Number (0-1) function located in the For Loop to observe the values generated. To do this, right-click on the output terminal or on the wire connected to this terminal and choose **Create>>Indicator** from the shortcut menu.

Place a Time Delay Express VI **(Functions>>Programming>>Timing>>Time Delay)** to delay the execution in order to have enough time to observe a current value. A configuration window is brought up for specifying the delay time in seconds. Enter the value 0.1 to wait 0.1 seconds at each iteration. Note that the Time Delay Express VI is shown as an icon in Figure L1.13 in order to have a more compact display.

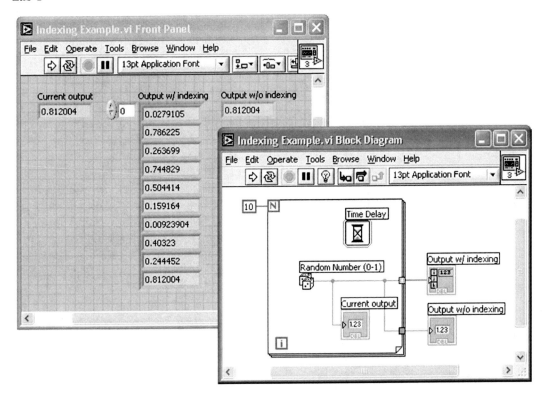

Figure L1.13: Creating array with indexing

Save the VI as **Indexing Example.vi** and run it to observe its functionality. From the output displayed on the FP, a new random number should get displayed every 0.1 seconds on the indicator residing inside the loop structure. However, no data will be available from the indicators outside the loop structure until the loop iterations end. An array of 10 elements should be generated from the indexing-enabled tunnel while only one output, the last element of the array, should be passed from the indexing-disabled tunnel.

L1.4 Debugging VIs: Probe Tool

The Probe tool is used to observe data that are being passed while a VI is running. A probe can be placed on a wire by using the Probe tool or by right-clicking on a wire and choosing **Probe** from the shortcut menu. Probes can also be placed while a VI is running.

Placing probes on wires create probe windows through which intermediate values can be observed. A probe window can be customized. For example, showing data of array data type via a graph makes debugging easier. To do this, right-click on the wire where an array is being passed and choose **Custom Probe>>Controls>>Modern>>Graph>>Waveform Graph** from the shortcut menu. As an example of using custom probes, a **Waveform Chart** is used here to track the scalar values at probe location 1, a **Waveform Graph** to monitor the array at probe location 2, and a regular probe window at probe location 3 to see the values of the Indexing Example.vi. These probes and their locations are illustrated in Figure L1.14.

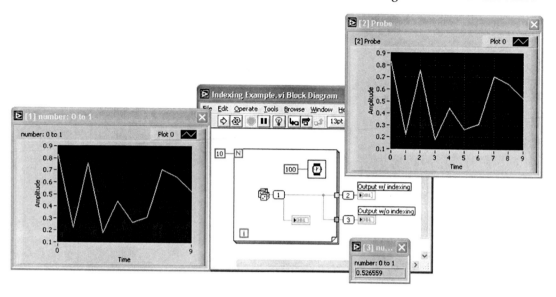

Figure L1.14: Probe tool

L1.5 Bibliography

[1] National Instruments, *LabVIEW User Manual*, Part Number 320999E-01, 2003.

L1.6 Lab Experiments

- Build a SubVI to compute the product, sum and difference of two given square matrices A and B.
- Build a SubVI to compute and display the roots of a quadratic equation $ax^2 + bx + c$ for given coefficients a, b and c.
- Build a SubVI to generate array of 10 random numbers, and find the minimum distance between any two.

Lab 2:

Building Signal Processing Systems in LabVIEW

Now that an initial familiarity with the LabVIEW programming environment has been acquired in Lab 1, this second lab shows how a DSP system can be built in LabVIEW. In addition, the hybrid programming approach is introduced.

L2.1 Express VIs vs. Regular VIs

A simple DSP system consisting of signal generation and amplification is covered here. The shape of the input signal (sine, square, triangle, or sawtooth) as well as its frequency and gain are altered by using appropriate FP controls. The system is built with Express VIs first, then the same system is built with regular VIs. This is done in order to illustrate the use of Express VIs versus regular VIs for building a system.

L2.1.1 Building a System VI with Express VIs

The use of Express VIs allows less wiring on a BD while providing an interactive user-interface by which parameter values can be adjusted on-the-fly. The BD of the signal generation system using Express VIs is shown in Figure L2.1.

To build this BD, locate the **Simulate Signal Express VI (Functions>>Express>>Input>>Simulate Signal)** to generate a signal source. This brings up a configuration dialog window as shown in Figure L2.2. Different types of signals including sine, square, triangle, sawtooth, or DC can be generated with this VI. Enter and adjust the parameters as indicated in Figure L2.2 to simulate a sinewave having a frequency of 200 Hz and amplitude swinging between -100 and 100. Set the sampling frequency to 8000 Hz. A total of 128 samples spanning a time duration of 15.875 milliseconds (ms) are generated. Note that when the parameters are changed, the modified signal gets displayed instantly in the **Result Preview** graph window.

Next, place a **Scaling and Mapping Express VI (Functions>>Express>>Arithmetic & Comparison>>Scaling and Mapping)** to amplify or scale this simulated input signal. When its configuration dialog is brought up (see Figure L2.3), choose **Linear (Y=mx+b)** and enter 5 in **Slope (m)** to scale the signal 5 times. Wire the **Sine** terminal of the Simulate Signal Express VI to the **Signals** terminal of the Scaling and Mapping Express VI. Note that a wire having a dynamic data type gets created.

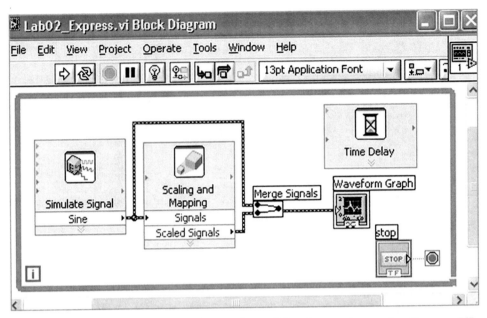

Figure L2.1: BD of signal generation and amplification system using Express VIs

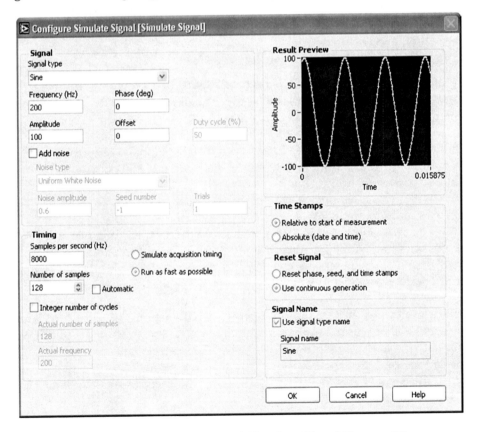

Figure L2.2: Configuration of Simulate Signal Express VI

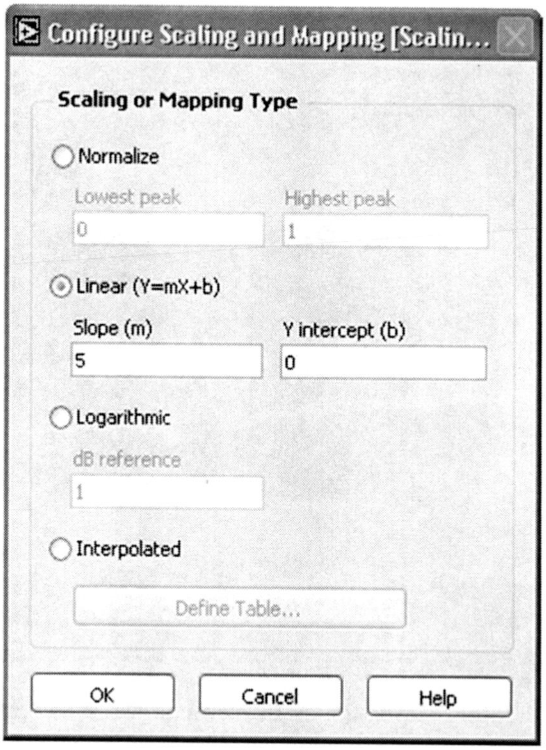

Figure L2.3: Configuration of Scaling and Mapping Express VI

To display the output signal, place a **Waveform Graph** (Controls>>Modern >>Graph>>Waveform Graph) on the FP. The Waveform Graph can also be created by right-clicking on the Scaled Signals terminal and choosing **Create>>Graph Indicator** from the shortcut menu.

Now, in order to observe the original and the scaled signal together in the same graph, wire the Sine terminal of the Simulate Signal Express VI to the Waveform Graph. This inserts a **Merge Signals** function on the wire automatically. An automatic insertion of the Merge Signals function occurs when a signal having a dynamic data type is wired to other signals having the same or other data types. The Merge Signals function combines multiple inputs, thus allowing two signals, consisting of the original and scaled signals, to be handled by one wire. Since both the original and scaled signals are displayed in the same graph, resize the plot legend to display the two labels and markers. The use of the dynamic data type sets the signal labels automatically.

To run the VI continuously, place a While Loop. Position the While Loop to enclose all the Express VIs and the graph. Now the VI is ready to be run. Run the VI and observe the Waveform Graph. The output should appear as shown in Figure L2.4. To extend the plot to the right-end of the plotting area, right-click on the Waveform Graph and choose **X Scale**, then uncheck **Loose Fit** from the shortcut menu. The graph shown in Figure L2.5 should appear.

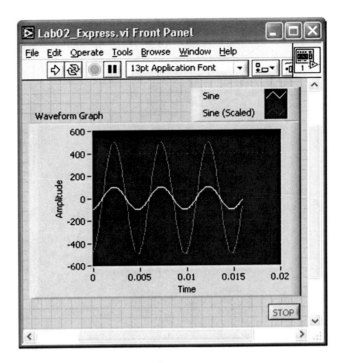

Figure L2.4: FP of signal generation and amplification system

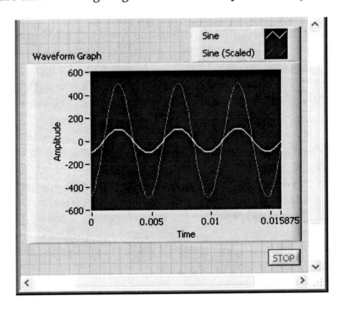

Figure L2.5: Plot with Loose Fit

If the plot runs too fast, a delay can be placed in the While Loop. To do this, place a **Time Delay Express VI (Functions>>Programming>> Timing>> Time Delay)** and set the delay time to 0.2 in the configuration window. This way, the loop execution is delayed by 0.2 seconds in the BD appearing in Figure L2.1.

Although this system runs successfully, no control of the signal frequency and gain is available during its execution since all the parameters are set in the configuration dialogs of the Express VIs. To gain such flexibility, some modifications need to be done.

To change the frequency at run time, place a **Vertical Pointer Slide** control (**Controls>>Modern>>Numeric>>Vertical Pointer Slide**) on the FP and wire it to the **Frequency** terminal of the Simulate Signal Express VI. The control is labeled as **Frequency**. The Express VI can be resized to show more terminals at the bottom of the expandable node. Resize the VI to show an additional terminal below the **Sine terminal**. Then, click on this new terminal, **error out** by default, to select **Frequency** from the list of the displayed terminals.

Next, replace the Scaling and Mapping Express VI with a **Multiply** function (**Functions>>Programming>>Numeric>>Multiply**). Place another **Vertical Pointer Slide** control and wire it to the **y** terminal of the Multiply function to adjust the gain. This control is labeled as Gain. These modifications are illustrated in Figure L2.6.

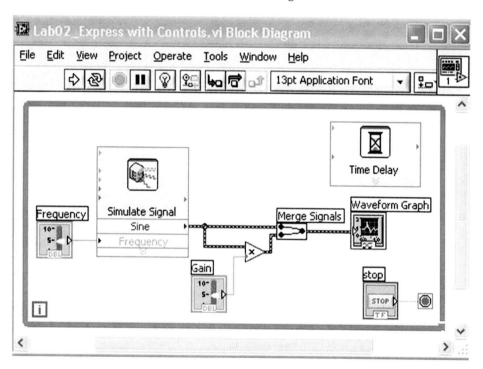

Figure L2.6: BD of signal generation and amplification system with controls

Now on the FP, set the maximum range and default values of each slide control to 1000 and 200 for the frequency control and 5 and 2 for the Gain control, respectively. By running this modified VI, it can be observed that the two signals get displayed with the same label since the source of these signals, i.e. the Sine terminal of the Simulate Signal Express VI, is the same. Also, due to the autoscale feature of the Waveform Graph, the scaled signal appears unchanged while the Y axis of the Waveform Graph changes appropriately. This is illustrated in Figure L2.7.

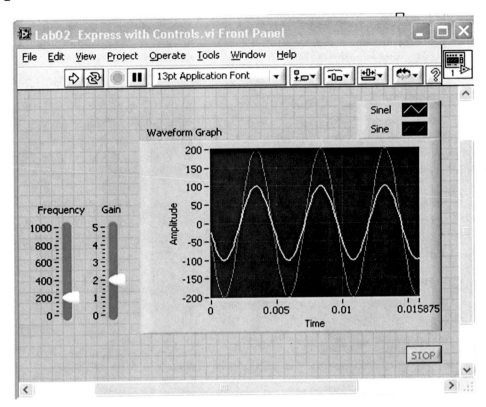

Figure L2.7: Autoscaled graph of two signals shown together

Let us now modify the property of the Waveform Graph. In order to disable the autoscale feature, right-click on the Waveform Graph and uncheck **Y Axis >>AutoScale Y**. The maximum and minimum scale can also be adjusted. In this example -600 and 600 are used as the minimum and maximum values, respectively. This is done by modifying the maximum and minimum scale values of the Y axis with the Labeling tool. If the automatic tool selection mode is enabled, just click on the maximum or minimum scale of the Y axis to enter any desired scale value. To modify the labels displayed in the plot legend, right-click and choose **Ignore Attributes**. Then, edit the labels to read **Original and Scaled** using the Labeling tool. The changing of the properties of the Waveform Graph can also be accomplished by using its properties dialog box. This box is brought up by right-clicking on the Waveform Graph and choosing **Properties** from the shortcut menu.

The completed FP is shown in Figure L2.8. With this version of the VI, the frequency of the input signal and the gain of the output signal can get adjusted using the controls on the FP.

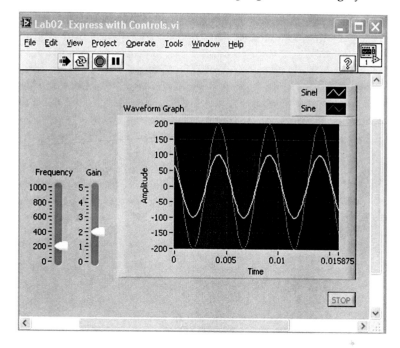

Figure L2.8: FP of signal generation and amplification system with controls

L2.1.2 Building a System with Regular VIs

In this section, the implementation of the same system discussed above is achieved by using regular VIs. After creating a blank VI, place a While Loop (**Functions>> Programming >>Structures>> While Loop**) on the BD, which may need to be resized later. To provide the signal source of the system, place a **Basic Function Generator VI** (**Functions>>Programming>>Waveform>>Analog Waveform>>Waveform Generation>>Basic Function Generator**) inside the While Loop. To configure the parameters of the signal, appropriate controls and constants need to be wired. To create a control for the signal type, right-click on the signal type terminal of the Basic Function Generator VI and choose **Create>> Control** from the shortcut menu. Note that an enumerated (Enum) type control for the signal gets located on the FP. Four items including sine, triangle, square, and sawtooth are listed in this control.

Next, right-click on the amplitude terminal, and choose **Create>>Constant** from the shortcut menu to create an amplitude constant. Enter 100 in the numeric constant box to set the amplitude of the signal. In order to configure the sampling frequency and the number of samples, create a constant on the sampling information terminal by right-clicking and choosing **Create>> Constant** from the shortcut menu. This creates a cluster constant which includes two numeric constants. The first element of the cluster shown in the upper box represents the sampling frequency and the second element shown in the lower box represents the number of samples. Enter 8000 for the sampling frequency and 128 for the number of samples. Note that the same parameters were used in the previous section.

Now, toggle to the FP by pressing <Ctrl-E> and place two Vertical Pointer Slide controls on the FP by choosing **Controls>>Modern>>Numeric>>Vertical Pointer Slide**. Rename the controls Frequency and Gain, respectively. Set the maximum scale values to 1000 for the Frequency control and 5 for the Gain control. The Vertical Pointer Slide controls create corresponding icons on the BD. Make sure that the icons are located inside the While Loop. If not, select the icons and drag them inside the While Loop. The Frequency control should be wired to the frequency terminal of the Basic Function Generator VI in order to be able to adjust the frequency at run time. The Gain control is used at a later stage.

The output of the Basic Function Generator VI appears in the waveform data type. The waveform data type is a special cluster which bundles three components (t0, dt, and Y) together. The component t0 represents the trigger time of the waveform, dt the time interval between two samples, and Y data values of the waveform.

Next, the generated signal needs to be scaled based on a gain factor. This is done by using a **Multiply** function (**Functions>>Programming>>Numeric>>Multiply**) and a second Vertical Pointer Slide control, named Gain. Wire the generated waveform out of the signal out terminal of the Basic Function Generator VI to the x terminal of the Multiply function. Also, wire the Gain control to the y terminal of the multiply function.

Recall that the Merge Signals function is used to combine two signals having dynamic data types into the same wire. To achieve the same outcome with regular VIs, place a **Build Array** function (**Functions>>Programming>>Array>>Build Array**) to build a 2D array, i.e. two rows (or columns) of one dimensional signal. Resize the Build Array function to have two input terminals. Wire the original signal to the upper terminal of the Build Array function, and the output of the Multiply function to the lower terminal. Remember that the Build Array function is used to concatenate arrays or build n-dimensional arrays. Since the Build Array function is used for comparing the two signals, make sure that the **Concatenate Inputs** option is unchecked from the shortcut menu. More details on the use of the Build Array function can be found in [1].

A Waveform Graph (**Controls>>Modern>>Graph>>Waveform Graph**) is then placed on the FP. Wire the output of the Build Array function to the input of the Waveform Graph. Resize the plot legend to display the labels and edit them. Similar to the example in the previous section, the **AutoScale** feature of the Y axis should be disabled and the **Loose Fit** option should be unchecked along the X axis.

Place a **Wait (ms)** function (**Functions>>Programming>>Timing>>Wait**) inside the While Loop to delay the execution in case the VI runs too fast. Right-click on the milliseconds to wait terminal and choose **Create>>Constant** from the shortcut menu to create and wire a Numeric Constant. Enter 200 in the box created.

Figure L2.9 and Figure L2.10 illustrate the BD and FP of the designed signal generation system, respectively. Save the VI as **Lab02_ Regular_Waveform.vi** and run it. Change the signal type, gain and frequency values to see the original and scaled signal in the Waveform Graph.

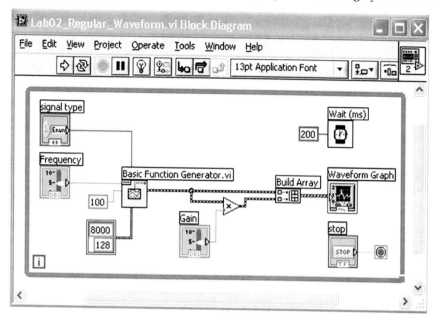

Figure L2.9: BD of signal generation and amplification system using regular VIs

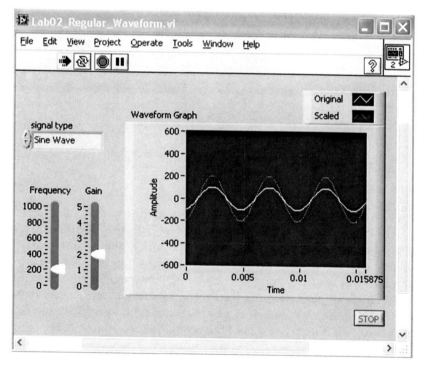

Figure L2.10: Original and scaled output signals

The waveform data type is not accepted by all the functions or subVIs. To cope with this issue, the Y component (data value) of the waveform data type is extracted to have the out-

41

put signal as an array of data samples. This is done by placing a **Get Waveform Components function (Functions >>Programming>>Waveform>>Get Waveform Components)**. Then, wire the signal out terminal of the Basic Function Generator VI to the waveform terminal of the Get Waveform Components function. Click on **t0**, the default terminal, of the Get Waveform Components function and choose **Y** as the output to extract data values from the waveform data type, see Figure L2.11. The remaining steps are the same as those done for the version shown in Figure L2.9. In this version, however, the processed signal is an array of double precision samples.

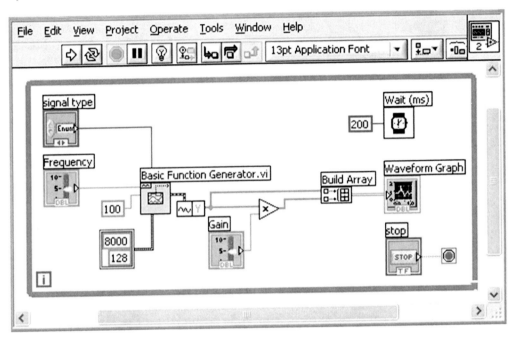

Figure L2.11: Matching data types

L2.3 Profile VI

The Profile tool is used to gather timing and memory usage information. Make sure the VI is stopped before setting up a Profile window. Select **Tools>>Profile>>Performance and Memory** to bring up a Profile window. Place a checkmark in the **Timing Statistics** checkbox to display timing statistics of the VI. The **Timing Details** option provides more detailed statistics of the VI such as drawing time. To profile memory usage as well as timing, check the **Memory Usage** checkbox after checking the **Profile Memory Usage** checkbox. Note that this option can slow down the execution of the VI. Start profiling by clicking the Start button on the profiler, then run the VI. A snapshot of the profiler information can be obtained by clicking on the Snapshot button. After viewing the timing information, click the Stop button. The profile statistics can be stored into a text file by clicking the Save button. An outcome of the profiler is exhibited in Figure L2.12 after running the Lab02_Regular VI. More details on the use of the Profile tool can be found in [2].

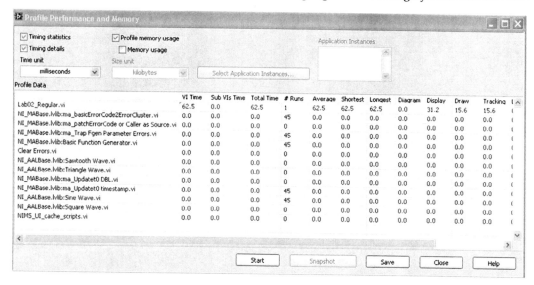

Figure L2.12: Profile window after running Lab02_Regular VI

L2.4 Bibliography

[1] National Instruments, *LabVIEW User Manual*, Part Number 320999E-01, 2003.

[2] National Instruments, *LabVIEW Performance and Memory Management*, Application Note 168, Part Number 342078B-01, 2004.

L2.5 Lab Experiments

- Build a VI to check whether a given positive integer 'n' is a prime number or not and display a warning message if it is not a prime number.
- Build a VI to generate the first 'n' prime numbers and store them using an indexing array. Display the outcome.
- Build a VI to generate two sinusoid signals with the frequencies f_1 Hz and f_2 Hz and the amplitudes A_1 and A_2, based on a sampling frequency of 8000 Hz with the number of samples being 256. Set the frequency ranges from 100 Hz to 400 Hz and set the amplitude ranges from 20 to 200. Generate a third signal with the frequency f_3 = (*mod* (*lcm* (f_1, f_2), 400) + 100) Hz, where *mod* and *lcm* denote the modulus and least common multiple operation, respectively, and the amplitude A_3 being the sum of the amplitudes A_1 and A_2. Use the same sampling frequency and number of samples as used for the first two signals. Display all the signals using the legend on the same waveform graph and label them accordingly.

CHAPTER 3

Introduction to VHDL for Signal Processing

In this chapter, an introduction to VHDL programming is provided. This introduction is not meant to be a substitute for many comprehensive VHDL programming guides that are available in the literature, e.g. [1-3]. It is merely included for the purpose of familiarizing signal processing students who normally do not have any prior VHDL programming experience. VHDL and Verilog constitute the most widely utilized FPGA programming languages. In this book, VHDL is considered due to its support by the LabVIEW FPGA Module. VHDL stands for Very High Speed Integrated Circuits Hardware Description Language.

3.1 VHDL Language Basic Elements

3.1.1 Entity and Architecture

Every VHDL program that defines or represents a digital system must have an entity and architecture. Entity defines the interface or the window with which the system interacts with the outside world, and architecture defines the internal behavior or structure of an entity.

3.1.1.1 Entity Declaration and Specification

The syntax for declaring an entity is:

```
entity entity_name is
port (signal_name: mode_type signal_type;
signal_name: mode_type signal_type;
:
signal_name: mode_type signal_type);
end [entity_ name] ;
```

Before going into the specifics of the syntax, note that keywords are shown in bold letters, and optional elements in square brackets. This convention is followed throughout the chapter. The above syntax has the following attributes:

- Every entity starts with the keyword **entity**, followed by the name of the entity entity_name.This name should match the name of the VHDL program, that is, if the entity name is 'abc', the VHDL file should be saved with the name 'abc.vhd'.

45

- The keyword **port** provides the interface to the outside world via so called signal, and has three parts:
    - signal_name specifies the name given to the signal. Signals carry information in VHDL. A signal_name inside a port declaration is always followed by a colon. Multiple signals, of the same type, can be declared inside an entity by specifying the corresponding names separated by commas.
    - mode_type specifies the mode or the manner in which a port is to be used. The available modes include in, out, inout and buffer, which correspond to a port that can be used as input (read only), output (write only), inout (write as well as read), and bidirectional, respectively.
    - signal_type specifies the type of signals, such as bit, boolean, etc.
- The keyword **end** specifies the end of an entity followed by the entity name used in its declaration. Mentioning of entity_name at the end is optional.
- A port specification is always enclosed in parentheses (). Each statement inside a port declaration ends with a semicolon, except the last one, where the semicolon is preceded by the closing parenthesis of the port declaration.

3.1.1.2 Architecture

Architecture defines the body of an entity with which it is associated. Its syntax is as follows:

Architecture architecture_name **of** entity_name **is**
architecture-declaration
begin
architecture-statement;
architecture-statement;

...
end [architecture_name] ;

The following items are to be noted in the architecture declaration:

- The keyword **Architecture** is followed by architecture_name that specifies the name for the architecture. This is followed by the keyword **of** and entity_name with which this architecture is associated. The entity name must match the name used in the entity declaration.
- architecture-declaration is used to declare the items used within the design entity. These items can be signals, constants, subprograms, types, etc. It is possible to have no declarative items within an architecture. This situation occurs when the only signals used in the architecture body are port signals.
- An architecture statement or body begins with the keyword **begin**, and ends with the keyword **end** and the optional architecture_name. It defines what the circuit or the system (that the entity represents) does. Thus, it gives a description of the architecture/organization of the entity.
- Every architecture must have an entity, and it is possible to have multiple architectures for the same entity.

It helps to go through an example code for better understanding of the entity and architecture concepts. Let us consider a simple VHDL code for an 'and gate' circuit. The code appears in Figure 3.1.

```
library ieee;
use ieee.std_logic_1164.all;
entity and_gt is
port( a,b : in bit ;
      c:out bit);
end entity;
architecture a1 of and_gt is
begin
c<= a and b;
end a1;
```

Figure 3.1: VHDL code for 'and gate'

Before examining the syntax, let us note two new elements of the language: library and package. A VHDL library is a collection of various subsystems or units such as entities. The benefit gained by using libraries is that similar designs can be organized into a common structure that can be analyzed and tested independently. This concept is similar to the way one organizes similar VIs in a LabVIEW library. A VHDL package provides a means of combining functions, data types, etc., into a common structure.

In the example shown in Figure 3.1:

- The first line of the code uses the keyword library to define the name ieee for the design library. This makes the design units within the library ieee available to the entity and architecture declared later.
- The second line is a package use statement with the syntax

    use library_name.package_name.**all**

  This causes the definitions given in the package, defined by package_name in the library library_name, to become available in the current program or design. The clause **all** refers to all the definitions within the package. Thus, in the example, the package std_logic_1164 makes all its definitions in the library ieee available to the entity and_gt and its architecture a1.
- Next, the entity and_gt is declared with three port signals consisting of signals a and b as input signals of type bit and signal c as output signal of the same type.
- Next statement assigns the name a1 for the architecture.
- The body of the architecture has one statement, which ends with a semicolon, and implements the 'and' operation on signals a and b generating signal c. This is a signal assignment statement that computes a and b and assigns the result to c using the assignment operator <=. The syntax of a signal assignment is given by

target <= source expression

> where the right hand side is the source expression, which is computed and assigned to the left hand side.

- The last line ends the architecture a1.

It should be noted that the above program directly assigns the result of the and operation to the output port. If the signal c was to be used for a read operation within the architecture, a compilation error would be generated as this would correspond to reading an output port, which is not allowed. For example, Figure 3.2 below provides the syntax for creating a signal d that is obtained by an 'or' operation on c and a:

```
library ieee;
use ieee.std_logic_1164.all;
entity and_gt is
port (a,b : in bit ;
c:out bit);
end entity;
architecture a1 of and_gt is
signal d: bit;                    -- signal declaration
begin
c<= a and b;
d<= c or a;
end a1;
```

Figure 3.2: VHDL code giving compilation error

Note: In the above code and other codes that follow, VHDL comments are specified with double hyphens '—'. Also, comments are sometimes extended to the next line due to a lack of space. Second lines are not commented again with double hyphens. However, it should be noted that while running codes in ModelSim, such double-line comments should be written as one line.

As can be seen from the code given in Figure 3.2, the result of c or a is assigned to the signal d. This is not allowed since port c is regarded as an output port. To get around this problem, an intermediate signal x is created in the architecture, and it is or-ed with the signal a, and finally it is assigned to the signal d, see Figure 3.3. This way signals (not port signals) become visible within the architecture and can be used for both read and write operations.

A signal can be declared inside an architecture by using an architecture declarative port as follows:

signal signal_name: signal-type [:= expression];

The definition starts with the keyword **signal.** The type of the signal and its name are specified by signal-type and signal_name, respectively. A signal can be initialized to a value at the time of signal declaration by using expression. However, if expression is not used explicitly, the signal is assigned a value implicitly at the time of declaration, which is the left-most value in a scalar type. For example,

signal x: integer:= 4; - declares signal x, of type integer, and initialized to value 4.

signal x: std_logic; - declares signal x, of type std_logic, and due to its implicit assignment, it is assigned the leftmost value in the type std_logic which is 'U'.

Note that the signal declaration is the same as the port signal declaration, except that (1) port signals do not begin with the keyword **signal**, and (2) port signals must have their mode type specified.

3.1.2 Constants and Generics

Both constants and generics allow a mechanism for defining static information of various parameters such as bus width.

3.1.2.1 Constants

The general definition of a constant is

constant constant_name: constant_type := value ;

The definition starts with the keyword **constant**, followed by constant_name, which specifies the name of the constant. This is followed by a colon, and the type of constant given by constant_type (similar to signal types of bit and boolean). Then, the operator := follows and in the end value for the constant is specified. More than one constant of the same type and with the same value can be declared in a single statement by using commas, that is

constant constant_name1,constant_name2,...: constant_type:=value;

Constants can be declared inside an architecture. Once they are declared with a value assigned to them, the value cannot be changed. Here are two examples:

constant prop_delay: time := 5ns – declares a propagation delay constant of a gate which is of type time and value 5 nanosecond.

constant sqr_three: real := 1.732– declares constant square root of 3 of the type real number.

The advantage of using a constant in a program is for the ease of debugging and modification in the program. For example, if one wishes to change a particular value in multiple statements inside a VHDL program, that value can be defined as a constant. This way, if the value needs to be modified, the modification is done at one place. A constant may not be initialized at the time of declaration when it is a deferred constant, which is a constant already declared in a package.

```
library ieee;
use ieee.std_logic_1164.all;
entity and_gt is
port (a,b : in bit ;
      c:out bit);
end entity;
architecture a1 of and_gt is
signal d,x: bit;                    -- signal declaration
begin
x<= a and b;
d<= x or a;
c<= d;
end
```

**Figure 3.3: Modified VHDL code that uses an intermediate signal
to prevent output port read**

3.1.2.2 Generics

The basic generic type is a generic constant, whose value is set when the design is initialized and cannot be changed later in the design. It is defined as:

generic (generic_name**:** generic_type := value)**;**
This creates a generic constant of the name generic_name having its data type defined by generic_type, and initialized to a value. The difference with the declaration of the constant is that the keyword **generic** is used, instead of the keyword **constant**, and parentheses () are used in the definition. The initialization is not necessary. As in the case of constants, multiple generic constants may be declared using a comma separator.
Here is an example of declaring a generic

generic (memory_sz: integer:= 8); -- declares a memory of size 8 bits

Another use of generics is in package declaration for a package to be reused with different types and operations.

Both generics and constants can be used to control the size of buses or timing parameters, loop counters, physical characteristic of blocks. However, the difference between the two is that while constants are declared in the architecture, generics are declared in the entity. This means that constants are visible or localized to the architecture. Also, constants cannot be used to define the size of ports (unless they are declared in a package that the entity uses).

3.1.3 Concurrent Statements

The statements written inside an architecture represent concurrent statements that are executed together, independent of the order in which they occur in the program. Thus, signal assignments for concurrent statements cause target to be assigned to the source values whenever the condition changes. This models a combinatorial behavior. These statements can have various types of signal assignments that are covered in the next section.

3.1.3.1 Conditional Signal Assignments

In these signal assignments, based on a condition(s), one of the multiple source expressions is chosen and assigned to the target.

In the statement

> target **<=** expression 1 **when** condition true **else**
> expression 2;

a **when-else** clause is used, that is when the condition is true, the target is assigned expression 1; and when the condition is false, the target is assigned expression 2.

Synthesis interpretation: Conditional signal assignments get synthesized to multiplexers because the general behavior of a multiplexer closely matches the behavior of signal assignment statements (a multiplexer behaves by passing one of the inputs (source) to the output (target) based on a conditional signal). For example,

> z **<=** x when sel **='1'** else
> y;

is mapped to a multiplexer, with sel as the select terminal, x & y as data inputs, and z as the output. This is indicated in Figure 3.4.

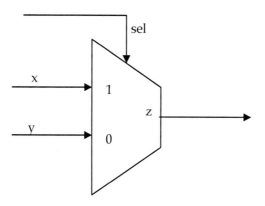

Figure 3.4: Single condition signal assignment synthesis

Similarly, one can have multiple conditions leading to a synthesis of cascaded multiplexers. That is,

target **<=** source expression 1 **when** condition1 true **else**
source expression 2 when condition 2 true **else** source expression 3;

This means that if the condition1 is true, then expression 1 is selected; if false, then either expression 2 or expression 3 is selected depending upon the truth status of condition 2. Thus, the synthesis resembles a two-step multiplexer. When condition 1 is true, multiplexer 1 will select expression 1 regardless of the condition 2, so condition 1 is synthesized to the select input of multiplexer 1, expression1 as its data input 1, and the target as its data output. Similarly, the remaining statements are synthesized to multiplexer 2 with condition 2 synthesized to its select input, expression 3 as its data input, and its data output is connected to the data input 2 of multiplexer 1. This is made more clear by considering the example below

z **<=** x when sel =′1′ else
y when sel1= '1' else
w;

In the above example, sel and sel1 are the select terminals of the two multiplexers with the inputs x and y. Figure 3.5 provides an illustration of this example.

The drawback of conditional signal assignment statements is that they are synthesized to cascaded or chained multiplexers, which are bound to get skewed, leading to delays as more and more multiplexers are added to synthesize multiple conditions. Also, with each statement being independent, in complex designs, it is possible that some redundant multiplexers are synthesized.

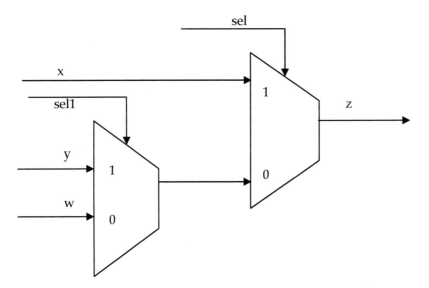

Figure 3.5: Multiple condition signal assignment synthesis

3.1.3.2 Selected Signal Assignments

These are similar to conditional signal assignments in the manner a condition causes the source expression(s) to be assigned to the target, but the difference lies in the fact that a single condition is used instead of multiple conditions in a conditional signal assignment. The definition uses the keyword **with** and **select** as follows:

> **with** condition **select**
> target **<=** expression 1 **when** choice 1;
>   expression 2 **when** choice 2;

Thus, based upon the condition being choice 1, expression 1 is selected; and if choice 2 occurs, expression 2 is selected. The important point to note here is that the condition type may have multiple values all of which must be covered in the branches or the choices. If this is not done and the keyword 'others' is used to describe all the choices that are not covered in the previous statements, incorrect results are obtained. For example, consider the following code:

> signal x: bit ;
> with x select
> y <= x when '0' ;
>   not(x) when others;

Here, if the signal x has value '0', y is assigned the value of x; while if x has the value others, y is assigned the value not(x). The type bit has only two values, 1 and 0, so others covers the value of 1 without any ambiguity. However, consider the same code with the type changed to std_logic, that is

> signal x: std_logic ;
> with x select
> y<= x when '0';
>   not(x) when others;

Now, the choice others creates a problem because std_logic has 8 other values besides '0', so y will be assigned not(x) when any of the choices occur (which is what others should cover). However, the actual circuit may not require this. For example, for alarm<= not (reset) with LED 'on' as the requirement, by using others, alarm is set even when LED have states other than 'on' or 'off', which is not correct. Thus, here others is not equivalent to 'on'.

3.1.4 Sequential Statements

These statements are executed sequentially in the order they appear in the program. They are written inside a VHDL construct called **process**, and are of many types. First, let us see what a VHDL process is.

3.1.4.1 Process

A process is a structure that includes statements that are executed one after another, similar to a normal program execution in MATLAB or C languages. A process is declared inside an architecture body (after the keyword **begin**) as follows:

> [name]:**process** [sensitivity list]
> declarations
> > **begin**
> > sequential statements
> **end process** [name];

A process definition starts with the keyword **process**, and can be preceded by an optional name of the process followed by a colon. The name should then match the name following the keyword **end process** at the end of the process block. The declarations part includes the declaration of various VHDL elements such as variables and functions. The sequential statements start after the keyword **begin**. The important aspect of a process is its sensitivity list. The sensitivity list includes the name of the signals (separated by commas) to which the process is sensitive. How a process with a sensitivity list starts, suspends, and then starts again, is explained in the following steps:

- An event on a signal denotes the change of its value. Whenever any of the signals defined in the sensitivity list of a process has an event placed on it, the simulation starts and the process is executed once, that is, the statements defined in the process body are executed (or placed in queue).
- During this execution time, the signals are not updated and they retain their old values.
- As the process reaches the last statement, it is suspended.
- The process remains in the suspended state until there is an event on any of the signals in its sensitivity list, at which time it resumes again. During this time, the signal assignments (that were placed in the queue) are executed, and the signals are updated.
- Alternatively, a process may not have a sensitivity list, but can have a **wait** statement at the end, just before the last statement (other forms are also possible).
- As the process encounters a wait statement, it is suspended.
- The process remains in the suspended state until the condition specified in the wait statement is met. As the wait condition is satisfied, the process resumes. In other words, the process statements shown in Figure 3.6 are equivalent.

3.1.4.2 Variable

The motivation to use variables as a means of carrying information within an architecture is governed by two facts. First, signals cannot be declared inside processes. Second, signals cannot store values which are older than the last signal assignment. This makes the storage of intermediate values via signals not possible. Variables provide a solution to these problems. Their definition is similar to signal definition and is given by:

> **variable** identifier list: subtype [:= expression];

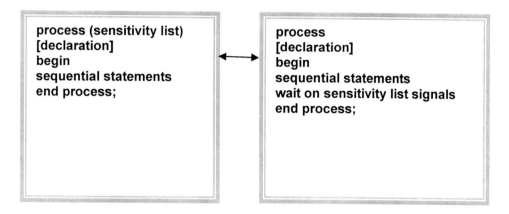

Figure 3.6: Equivalence of two types of process statements

The declaration uses the keyword **variable**, the list identifier list contains the variable names of the variables to be declared, and the type is given by subtype. Similar to signals, variables can be assigned values at the time of declaration by using expression. Variable assignment is similar to signal assignment except that the assignment is made with the symbol ':=' instead of '<=' via

target := source expression;

Note that the target is a variable, and the source expression can involve variables only, signals only, or mix of the two. The real power of variables lies in the fact that they are assigned values immediately (without delay), thus making the storage of intermediate values possible. For example, Figure 3.7 describes a 1-bit full adder circuit using two different approaches.

The program on the left side of Figure 3.7 uses the port signals directly in the architecture, while in the one on the right, a variable d is created to hold the result of the xor operation, and then it is used in the source side of the signal assignments. These assignments contain the source expression which is a mix of a signal and a variable. Since a variable has no scope outside a process, the last statement in a process must be a signal assignment in order to carry the values held by the variables to outside the process. Table 3.1 summarizes the difference between a signal and a variable.

Similar to concurrent statements, there are two types of signal assignments in the sequential statements that are covered in the next section.

```
library ieee;
use ieee.std_logic_1164.all;

entity full_add is
port (a,b,cin:in bit;
        s,cout: out bit
        );
end entity;

architecture a1 of full_add is
begin
        s <= a xor b xor cin;
        cout <= (a and b) or
        (cin and (a xor b))

end a1;
```

```
library ieee;
use ieee.std_logic_1164.all;

entity full_add is
port (a,b,cin:in bit;
        s,cout: out bit
        );
end entity;

architecture a1 of full_add is
begin

process (a,b,cin)
variable d: bit;
    begin
            d:= a xor b;
            s <= d xor cin;
            cout <= (a and b) or
            (cin and  d);
    end process;
end a1;
```

Figure 3.7: VHDL codes for 1-bit adder

SIGNAL	VARIABLE
Defined inside architecture	Defined in a process
Signal assignment can appear anywhere inside an architecture, including process	Variable assignment cannot be done outside the process, within architecture
Signals cannot hold intermediate values	Variables can hold intermediate values
Signals assignment within a process takes place when a process is suspended	Variable assignment within a process is immediate

Table 3.1: Differences between signals and variables

3.1.4.3 Conditional Signal Assignments

The conditional signal assignments, within the framework of sequential statements, use **if** clause similar to **when** clause in the concurrent signal assignment. The format is given by

```
if condition then
statements set a;
else
statements set b;
end if;
```

If the condition is true, statements set a is executed. If the condition is false, statements set b is executed.

Synthesis interpretation: The above **if-else-end if** clause is synthesized to a multiplexer with condition mapped to the select input of the multiplexer (as illustrated in Figure 3.4). The other type of if statement is **if-elsif-else-endif** with the format of:

> **if** condition 1 **then**
> statement set a;
> **elsif** condition 2 **then**
> statement set b;
> else
> statement set c;
> **end if**;

If condition 1 is true, statement set a is executed, and the other branches are ignored. If condition 2 is true, statement set b is executed, and the last branch is ignored. If both the conditions are false, then statement set c is executed.

Synthesis Interpretation: The above code gets synthesized to a two-level multiplexer, with condition 1 mapped to the select input of the first multiplexer and condition 2 mapped to the select input of the second multiplexer (see Figure 3.5).

3.1.4.4 Selected Signal Assignments

Similar to a concurrent selected signal assignment that uses the clause **with-select-when**, a selected signal assignment inside a process uses the clause **case-when**. The format is

> **case** expression **is**
> **when** choices **=>** statements;
> **end case;**

The statement starts with the keyword **case** followed by expression, which can be an enumeration type, integer type, or a character array type (these are discussed later). Depending upon the value of the expression, as declared by choices, only one of the statements is executed. Finally, **end case** specifies the end of the case statement.

There exists a difference between multiple **if** statements and **case** statements with multiple choices. Multiple **if** statements usually have conditions that depend upon expressions involving different signals/variables while **case** statements have a single expression involving a single variable or signal. Also, if the expression in a case statement involves integer types, then the clause other is needed to cover all the types that are not covered by **when** choices.

3.1.5 For Loop

Similar to a for loop in any other language, a **for loop** repeats a section of the VHDL code a specified number of times. The format is

> **for** loop_parameter **in** discrete_range **loop**
> sequential statements;
> **end loop;**

The first line of the code starts with the keyword **for,** followed by loop_parameter, which takes a discrete set of values specified by discrete_range after the keyword **in.** The statement ends with the keyword **loop.** The for loop ends with the keyword **end loop.** The statements within the first and the last line constitute sequential statements to be repeated. This means that the sequential statements are repeated a number of times depending upon discrete_range taken by loop_parameter. loop_parameter has the same type as the base type of discrete_range. A point worth nothing here is that the loop parameter cannot be used as the target of signal assignment statements.

Let us consider an example here. Suppose, a signal x has been declared of the type integer and it is desired that whenever this signal changes, the value of the signal sum is incremented by 4 and then gets assigned to the signal y. This can be achieved by the code provided in Figure 3.8 (shown after the process statement).

As seen on line 4 in Figure 3.8, the signal x is assigned to the variable sum of the type integer. Then, the variable sum is incremented four times as the loop index k goes from three to zero (the keyword downto indicates a descending count, more on this discussed later in the chapter). Finally, the variable sum is assigned to the signal y.

3.1.6 While Loop

The syntax of While Loop is given by

> **while** condition **loop**
> sequential statements;
> **end loop;**

This loop repeats sequential statements a number of times as long as condition is true. As soon as condition becomes false, the loop is exited.

3.1.7 Exit Statement

The syntax is

> **exit when** exit_condition

This statement allows for the stoppage of a for loop execution before the required number of iterations are completed. This statement is placed within the for loop body and is executed once the condition exit_condition is satisfied, leading to the termination of the for loop.

```
process(x)
  variable sum: integer;
  begin                     -- process starts
  sum:= x;                  -- signal to variable assignment
  for k in 3 downto 0 loop  -- k=3,2,1,0
          sum:= sum+1;
  end loop;
  y<=sum;                   -- variable to signal assignment
end process;
```

Figure 3.8: Using a For Loop in VHDL

3.1.8 Next Statement

The syntax is

next when condition;

Within a for loop or a while loop, when this statement is reached, condition is evaluated. If true, the rest of the sequential statements are executed, and the next iteration starts; if false, the statements following **next** are ignored, and the loop jumps to the next iteration.

3.1.9 Wait Statements

The following wait statements are commonly seen in VHDL:

- **wait on sensitivity list:** This is the most common form of the wait statement, where based upon the events that get generated on the sensitivity list, the process resumes. As discussed before, a process with a sensitivity list is equivalent to a process with no sensitivity list and a wait on sensitivity list at the end. However, the wait statement can also appear at the start of a process (just after begin, see Figure 3.9). In this figure, two statements are equivalent from the synthesis point of view, but not from the simulation point of view. For a detailed discussion of the differences in the simulation, the interested reader is referred to reference [4].

- **wait for [time]:** When this statement appears inside a process, the process is suspended for the specified time.
- **wait until condition:** This causes the executing process to suspend until the condition becomes true.

```
process
begin
wait on x;
statements;
end process;
```
```
process
begin
statements;
wait on x;
end process;
```

Figure 3.9: Equivalent set of wait statements

Having gone through both the concurrent and sequential signal assignment statements, let us now go through the examples exhibited in Figure 3.10 that clearly distinguish between the two. All the codes are synthesized to a multiplexer.

3.1.10 D Latch

A D latch or a transparent latch is the one that passes the data from input to output in a transparent fashion when an enable signal is high. A simple implementation of a D latch in VHDL uses the clause **if-endif** as shown in Figure 3.11. The latch passes the input data to the output when the enable signal en is high. With the signal en at level high, as the input data changes, the process is executed and the output is assigned the new value of the input through the signal assignment q<=d. Thus, the output follows the input. When en is low, the latch output remains in the previous state. It is important to use both d and en as the sensitivity list parameters. For example, if the statement process(en) is used, the process will resume when the enable signal goes high; but as long as the enable signal remains high, the assignment q<=d will not get executed even if d changes, as d is not in the sensitivity list. Figure 3.12 illustrates the simulation of the behavior of a D latch.

3.1.11 Register or D Flip Flop

A register is a combination of storage elements. Both flip flops and latches are considered to be registers. However, strictly speaking, a flip flop has its output changing only on clock pulses, while a latch has its output changing when enable is high, independent of any clock. Here, let us follow the convention of using the term register as a group of flip flops. The D flip flop or edge triggered D flip Flop passes data from its input to output at every clock cycle. Figure 3.13 shows differing coding styles for implementing a D flip flop. Earlier, it was seen that if-endif without an else synthesizes to a latch, however, in two of the stated codes appearing in Figure 3.14, the clause if clk'event and clk='1' then is used, which causes a D flip flop synthesis. Let us consider the first code. The process statement has clk or the clock signal in its sensitivity list. The statement if clk'event and clk='1' specifies an event on the clock signal, which is the clock going from low to high or the rising edge of the clock. When this event occurs, the process executes. The next statement q<= d is queued up for the next delta cycle. Then, the process suspends, the signal assignment is executed causing the signal q to be updated on that delta cycle, or the same simulation time as the clock rising edge. The signal q holds this value till a new event is generated at the next clock edge, and the process resumes operation. Thus, one can see that the D flip flop holds the data within clock periods (or at clock levels) and updates it over clock edges.

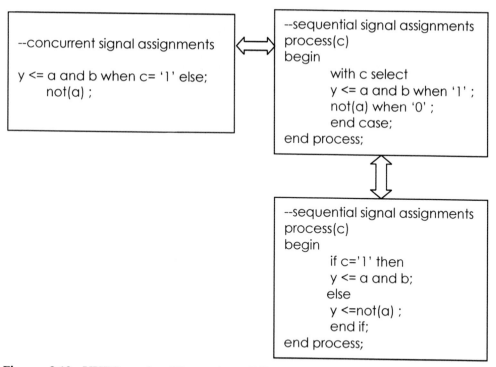

```
--concurrent signal assignments

y <= a and b when c= '1' else;
    not(a) ;
```

```
--sequential signal assignments
process(c)
begin
        with c select
        y <= a and b when '1' ;
        not(a) when '0' ;
        end case;
end process;
```

```
--sequential signal assignments
process(c)
begin
        if c='1' then
        y <= a and b;
        else
        y <=not(a) ;
        end if;
end process;
```

Figure 3.10: VHDL codes illustrating difference in syntax between concurrent and sequential signal assignments having the same synthesis interpretation

```
library ieee;
use ieee.std_logic_1164.all;

entity Dlatch is
  port(d, en: in std_logic;
       q:out std_logic);
end Dlatch;

architecture behav of Dlatch  is
begin
  process (d,en)
    begin
     if (en='1') then
       q <= d;
     end if;
   end process;
end behav;
```

Figure 3.11: D latch VHDL code

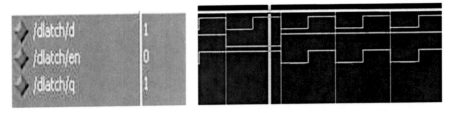

Figure 3.12: Simulation result for a D latch

3.1.12 Synchronous and Asynchronous Resets

The purpose of reset signals in a register is to put the register in a state which is not an arbitrary one, but a state which is well defined, thus allowing simulation. While the asynchronous mode resets the register independent of the clock, the synchronous mode does the reset with respect to the clock edge. While a detailed discussion is beyond the scope of this introduction, some of the differences between the two are listed in Table 3.2

Figure 3.14 shows the VHDL code for a D Flip flop with an asynchronous reset. Line 3 of the code checks for the level of the reset signal rst. If this is high, the output q of the register is set to 0. Note that this takes place irrespective of how the clock is applied (as the reset statement comes before the clock statement), hence the asynchronous behavior. If the reset is low, then the register responds to the data input on the active edge of the clock, as shown in Figure 3.14. Figure 3.15 implements a D flip flop with a synchronous reset.

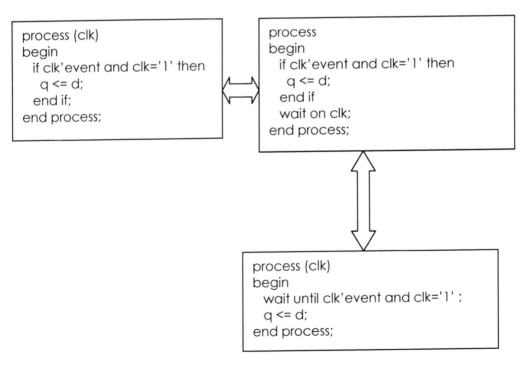

Figure 3.13: Equivalent VHDL coding styles for D flip flop

Synchronous reset	Asynchronous reset
requires more gates to implement	requires less gates to implement
requires a clock to be effective	works independent of the clock
provides glitch free reset (unless the glitch occurs exactly at the clock edge)	suffers from metastability problems
slow	fast

Table 3.2: Differences between synchronous and asynchronous resets

```
process (clk,rst)
begin
If rst='1' then
  Q <= '0';
elsif
  clk'event and clk='1' then
  q <=d;
end if;
end process;
```

Figure 3.14: VHDL code for D flip flop with asynchronous reset

```
process(clk)
begin
 If clk'event and clk='1' then
   If rst ='1' then
     q<='0';
   else
     q<=d;
   end if;
 end if;
end process;
```

Figure 3.15: VHDL code for D flip flop with synchronous reset

One can immediately see that the test for the event on the clock takes place before the test for the reset, hence the reset is synchronized with the clock. For further discussion on this topic, the interested reader is referred to reference [5].

3.1.13 Component

In VHDL, a system can be designed as an interconnection of components with information flowing between the components through signals. Each component itself can have a behavioral or a structural description (examples of these will be seen shortly).

To better understand this modeling style, the first step involves declaring a component. It can be declared as a separate structure with its own entity and architecture, or it can be declared inside an architecture beginning with the keyword **component**. The latter style is shown here:

```
component comp_name is
[generics]port definition
end component;
```

The component name is specified by comp_name, followed by the keyword **is**; generics are optional (as in an entity definition), followed by port definition. The end of the component structure is specified by **end component**. Here is an example of a D flip flop defined as a separate component

```
component Dff is
port(d, clk: in std_logic;
q:out std_logic);
end component;
```

So, Dff is the name of the component, which is also an entity with the ports d, clk, and q. Next, let us create a component instantiation. In simple terms, one can think of component instantiation as replacing a component within the main or enclosing architecture by the internal details of the component, and then connecting what goes in or out of the component with what goes in and comes out of the main entity (signals). This is done through **port map**, which specifies the connection between the ports of the entity and the signals of the enclosing architecture body. VHDL allows for positional or named association.

In positional association, the order in which signals are listed is the same as that of ports (local to the component) that are declared in the component entity declaration. For example, suppose a component is declared with ports a and b. These are mapped to the internal signals c and d of the main entity. The syntax is

```
port map (c,d);
```

Positional association has the format

```
[association_name]: component_name port map(signal list);
```

In named association, the association between the signals and the component ports is shown explicitly by using the symbol **=>**. The above example can be written as a named association as follows:

port map (a=>c, b=>d);

with the format

[association_name]: **component**_name **port map** (port name=> signal name, ...);

The named association has the advantage that it clearly shows the relationship between the component ports and the signals, while the positional association does not convey which signal is connected to which port.

3.1.14 Generate Statement

This statement provides a method by which one can repeat certain sections of a code. Thus, functionally, it is similar to a for loop with the difference that when a for loop is used inside a process as a sequential statement, all the statements are concurrent statements. Hence, they can also be used to repeat or replicate hardware or components to form larger systems. There are two types of generate statements, **for generate** and **if generate**. The first one is stated as

[label]: **for** gen_parameter **in** discrete_range **generate**
statements;
end generate

The **label** specifies the name for replicating the code and is optional. The parameter gen_parameter is similar to loop_parameter in for loop. The statements within the clause **for-end-generate** are executed a number of times depending upon discrete_range that loop_parameter goes through. This range can be an ascending or descending range. The keyword **generate** ends the first line of the loop, and **end generate** ends the statement. For example, suppose a D flip flop is implemented as a single bit system (see Figure 3.14). Let us make an 8-bit serial in parallel out (SIPO) right shift register using the statements component and generate. A right shift SIPO takes serial data from the left and shifts it bit by bit on each clock signal to the right whereby the output bits are available as parallel outputs.

The code is provided in Figure 3.16 and the same can be inferred diagrammatically as shown in Figure 3.17 (drawn for a 4-bit register for a better understanding). In this code, first, the main entity nbitsiso_rs_usingcomp is created with its port signals as din, clock, qout corresponding to the data input, clock signal, and data output terminals. The signal din is single bit of the type std_logic to read 1 bit data, and qout is 8 bits of the same type to provide the 8-bit parallel output.

Next, the component Dff is created, which is the same as the D flip flop of Figure 3.14. However, in this code, it is being used as a component with its port signals being d, clk, q corresponding to the data read, clock and data out terminals. Then, an 8-bit signal sig is created, followed by the positional component instantiation via the statement

MSBmapping: Dff port map (din, clock, sig(n-1));

This way, the input d of the flip flop 3 is mapped to din, the input clk of the flip flop 3 is mapped to the system clock clock, and the output q of the flip flop 3 is mapped to sig(3) (as n=4, n-1=3).

Next, the generate statement is used to create three copies of the D flip flop. To run the loop three times, the index i is used from 2 to 0. Within the generate statement, the signals and ports are mapped in a loop.

When i=2, the following mapping is obtained

port map (sig (3), clock, sig (2)) which means
d=> sig(3), clk=>clock, q=>sig(2) for D flip flop 2.

Similarly, for i=1 and 0, the associations shown in Figure 3.18 are obtained. Finally, the signal sig is assigned to the output port qout. The simulation result for the above code is shown in Figure 3.19 for an 8-bit register.

3.2 VHDL Types

VHDL has a number of types which are defined in the standard package. However, not all of these types are synthesizable. The synthesizable ones are included in the package std_logic_1164. Each VHDL type has a value and certain operations associated with it. These operators are used on various signals and/or variables having different types, combined together in expressions. The VHDL 87 standard defines four types: (1) Scalar types that include integers, physical types, floating-point types and enumeration types; (2) Composite types that include array types and record types; (3) Access types; and (4) File types.

Let us go through these types one by one.

- **Bit:** This is the logical type with two values of 0 and 1. It is defined as

    type bit **is** ('0','1');

    The values of the type bit must be enclosed in single quotes to signify that they should not be treated as numeric. Each of these values is synthesized to a single wire, representing a logic 0 or a logic 1.

```
library ieee ;
use ieee.std_logic_1164.all;

entity nbitsiso_rs_usingcomp is
generic (n:natural :=4; -- width ff 3to ff0
port(din, clock: in std_logic;
     qout: out std_logic_vector(n-1 downto 0));
end nbitsiso_rs_usingcomp;

architecture a1 of nbitsiso_rs_usingcomp is

component Dff is
    port (d ,clk: in std_logic;
    q:out std_logic);
end component;

signal sig: std_logic_vector (n-1 downto 0); -- sig3-sig0

begin
  MSBmapping: Dff port map (din, clock, sig(n-1)); -- d3 to din, q3 to sig(3))

  gen: for i in (n-2) downto 0 generate     -- FF 2 to 0, right shift, msb first
                                 -- so, 2 downto 0

  restofFFmapping: Dff port map (sig(i+1),clock,sig(i));
                                        --(d2 to sig(3)),q3 to sig(3))
end generate;

qout<=sig;   -- signal to ouput mapping
end a1;
```

Figure 3.16: VHDL code for an n-bit serial in parallel out shift register

- **Boolean:** It is defined as

    type Boolean **is** (false, true);

Boolean type has only two values of false and true. When two or more expressions are compared, the result is either true or false, which is a Boolean type. Note that in conventional logic, one does not differentiate between a bit value or a Boolean value. Thus, if a variable has a value 0 or 1, which are bit values, they are interpreted as false and true. In other words, in VHDL, the result of a comparison is true or false, and not 0 or 1. However, for synthesis, the Boolean signal is mapped to a single wire (as for bit), where the Boolean values true and false represent 1 and 0, respectively.

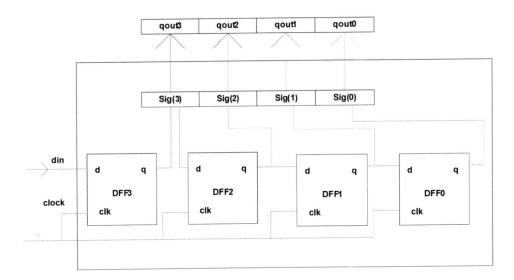

Figure 3.17: Structural description for an n-bit serial in parallel out shift register

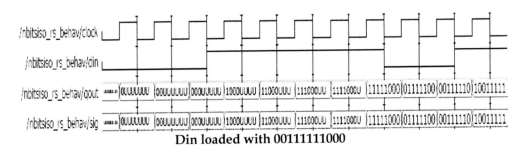

Din loaded with 00111111000

Figure 3.18: Simulation result for an 8-bit right shift register

- **Integer:** The definition is

    type integer **is range** -2147487483648 **to** +2147487483647

This definition shows that integer types have a range corresponding to 32-bit two's complement signed representation (-2^{32} to $2^{32}-1$). Thus, the range represents a series of different integers that can be used in the current design. Although the range can go beyond this and is implementation dependent, it may not be possible to simulate such integer types. The range can be an ascending range (from low value to high) or a descending range (from high value to low). For example, the above example is an ascending range. Also, it is possible to define new integer types by defining a new range which must be a subset of the integer range as follows:

    type new_type **is range** range'low to range'high –
    defines an ascending range.
    type new_type **is range** range'high downto range'low –
    defines a descending range.

A descending range is specified by the keyword downto and an ascending range by the keyword to.

- **Integer Subtypes:** The VHDL language is flexible in the sense that it allows not only creating new types, but also subtypes, where a subtype has a range of values that are a subset of the range of type (named basetype). For example, natural numbers are a subset of integers as they include positive integers. The **subtype** natural can be defined with a range that is given by

subtype natural **is integer range** 0 to integer 'high

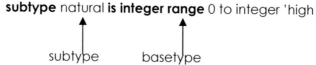

subtype   basetype

This shows that the subtype natural has a range 0 to the highest value (specified by the attribute integer'high) in the range defined by the basetype **integer**. The integer types or its subtypes are synthesized to buses or wires with the size of the bus equal to the number of bits needed to represent the range of the subtype.

- **Floating-Point Type:** VHDL represents real numbers by the type floating-point. The predefined floating-point in the standard is the type real, which conforms to the IEEE 64 bit double precision representation for floating-point numbers and is defined as

type real **is range** -1E38 to +1E38

Again, as in the case of integers, the range of real is dependent upon implementation, but will be a subset of the maximum range shown in the above definition. Similar to integers, new floating-point types can be defined. For example, temperature in Dallas can be defined as a new floating-point type with a range varying between 30 to 100 degrees,

type temp_dallas is range 30.0 to 100.0

To be able to use floating-point arithmetic, the package float_generic_pkg should be included in the design. Floating-point types are not synthesizable, that is to say there is no hardware mapping of floating-point types.

- **Physical Types:** These represent some real-world physical measurement values. The physical type definition includes a primary or base unit of measurement and may include a secondary unit(s) where one can express the primary unit as having an integral number of secondary units. A simplified definition is given by

type physical_type_name **is range** range_expression
units
base unit declaration;
[secondary unit declaration];
end units;

In this definition, base unit declaration corresponds to the smallest representable unit, range_expression defines the range of physical_type_name, while secondary unit declaration is some integral multiples of the base unit. For example, suppose the physical type voltage is defined as

```
Type voltage is range 0 to 250
units
volt;
end units;
```

Notice that volt means 1 volt. The range is a multiple of the base unit of 1 volt, so it is equivalent to 1volt * 0 to 1volt * 250.
One may also define secondary units on this base unit as follows:

```
type voltage is range 0 to 250
units
volt;
kVolt=1000 Volt;  --
secondary unit Kvolt, as multiple of base unit volt
end units;
```

- **Physical Type Time:** This type is the predefined physical type in VHDL. Again, the range is implementation dependent but is guaranteed to include the range -2147483647 to +2147483467. All the physical types are non-synthesizable

- **Enumeration Types:** Enumeration types are those types which are given common names to represent the values they possess for a simple and easy to use representation. Literal values are enumerated in a list, hence the name enumeration type. For example, the climate in a region can be defined as an enumerated type as follows:
  ```
  type climate is (warm, cold, hot, snowy, rainy);
  ```
  Bit, Boolean and std_logic are also of the enumeration type.

- **Multi-valued Logic:** Ordinarily, two valued logic should be sufficient to model many digital systems. However, there are situations where one needs to go beyond this. For example, one might need to model don't care or tristate. This is achieved through multi-valued logic. The predefined multi-valued logic is std_ulogic. The definition is given by

```
type std_ulogic Is ('U',  -- Uninitialized
'X',  -- Forcing Unknown
'0',  -- Forcing 0
'1',  -- Forcing 1
'Z',  -- High Impedance
'W',  -- Weak    Unknown
'L',  -- Weak   0
'H',  -- Weak   1
```

```
'-'  -- Don't care
);
```

In other words, it is a 9 valued logic with the 9 values defined as enumeration values. The logic values mean the following:

- U: The signal is not initialized yet. During simulations, if any signal is not initialized during declaration, before starting simulations, the simulator will show the signal as U.
- X: The logic type value cannot be determined.
- 0: Logic 0
- 1: Logic 1
- Z: High impedance; This models resistive pull ups and downs.
- W: It is weak signal, so not possible to determine whether it is 0 or 1.
- L: Weak signal that could go to 0.
- H: Weak signal that could go to 1.
- -: Don't care. Note that this representation is different from the don't care representation (X) as seen in the logic type domain outside VHDL.

The type std_ulogic has the major shortcoming of not being able to resolve multiple drivers. This means that if there are simultaneously active inputs or drivers of the type std_ulogic driving a bus, the bus output is not determined correctly as the target of multiple input signal assignments. This shortcoming is overcome in the type std_logic which is a subtype of std_ulogic.

- **Std_logic:** It is a resolved type with a resolution function associated with it. The definition of this type contains the resolution function, which specifies how to compute the output signal value when more than one driver or source is active. The following gives the definition of std_logic and the resolution function resolved:

      subtype std_logic **is resolved** std_ulogic;
      type stdlogic_table **is** array (std_ulogic,std_ulogic) **of** std_ulogic;

The resolution table specifies the output when multiple inputs are active. The following points hold about this table:

- The forcing value and a weak driver when active together, get resolved to a forcing value. For example, column seven shows that L mixed with 0 or 1 results in a 0 or 1, respectively.
- If both drivers drive the same strength values, which are different, the result is an unknown value of that strength.
- If there are two drivers with values U and any other nine std_ulogic values, the output will be resolved as U or unknown.

Finally, each of the types std_ulogic and std_logic have corresponding array types, std_ulogic_vector and std_logic_vector, having elements of these types. These types are covered next.

- **Arrays:** As in any other programming language such as C or Matlab, a VHDL array is a collection of elements of the same type. The array index specifies the position of the array elements in the array and is a scalar type. More often, the index is defined to be of the type natural to exclude negative values. So, an array definition requires defining the type or the subtype of its elements, and the type and range of its indices. One dimensional arrays have only one index, while multidimensional arrays have more indices. For example, a basic definition is given by

  type array_name **is** array (index range) **of** element_subtype

This creates an array by the name array_name with the range specified by index range having elements of the type element_subtype. The range index range can be an ascending or descending range, as discussed earlier. With a descending range, if the elements are of type bit, the leftmost element becomes the Most Significant Bit (MSB) and the rightmost element becomes the Least Significant Bit (LSB). On the other hand, with an ascending range, the LSB is to the left and the MSB to the right. For example, let us create one byte wide memory array as follows:

  type byte **is** array (7 downto 0) **of** bit; -- descending range

This creates the array byte as [byte(7) , byte(6),….byte(0)].

$$\text{MSB} \qquad\qquad \text{LSB}$$

Let us now create the same array with an ascending range,

  type byte **is** array (0 to 7) **of** bit; -- ascending range

This creates the array byte as [byte(0) , byte(1),….byte(7)].

$$\text{MSB} \qquad\qquad \text{LSB}$$

Ascending range is particularly appealing when working with DSP systems, as other commonly used software such as Matlab and C use ascending range for arrays. Hence, this would make easy porting of a C or Matlab code into the VHDL environment.

Once an array is declared, objects of that type including signals and variables can be defined. For example,

  signal mem: byte; -- creates a signal mem of type byte.

The arrays in VHDL can be defined in two ways: constrained and unconstrained. Constrained arrays have size predefined by specifying the range of the index and type. Unconstrained arrays have the type of the elements specified, but not the range, so the size is not specified. The symbol **<>** in an index subtype definition of an unconstrained array specifying an undefined range. For example, std_logic_vector is defined as an unconstrained array of the std_logic elements as follows:

type std_logic_vector **is** array (natural range <>) **of** std_logic;

As far as synthesis is concerned, an array with range 0 to n-1 or n-1 downto 0 having n memory locations, is synthesized to an n-bit bus. Thus, for synthesis, either constrained arrays should be declared, or unconstrained arrays should be declared, with their range constrained at the time of object creation. For example,

signal a: std_logic_vector (7 downto 0)

ensures that the array a, which is of the type std_logic_vector, is constrained in its range to 8 locations in the signal declaration. Another possible approach is to first create an unconstrained array. Then, create an array having a constrained range and type, which is a subtype of the unconstrained array. So, the above is restated as:

subtype b is std_logic_vector (7 downto 0);
-- subtype 'b' of type std_logic_vector
signal a: b; -- signal 'a' declared of the subtype 'b'

With respect to arrays, the following points and examples are worth noting:
- **Array slicing:** It refers to extracting a portion of an array. For example, suppose signal a is declared of the type std_logic_vector with 8 elements and a descending range. Then, most significant nibble out of it can be obtained by
  b<= a (7 downto 4);

- **Array concatenation:** It refers to combining elements of various arrays. For example, the least significant nibble from array a can be obtained by
  c<= a(3 downto 0);
  Now, these two slices can be concatenated into an array by
  d<= b & c -- & is concatenation operator
  Note that the order of b and c is important for correctly identifying the MSB.

- **Array attributes:** Attributes, in general, can be used to get information about scalar types. The references at the end of the chapter provide more information about the attributes that can be used with scalar types. Arrays have attributes, which give information about the size, range and indexing of an array. Assuming a one-dimensional array signal x of the type std_logic_vector, and range 3 downto 0, the following attributes can be defined:
  - **x'left:** Index of the leftmost element of the array, that is, index of the MSB if descending range and the LSB if ascending range.
  - **x'right:** Index of the rightmost element of the array, that is, index of the LSB if descending range and the MSB if ascending range.

- **x'range**: Index range of the array.
- **x'reverse_range**: Reverse of the index range of the array.
- **x'length**: Length of the array, that is, number of elements in the array.
- **x'element**: The subtype of the elements of the array.

For example, for the array x, these attributes are: x'left = 3 , x'right =0, x'range = 3 downto 0 , x'reverse_range = 0 to 3 , x'length = 4, x'element = std_logic. These attributes should be used to prevent errors and make programs more generic. For example, suppose it is desired to normalize an unconventional array range (that does not start from 0). That is, the array range is a to b which needs to be changed to 0 to c. One way to do this is

```
signal x: std_logic_vector (a to b);
signal y: std_logic_vector (0 to x'length-1);
```

The second statement creates a signal y of the type std_logic_vector with the range 0 to 0 to x'length-1, that is, 0 to b-a.

- **Other VHDL Types**: As stated earlier, there are other VHDL types such as records, access types and file types. Their discussion is beyond the scope of this introduction, and thus the interested reader is referred to the references at the end of the chapter for further explanation.

3.3 VHDL Operators

VHDL has a number of operators that are summarized in Table 3.3. Note that the original VHDL 87 release did not have the operators xnor, shift and rotate. They have been added in subsequent releases. If we try to run VHDL code having these operators based on the VHDL 87 tools, we would get errors.

For a detailed discussion of these operators, and of synthesis interpretation/limitations, refer to the references at the end of the chapter. However, to give a general understanding of these operators, the following examples are useful to go through.

OPERATOR	TYPE	USE/FUNCTION
+	Arithmetic	**addition**: add the operands
-	Arithmetic	**subtraction**: subtract the operands
*	Arithmetic	**multiplication**: multiply the operands
/	Arithmetic	**division**: divide the operands
abs	Arithmetic	**absolute**: absolute value of the operand
mod	Arithmetic	**modulo**: modulo operator
rem	Arithmetic	**remainder after division**: gives the remainder after dividing the operands
**	Arithmetic	**exponentiation**: result of raising one operand to the power of the other
not	Boolean	**not** operation on operands

and	Boolean	and operation on operands
nand	Boolean	nand operation on operands
nor	Boolean	nor operation on operands
or	Boolean	or operation on operands
xor	Boolean	xor operation on operands
xnor	Boolean	xnor operation on operands
=	Comparison	**equality:** test whether the operands are equal
/=	Comparison	**inequality:** test whether the operands are unequal
>=	Comparison	**greater than or equal:** test whether one operand (left) is greater than or equal to the other (right)
<=	Comparison	**less than or equal:** test whether one operand (left) is less than or equal to the other (right)
>	Comparison	**greater than:** test whether one operand (left) is greater than the other (right)
<	Comparison	**less than:** test whether one operand (left) is less than the other (right)
sll	Shifting	**shift left logical:** shift the operand to left, filling the right positions with zeroes
srl	Shifting	**shift right logical:** shift the operand to right, filling the left positions with zeroes
sla	Shifting	**shift left arithmetic:** shift the operand to left, copying the rightmost element of the operand to the left, treating it as sign element
sra	Shifting	**shift right arithmetic:** shift the operand to right, copying the leftmost element of the operand to the right, treating it as sign element
rol	Shifting	**rotate left:** rotate the operand to the left
ror	Shifting	**rotate left:** rotate the operand to the right
&		**concatenation:** concatenate the operands

Table 3.3: Standard VHDL operators

```
signal x,y,z,w: std_logic_vector(3 downto 0);
signal a,b,c: integer
x<= ('1','0','1','0');
w<= ('1','0','0','1');
a<=2;
b<=3;
a=b; -- false
a>b; -- false
c<= a+b ;              -- c= 5
c<= a – b;             -- c=-1
c<= a* b;              -- c= 6
c<= 2/3 --c=0          (note)
c<= abs(a – b)         – c= 1
c<= b mod a;           -- c= 1
```

```
c<=b rem a;            -- c= 1, 3/2 has remainder 1
c<= b**a ;             -- c= 3^2=9
y<= not(x);            -- y= [0 1 0 1];
z<= x and y;           -- z= [1 0 1 0] and [0 1 0 1] =[0 0 0 0];
z<= x or y;            -- z= [1 1 1 1];
z<= x xor y;           -- z= [1 1 1 1];
z<= w sll 2;           -- z= [0 1 0 0];
z<= w sla 2;           -- z= [0 1 1 1];
z<= w srl 2;           -- z= [0 0 1 0];
z<= w sra 2;           -- z= [1 1 1 0];
z<= w ror 2;           -- z= [0 1 1 0];
z<= w rol 2;           -- z= [0 1 1 0];
```

Table 3.4 gives a list of the types, upon which these operators can be applied based on the assumption that the types and the operators are defined in the package std_logic_1164.all.

VHDL TYPE	USED WITH
Boolean	Boolean, comparison
bit	Boolean, comparison
integer	comparison, arithmetic
arrays	comparison, concatenation
std_logic_vector	comparison, concatenation, shifting, Boolean

Table 3.4: VHDL operators

3.3.1 Operator Precedence

It refers to the order in which operators are processed in an expression containing multiple operators. Operators can also be classified as per precedence, so that operators of the same class have the same precedence among themselves. For a sequence or chain of operators with the same precedence level, the operators are associated with their operands in textual order, from left to right. For the precedence table showing the precedence among various operators, refer to the IEEE VHDL standard reference. The following points are worth noting:

- Logical operators have the lowest precedence.
- Logical operators should not be mixed, unless brackets are used. For example,
        a and b or c
  is incorrect while
        (a and b) or c
  is correct.
- The operators that can be chained or used in expressions involving multiple operators are the addition operators (+,-), concatenation operator, multiplication operators (*, /, mod, rem), logical operators (and, or, xor, xnor).
- For the exponentiation operator **, the only form allowed for synthesis is raise to power 2, that is $x**2$, which means square of the operand.

3.4 VHDL for Signal Processing

Up to this point, the packages with operator definitions are discussed that can be applied to various VHDL types for simulating and synthesizing digital circuits not involving many mathematical operations. As seen earlier, mathematical operators can be applied to few VHDL types. However, when synthesizing digital signal processing algorithms, it is desired to have operators that can be applied to many types with no limitation on their usage. For this purpose, a number of packages have been defined that are suited for numeric computations. Two of such common packages include IEEE and SYNOPSIS as indicated in Table 3.5.

IEEE Packages	Synopsis Packages
numeric_bit: It has VHDL types defined in terms of the type **bit** from the package **standard** in the library **std**.	**std_logic_arith**
numeric_std: It has types defined in terms of the subtype **std_logic** from the package **std_logic_1164**, in the library **ieee**.	**std_logic_signed**
	std_logic_unsigned

Table 3.5: Packages for arithmetic operations in VHDL

Before stating the major differences between these two packages, the following points are worth noting:

- std_logic_signed and std_logic_unsigned are relatively obsolete packages, and so should be avoided in the design. It is better to use the packages numeric_std, numeric_bit and std_logic_arith.
- Both std_logic_arith and numeric_std represent numeric values or integers as unconstrained arrays of the subtype signed and unsigned, which are subtypes of the type std_logic.
- The type signed represents binary equivalent of a 2's complement number, while unsigned represents a positive binary number in sign magnitude form. For example, the package numeric_std is defined as

  library ieee;
  use ieee,std_logic_1164.**all**;
  package numeric_std **is**
  type unsigned **is** array (natural range <>) **of** std_logic;
  type signed **is** array (natural range <>) **of** std_logic;
- Since integers are treated as arrays, computations are more time consuming when bitwise operations are performed.
- To use these packages effectively, create signals and variable of the type signed and unsigned.
- Both of these packages use the concept of operator overloading which allows mixing vectors and integer types in expressions without an error.

The major differences between the packages numeric_std and std_logic_arith are listed in Table 3.6.

NUMERIC_STD	STD_LOGIC_ARITH
Type conversion: keyword **to** is used. Syntax is **to_type(signal,size)** where type conversions follow Table 3.7.	**Type conversion:** keyword **conv** is used. Syntax is **conv_type(signal,size)** where type conversions follow Table 3.8.
No type conversion allowed between **std_logic_vector** and **signed** and **unsigned** types.	No **conv** clause needed for conversion between **std_logic_vector** and **signed** and **unsigned** types, as both are subtypes of **std_logic_type**. Thus, for example, **a< std_logic_vector(b)** converts unsigned signal **b** to **std_logic_vector** signal **a** automatically.
Shift operations: These may not follow the normal conventions for shift operations. Four types of shifts are used with the keyword representing the shift / rotate direction • shift_left (equivalent to **slr** in **std_logic_1164 package**) • shift_right (equivalent to **srl**) • rotate_left (equivalent to **rol**) • rotate_right (equivalent to **ror**) The syntax is: **keyword(signal,distance);** where distance is an integer type. For example, **y<=shift_left(x,4)** shifts signal x 4 bits to the left. Note: For synthesis, shift distance must be a **constant** type.	**Shift operations:** These operations function as defined. However, there are no rotate operations defined, but only shift operations with the following keywords: • shl (shift left) • shr (shift right) The syntax is **Keyword(signal, distance) ;** where distance is an unsigned type For example, **y<= shl(x, "100");** shifts signal x 4 bits ($[100]_{10}=4$) to the left. Note: For synthesis, shift distance must be a **constant** type.

Boolean operators: These are supported in the package. For example, the following code shows 'or' operation on unsigned types.
 	
Architecture a1 of and_gt is
signal x,y,z: unsigned (3 downto 0);
begin
z<= x or y;
end;
	Boolean operators: These are not supported in the package. Therefore, type conversions are needed when such need arises. Thus, the example to the left can be implemented as:
	
	Architecture a1 of and_gt is
	signal x,y,z: unsigned (3 downto 0);
	signal x1,y1,z1:
	std_logic_vector (3 downto 0);

	begin
	x1<= std_logic_vector (x);
	y1<= std_logic_vector (y);
	z1<= x1 or y1;
	z<=unsigned(z1);
	end;
	
	That is, **x**, **y** are first converted to **std_logic_vector** types, Boolean operator is applied on them, and the outcome is converted back to unsigned type. As pointed out earlier, between std_logic_vector and signed/unsigned types, no explicit conversion is needed. Also, the above could be done in a short way as
	z<= unsigned ((std_logic_vector(x)) or (std_logic_vector(y)));
	This reduces the signal count in the design.

Arithmetic operators: The operators that are allowed are **sign-, abs, +, -, *, mod, rem**	**Arithmetic operators:** The operators that are allowed are **sign-, abs, +, -, ***
Resize: The purpose of resizing is to either increase or decrease a signal of the type signed or unsigned. The package provides for the function **resize** whose syntax is **Target <= resize(source, size)** This function resizes the source signal to 'size' number of bits. **Size** must be of the type **natural** and **constant** to be synthesizable. For example, if x is **signed** type of length 8 bits, it can be resized to 16 bits by **y <= resize (x, 16);** where sign extension is used for resizing. Thus, for signed signals, **1** would be filled up in the extra Most Significant Bit position and **0** for unsigned signals. For resizing to smaller size, the unsigned signals are truncated by discarding the extra MSB bits, which is a normal behavior. For example, if x=" 0110", if its resized as **y <= resize (x, 2);**	**Resize:** Here, resizing is done using two functions **conv_signed** and **conv_unsigned** for signed and unsigned conversion, respectively. The syntax is **Target <=conv_signed(source, size)** **Target <=conv_unsigned(source, size)** These functions resize the source signal to **size** number of bits. **Size** must be of the type **natural** and **constant** to be synthesizable. For example, if **x** is a **signed** of the length 8 bits, it can be resized to 16 bits by **y <=conv_signed (x, 16);** where sign extension is used for resizing. Thus, for signed signals, **1** would be filled up in the extra MSB positions and **0** for unsigned signals. However, for resizing down, both conv_signed and conv_unsigned discard the extra Most Significant Bits. For example, if x="100001"

the result is y="10". However, signed resizing to a smaller size does not follow the expected discarding of Most Significant Bits but keeps the MSB and discards the other bits. For example, if x="100001". **y<= resize (x, 3);** gives y= " 101". Also, while resizing up, a better way is **Target <= resize(source, target' length)** to prevent any length mismatch.	**y<= conv_signed (x, 3);** gives y= "001".

Table 3.6: Differences between numeric_std and std_logic_arith packages

TO_TYPE	SOURCE_TYPE	TARGET TYPE
to_integer	signed	integer
to_integer	unsigned	integer
to_signed	integer	signed
to_unsigned	integer	unsigned

Table 3.7: Type conversion between various types in the numeric_std package

CONV_TYPE	SOURCE_TYPE	TARGET TYPE
conv _integer	signed	integer
conv _integer	unsigned	integer
conv _ integer	std_ulogic	integer
conv _ integer	integer	integer
conv _unsigned	signed	unsigned
conv _ unsigned	unsigned	unsigned
conv _ unsigned	std_ulogic	unsigned
conv _unsigned	unsigned	signed
conv _signed	signed	signed
conv _ signed	unsigned	unsigned
conv _ signed	std_ulogic	signed
conv _signed	signed	signed
conv_std_logic_vector	unsigned	std_logic_vector
conv_std_logic_vector	std_ulogic	std_logic_vector
conv_std_logic_vector	unsigned	std_logic_vector
conv_std_logic_vector		std_logic_vector

Table 3.8: Type conversion between various types in the std_logic_arith package

Table 3.9 provides the data type conversions for all the arithmetic operators (z<= x (operator) y or z:= x(operator) y) for the package std_logic_arith.

X TYPE	Y TYPE	Z TYPE
unsigned	unsigned	unsigned
signed	signed	signed
unsigned	signed	signed
signed	unsigned	signed

Table 3.9: Type conversions in the package std_logic_arith

For the addition operator '+' and the subtraction operator '-', various operator overloadings are provided which are listed in Table 3.10.

X TYPE	Y TYPE	Z TYPE
unsigned	unsigned	unsigned
signed	signed	signed
unsigned	signed	signed
signed	unsigned	signed
unsigned	integer	unsigned
integer	unsigned	unsigned
signed	integer	signed
integer	signed	signed
unsigned	std_ulogic	unsigned
std_ulogic	unsigned	unsigned
signed	std_ulogic	signed
std_ulogic	signed	signed

Table 3.10: Operator overloading

Similarly, Table 3.11 provides all the arithmetic operators (z<= x (operator) y or z: = x (operator) y) for the package numeric_std.

X TYPE	Y TYPE	Z TYPE
unsigned	unsigned	unsigned
signed	signed	signed
unsigned	natural	unsigned
natural	unsigned	unsigned
signed	integer	signed
signed	integer	signed

Table 3.11: Type conversions for the package numeric_std

From Table 3.11, note that the unsigned type is overloaded with the type natural not integer. For a detailed description of the operators, and the overloading used, refer to the respective package documentation. As a final note, either use numeric_std package or std_logic_arith in the design, but never use both.

Lab 3:

VHDL Example Codes

L3.1 Example Codes

For further understanding of the above definitions and concepts, it helps to go though some examples.

L3.1.1 8-bit Adder

Let us first implement an 8-bit adder using the package std_logic_arith. The code is given in Figure L3.1. Generic clause is used to specify the number of bits for the addition of the two numbers a and b. Both the numbers are of the type signed. When adding two n-bit signed numbers, the result should be of the length n+1 to prevent overflow. So, both a and b are resized to n+1 using conv_signed and then are added together.

```
library ieee;
use ieee.std_logic_1164.all;
use ieee.std_logic_arith.all;
entity adder_nbit is
generic (n:natural:=8);

port (a,b:in signed(n-1 downto 0);
   sum:out signed (n downto 0));
end adder_nbit;

architecture a1 of adder_nbit is
signal d:signed (n downto 0);
begin
    d<= conv_signed (a,n+1)+conv_signed(b,n+1);
    sum<=d;
end a1;
```

Figure L3.1: VHDL code for an 8-bit adder

L3.1.2 Adder for Numeric Addition

Let us now create an adder that adds two integer arrays. This would be useful in many cases including convolution. There are two methods for creating a type. A type can be declared either within a package, where the package itself can be declared in the same program, or in a separate VHDL file. In both cases, the package name appears in the library work, so any program can use it by the clause use. work.[package name].all. Also, if a package has just the type definition and no functions, and is declared separately in a VHDL file, it will not need an entity and an architecture.

```
--*************using package from separate file******************
--File name: pack_int_array.vhd (note: name can be different from
 --package name)
library ieee;
use ieee.std_logic_1164.all;
package int_array is
    type integer_vector is array (natural range <>) of  integer;
end int_array;
 -- ********************************************************************
library ieee;
use ieee.std_logic_1164.all;
use ieee.std_logic_arith.all;

use work.int_array.all;
entity add_intarr_wo_carry is

  generic (n:natural :=4);
  port (a,b:in integer_vector (n-1 downto 0);
       sum: out integer_vector (n-1 downto 0));
end add_intarr_wo_carry;

architecture a1 of add_intarr_wo_carry is
    begin
    process(a,b)
      begin
       for i in n-1 downto 0 loop
         sum(i)<= a(i)+b(i); -- must add arrays,  --element by element add
       end loop;
    end process;
  end a1;
```

Figure L3.2: VHDL code for an integer array adder

Refer to Figure L3.2. First, create a VHDL file pack_int_array.vhd, where the package definition for the package int_array is provided, and then define a new type integer_vector as an unconstrained array of the element type integer. Next, create a VHDL file add_int_arr_wo_carry. Constrain the range of integer_vector to 4 by using the generic clause. Here, two port signals a and b of the type integer_vector are created, then added element by element and the result is assigned to a 4 element integer array sum. Note that the sum array needs to be 4 elements wide (as these are integers not bits). The simulation result is shown in Figure L3.3.

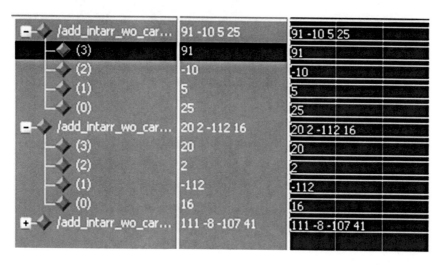

Figure L3.3: Simulation result for an integer array sum

It can be verified that [91, -10, 5, 25] + [20, 2, -112, 16] = [111, -8, -107, 41].

L3.1.3 Full Adder (1-bit)

Consider the VHDL code shown in Figure L3.4 that implements a one bit full adder.

```
library ieee;
use ieee.std_logic_1164.all;
entity fulladder_1bit is
  port (a,b,cin:in bit;
       cout,sum:out bit);
end fulladder_1bit;
architecture dataf of fulladder_1bit is
     signal xr:bit;
     begin
       xr <= a xor b;
       sum <= xr xor cin;
       cout <= (xr and cin)or (a and b);
end dataf;
```

Figure L3.4: VHDL code for 1-bit full adder

The Boolean equations that implement a full adder are used in this code through signal assignments. The synthesis includes xor, or, and gates. Figure L3.5 shows the simulation result.

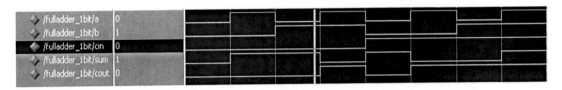

Figure L3.5: Simulation for 1-bit full adder

L3.1.4 N-Bit Full Adder

Next, a generic full adder is designed using the above designed 1-bit adder in two different ways. These are:

- Perform a bit by bit operations as per the truth table for 1-bit adder a specific number of times by using a generate statement, and then by using n as generic in the entity.
- Use a generate statement to copy the 1-bit adder component n times, where n is used as generic in the entity.

These schemes appear in Figure L3.6 and L3.7, respectively.

```
library ieee;
use ieee.std_logic_1164.all;

entity nbitadder_wo_comp is
  generic (n:natural :=4); -- n can be positive only
  port (a, b : in std_logic_vector (n-1 downto 0);
       cin: in std_logic;
       sum:out std_logic_vector(n-1 downto 0);
       cout: out std_logic);
end nbitadder_wo_comp;

architecture a1 of nbitadder_wo_comp  is
  signal c: std_logic_vector(n downto 0); -- create vector of length
                                          -- ci+cout+c1 to cn-1

  begin
    c(0)<= cin;
    gen:for i in 0 to n-1 generate

        sum(i) <= (a(i) xor b(i))xor c(i);
        c(i+1) <= ((a(i) xor b(i))and c(i) )or (a(i)and b(i));

    end generate;
    cout <= c(n);-- c(i)is also c(4) but 'i' cant be used outside loop
end a1;
```

Figure L3.6: N-bit adder using generic and generate, without component

```
library ieee;
use ieee.std_logic_1164.all;

entity nbitfulladd_usingcomp is
    generic (n:natural:=8); --8 bit adder
    port (a,b:in bit_vector(n-1 downto 0);  -- 4 bit input vectors
          s: out bit_vector(n-1 downto 0);  -- 4 bit sum
          ci: in bit;                        -- carry in
          co:out bit);                       -- carry  out
end nbitfulladd_usingcomp ;

architecture behav of nbitfulladd_usingcomp is

-- component declaration
 component fulladder_1bit
      port (a,b,cin: in bit; -- same name declared as in entity
                         --fulladder_1bit
           sum,cout: out bit);
 end component;

signal carry:bit_vector (n downto 0);

begin
   carry(0) <= ci;
   gen:for i in 0 to n-1 generate

   addern:fulladder_1bitport map(a(i),b(i),carry(i),s(i),carry(i+1));

      --a>= a(i),b=>b(i),cin=>carry(i),sum=>s(i),cout=>carry(i+1)
   end generate;
   co <= carry(n);
end behav;
```

Figure L3.7: N-bit adder using generic and generate, with component

L3.1.5 N (4)-Bit Counter

Let us now design a 16-bit counter that has an asynchronous reset. The code is given in Figure L3.8. Here, the count is captured in the port signal q as a 4-bit unsigned array. Thus, q would take values from 0 to 15, or 16 values, which is the Mod number of the counter. Thus, at each clock rising edge, the counter increments by one till it reaches 15, and then resets to 0. Also, the signal arst is used as an asynchronus reset signal. Consequently, in the simulation diagrams, it is seen that the counter counts 16 values (see Figure L3.10), but when arst is asserted in the middle of the clock, the counter resets to 0, irrespective of the clock edge (see Figure L3.9).

```
library ieee;
use ieee.numeric_std.all;
use ieee.std_logic_1164.all;

entity nbitcntr_acl_behav is
generic (n: natural :=4);
port (clk,arst:in bit;
      q: out unsigned(n-1 downto 0));
end nbitcntr_acl_behav ;

architecture behav of nbitcntr_acl_behav is
signal count:unsigned(n-1 downto 0);
begin
  process(clk,arst)
  begin
        if arst='1'then
        count<= (others => '1');
        elsif (clk'event and clk='1')then
        count<= count+1;
        end if;
  end process;
  q<=count;
end behav;
```

Figure L3.8: VHDL code for 4-bit counter

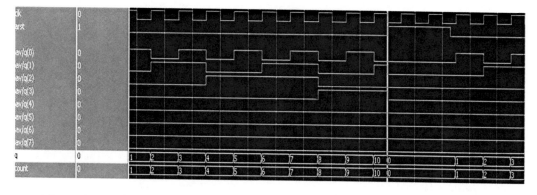

Figure L3.9: Simulation for 16-bit counter, with reset asserted asynchronously

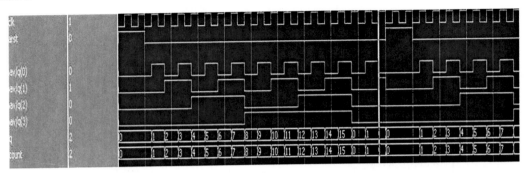

Figure L3.10: Simulation for 16-bit counter, with reset deasserted

L3.2 Bibliography

[1] Z. Navabi, *VHDL –Analysis and Modeling of Digital Systems*, McGraw-Hill, 1992.

[2] P. Ashenden, *The Designers Guide to VHDL*, Morgan Kaufmann, 1999.

[3] A.Rushton, *VHDL for Logic Synthesis*, Wiley, 1998.

[4] Language Reference Model, IEEE Std 1076, 1987.

[5] C. Cummings, "SynchronousResets?Asynchronous Resets?", http://www.sunburst-design.com/papers

L3.3 Lab Experiments

- As in example L3.1.1, write a VHDL code for multiplying two *n* (4) bit numbers using the package std_logic_arith.
- Write a VHDL code for multiplying two intger numbers (8 digits) using the package that declares integer arrays in a separate file.
- Similar to the example in the text for a shift register that shifts in one direction, write a VHDL code for a universal shift register with an asynchronus reset. Use 'dir' as an input line to control shifting the register left or right.

Appendix L3:
An Introduction to ModelSim

ModelSim is widely used for simulation of VHDL or VeriLog codes. For our simulations here, we have used a student version of ModelSim called ModelSim PE student edition 6.5d. This software is provided by Mentor Graphics. To install this software, follow the steps below:

- Go to www.Model.com.
- Select ModelSim PE Student Edition.
- Select Downloads.
- Fill the license agreement and a personal information form.
- Download the software to a desired folder.
- Run the executable file. The software gets installed.

Although a detailed discussion is beyond the scope of this book, a quick tutorial is provided here.

1. Go to **Start>> All programs>>ModelSIM PE>> ModelSim** as shown in Figure L3.A.1.

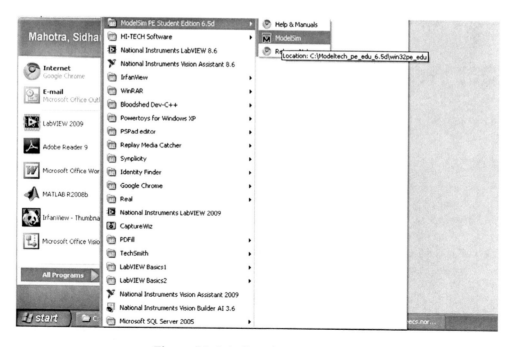

Figure L3.A.1: Opening ModelSIM

2. As shown in Figure L3.A.2, the ModelSIM main window opens up.

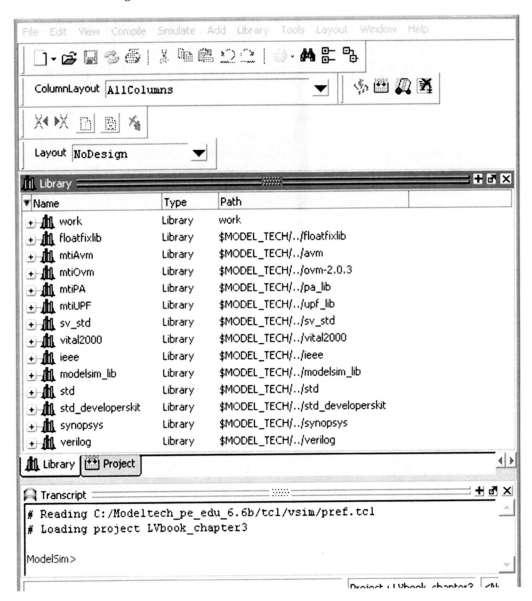

Figure L3.A.2: Main window of ModelSim

3. Select **File>>new>>Project** and name the project as **ADSP_VHDL**. The default library is **work** (Figure L3.A.3).

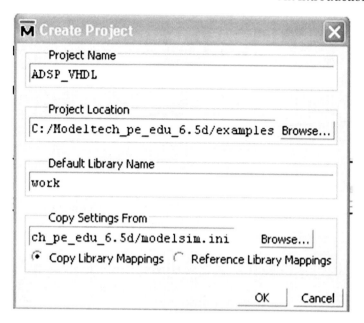

Figure L3.A.3: Naming a project in ModelSIM

Choose the project location. For example, as shown in Figure L3.A.4, it is set to **C:\NIFPGA2009\Ipnode_projects\vhdl files**.

Figure L3.A.4: Changing the project location

4. The dialog box **Add items to Project** opens up (Figure L3.A.5).

Figure L3.A.5: Add items dialog box

At this point, one can see two options:

- Creating new VHDL files from scratch and then adding them to the project, or
- Adding already existing VHDL files to the project.

Suppose a new VHDL file is to be created.

5. Select **Create New File.** The dialog box **Create project file** opens up (Figure L3.A.6).

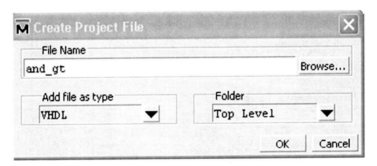

Figure L3.A.6: Create project file dialog box

Name the file **and_gt** and choose the file type **VHDL,** then press **ok.**

6. On the workspace section of the Main Window (Figure L3.A.7), double-click on the file just created. Type the code in the window. Save the file.

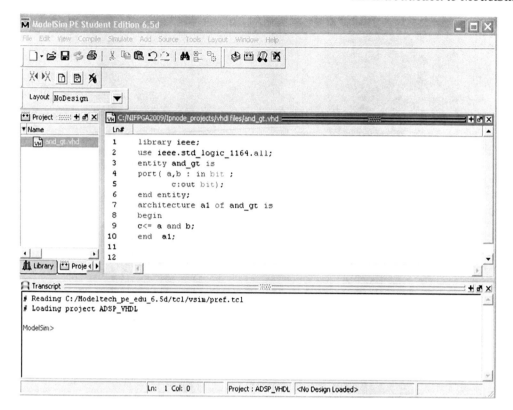

Figure L3.A.7: Typing the VHDL code

7. Compile the file by choosing **compile>> compile selected** or press the compile button indicated in Figure 3.A.8.

Figure L3.A.8: Toolbar showing Compile button

8. Examine the compilation result (Figure L3.A.9) for the successful compilation of the file file_name.vhd.

9. Next, to simulate the design, click on the Library tab of the main window and then click on the + sign next to the work library. One should be able to see the name of the entity of the code just compiled "and_gt". Right click on **entity name>> simulate** (Figure L3.A.10).

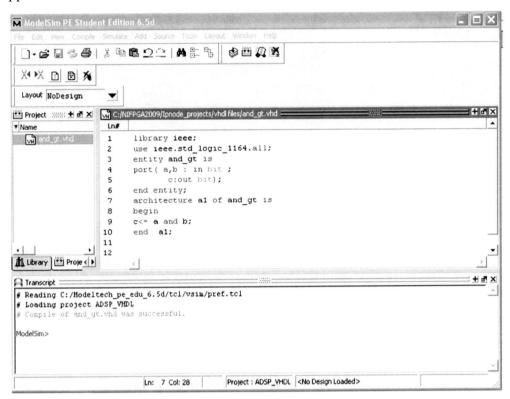

Figure L3.A.9: Successful code compilation

10. Within the **sim tab,** right click on the entity name and select **Add>>To wave>> All items in the region** (see Figure L3.A.12). This adds all the signals/variables (if used) to the wave file for simulation purposes.

11. The simulation wave window opens up, as shown in Figure L3.A.13, the signals **a**, **b** and **c,** and the total simulation time appear as 100 ns. Change the simulation time to 400 ns. Right click on the signal **a>>force selected signal** (Figure L3.A.14). Here, one can assign simulation values to the signal **a**. Assign value **0** for **200 ns**, assign value **1** for next **200 ns** (Figure L3.A.15). Similarly, assign the values 0, 1, 01, to the signal b for 0-100, 100-200, 200-300, 300-400 ns. Alternatively, since this bit stream looks like a clock signal, one can right click on the signal b, choose the clock and set the clock parameters as shown in Figure L3.A.16.

Run the simulation. Zoom for better display of the results.

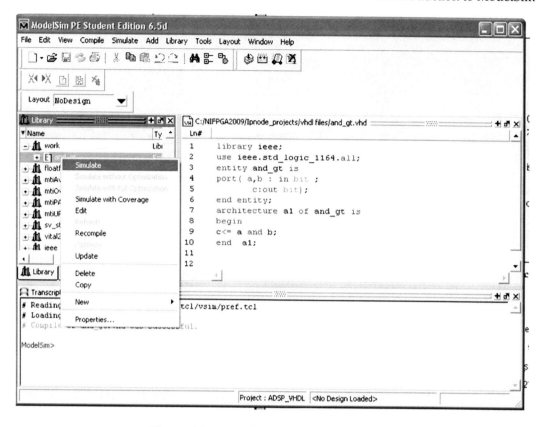

Figure L3.A.10: Choosing simulate design

This would open a third tab "sim" in the main window as shown in Figure L3.A.11.

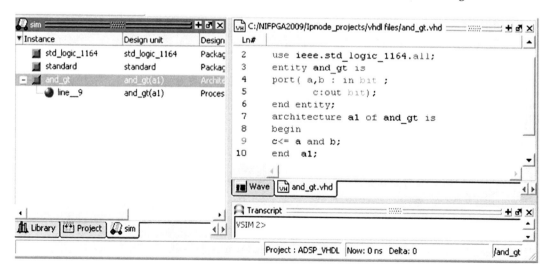

Figure L3.A.11: Simulate tab

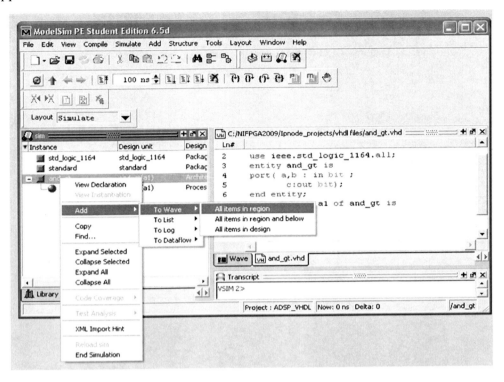

Figure L3.A.12: Adding signals/variables for simulation

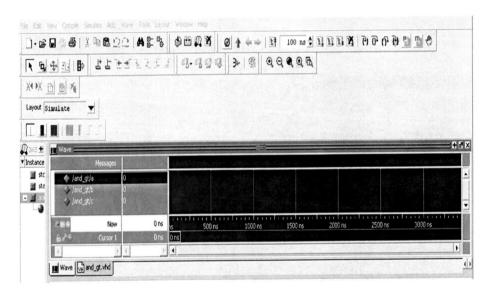

Figure L3.A.13: Simulation window

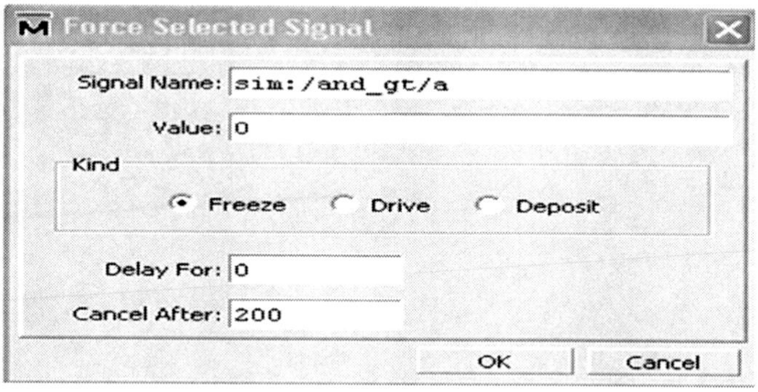

Figure L3.A.14: Changing signal values for signal 'a' in first duration

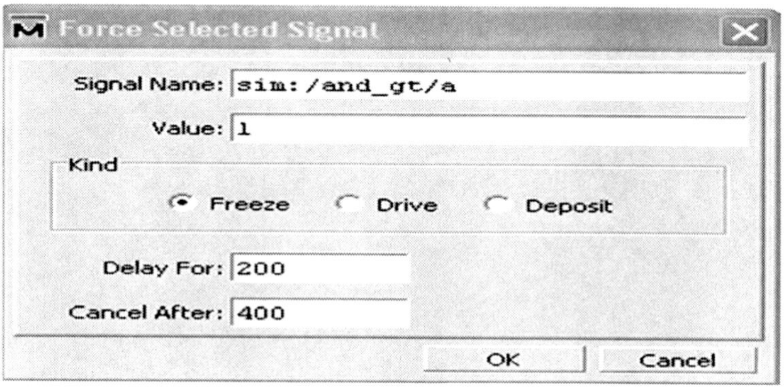

Figure L3.A.15: Changing signal values for signal 'a'

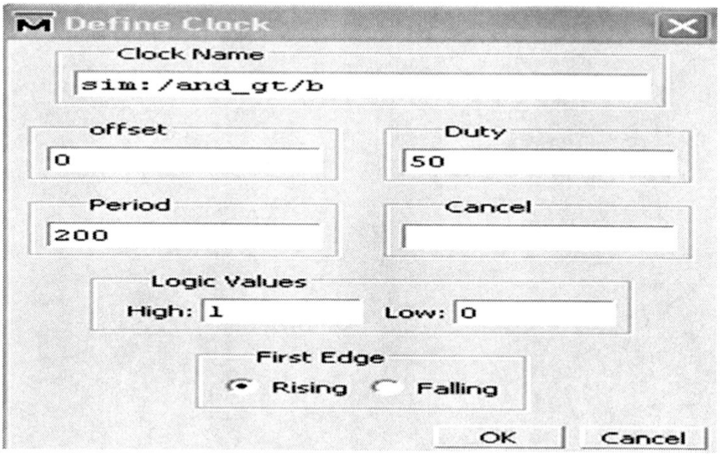

Figure L3.A.16: Changing signal values for signal 'b'

CHAPTER 4

Hybrid Programming: LabVIEW FPGA and IP Integration Node

This chapter discusses how to use the LabVIEW FPGA Module to create VIs that would run on FPGA targets. This module deploys the same graphical approach as used for designing standard VIs. In addition, this chapter covers LabVIEW IP (Intellectual Property) Integration Node which is a tool for running third party IP VHDL codes within the LabVIEW environment. By using this node, one can make use of extensive HDL codes that have already been written for digital signal processing. The combination of textual HDL through IP Integration Node and graphical FPGA Module allows a hybrid approach towards FPGA implementation of signal processing algorithms leading to design cycle efficiency.

4.1 Types of VIs in LabVIEW FPGA

Two types of VIs can be created using LabVIEW FPGA: FPGA VIs that run on FPGA targets and Host VIs that run on the host PC. These VIs can be made to interact with each other using various modes that are covered in this chapter. However, it should be noted that an FPGA VI differs from a standard VI in a number of ways that are listed below:

- Even if a simulation run of an FPGA VI is done, care must be taken to use only the subVIs and functions which the target supports as part of the FPGA VI.
- **Array functions**: Arrays on an FPGA VI must be of fixed size, and one-dimensional only. In other words, in an FPGA VI, one should not use array functions that accept or return variable-size arrays.
- **High throughput math functions**: These functions can be used to perform similar operations as numeric functions in a standard VI, but when used in an FPGA VI, they allow high throughput data rates and automatic pipelining.
- **Single Cycle Timed Loop (SCTL):** This loop is similar to a while loop in a standard VI with the difference that the code within this loop is highly optimized and utilizes less space on an FPGA target as compared to a conventional while loop. Therefore, a code when placed inside a SCTL would run within one clock cycle provided that the clock cycle is long enough. For example, a 16-bit by 16-bit multiply and accumulate when executed inside a SCTL would only take one clock cycle to run. The way SCTL achieves this speed and memory saving is by removing registers from the code inside the loop making the logic combinatorial.

4.2 FPGA I/O

The interfacing between an FPGA board and the outside world is done through inputs and outputs (I/Os) specified on the FPGA target of interest. FPGA I/Os, available as functions in the Block Diagram of an FPGA VI, determine the I/O resources on the FPGA target. The following points are to be noted about FPGA I/O resources:

- These resources correspond to digital or analog I/O.
- An FPGA target can have multiple I/O resources of the same or different types (both analog and digital lines).
- An I/O resource can have one or more terminals for receiving or generating a physical quantity.
- I/O resources can be renamed to correspond to physical quantities they represent.
- When an FPGA VI runs on an FPGA target, I/O operations are performed on actual hardware, and so the obvious advantage gained is speed and fixed/deterministic behavior.

Let us go through some of the I/O types in LabVIEW FPGA.

4.2.1 Digital I/O

Digital I/O lines on an FPGA target, organized as individual lines or ports, constitute Digital I/O resources in the LabVIEW FPGA. These resources are of three types:

- **Digital Input resources**: These can be used to read or monitor the logic outside an FPGA target (or simulated target), and are read by using the **FPGA I/O Node**.
- **Digital Output resources**: These can be used to control the logic outside an FPGA target (or simulated target), and are written into by using the **FPGA I/O Node**.

Figure 4.1 helps to see the above two definitions.

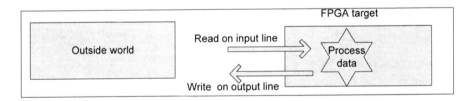

Figure 4.1: Simplified diagram of FPGA IO

- **Digital Input and Output resources:** These are tristate or bidirectional resources that are available for both read and write operations by using the **FPGA I/O Node.**

4.2.2 Analog I/O

These are of the following types:

- **Analog Input resources:** These can be read by using the **FPGA IO Node**, thus giving the binary representation of the physical quantity measured.
- **Analog Output resources:** These can be written into by using the **FPGA IO Node** in terms of analog voltage.

4.3 SubVIs in LabVIEW FPGA

SubVIs can be used in the LabVIEW FPGA configured as **reentrant** or **non-reentrant** mode. The difference between these are listed in Table 4.1

SUBVI TYPE	STRUCTURE	SPEED	SPACE /RESOURCE UTILIZATION
Reentrant	Each instance of subVI is mapped to a separate hardware, and so each caller VI calls/uses separate hardware resource	Faster as all instances of subVI run in parallel	Higher due to dedicated hardware mapping of each instance. However, if the same subVI is being called by many VIs, each subVI runs independently so that no arbitration is required for handling them. It does not thus use FPGA resources for arbitration.
Non-reentrant	All callers share a single copy of the subVI, mapped to a single hardware resource	Slower, as each call to the sub-VI must wait until the previous call ends	Theoretically, it is less as only one instance of the subVI is mapped to the hardware resource. However, in multiple calls to the subVI, there is a need for arbitration, which utilizes FPGA resources so that under certain circumstances, the arbitration resources might be comparable to the actual resource utilization for the subVI, nullifying the space savings.

Table 4.1: Differences between reentrant and non-reentrant subVIs

Figure 4.2 shows a simplified diagram that illustrates the difference between the two.

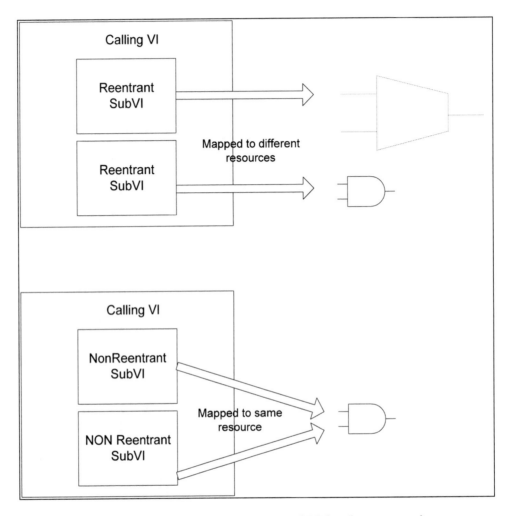

Figure 4.2: Reentrant and non-reentrant subVI hardware mapping

By default, VIs are not reentrant. To make a VI reentrant, select **File>>VI Properties>>Execution**, and place a checkmark in the **Reentrant execution** checkbox as shown in Figure 4.3. The diagram also displays a **clone** entry. A clone is a copy of the reentrant subVI, which opens when one clicks on the subVI from the Block Diagram of the caller VI. Since it is not the actual subVI, but it is a copy, a clone cannot be edited from within the main VI or caller VI. The title bar of the VI contains the name to indicate that it is a clone of the source VI. To view the original Front Panel of a reentrant VI (for editing) from a clone of the reentrant VI, select **View>>Browse Relationships>>Reentrant Original.**

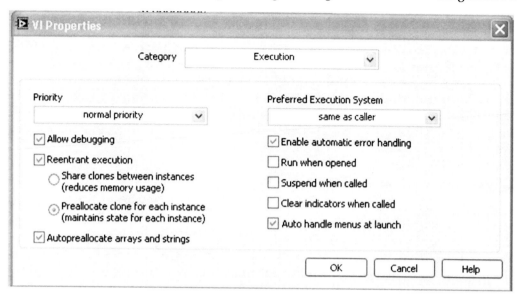

Figure 4.3: VI properties for reentrant subVI

4.4 Different Ways of Creating Test Applications in LabVIEW FPGA

There are many ways in which one could create VIs for test purposes. These are shown in Figure 4.4 and explained next.

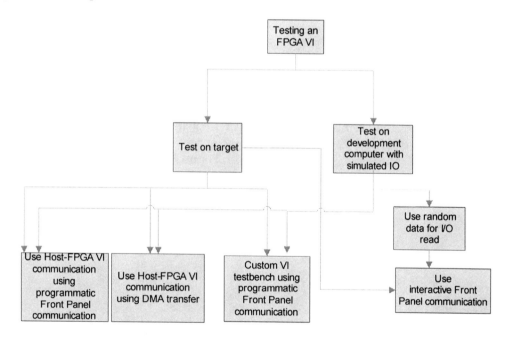

Figure 4.4: Different methods for testing an FPGA VI

4.4.1 Testing on FPGA Target via Interactive Front Panel Communication

Here, an FPGA VI is created corresponding to an FPGA target. The Front Panel of the FPGA VI acts as the host and communicates with the Block Diagram of the FPGA VI which represents the FPGA target. Front Panel controls and indicators are memory mapped onto the actual FPGA target. This way, an interactive Front Panel communication takes place between the host computer displaying the Front Panel window as well as the FPGA target executing the FPGA Block Diagram. The host updates the state of controls and indicators to correspond to the actual states of the FPGA target, but with some delay.

4.4.2 Testing on PC with Simulated I/O

Here, the Block Diagram of the FPGA VI executes as a simulated FPGA target (for which the support is installed on the PC or development computer), with the Front Panel controls and indicators getting values as Random Data for the FPGA I/O read. Thus, through this option, random data (random not in value but structure) is provided as input to the simulated FPGA target using Front Panel controls, and then the results are read back as random output data using Front Panel indicators.

4.4.3 Testing on PC/FPGA Target with Host-FPGA VI Communication

A host VI may need to communicate with an FPGA VI for the following reasons:

- More data processing might be needed than what can fit on the FPGA, e.g., implementing a part of an algorithm on the host VI and the rest on the FPGA target/simulated target.
- There may be a need to perform operations not available on the FPGA target such as floating-point operations.
- Data logging may be required which can be achieved with FIFOs.
- There may be a need to control the timing and sequencing of data transfer.

For example, this method can be used to send or write fixed-point data, convert from floating-point format in the host VI to the FPGA VI memory, process it, and then read back the result from the FPGA memory. This is illustrated in Figure 4.5.

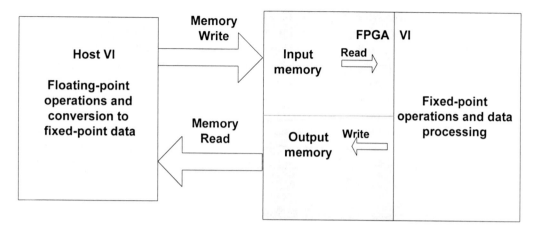

Figure 4.5: Method of transferring data between FPGA VI and host VI using
Block Memory

Programmatic Front Panel Communication: For better debugging and control over the algorithm running on an FPGA VI, a standalone host VI can be created (project>>my computer, similar to a standard VI) to run on the host computer, and get programmatically controlled and monitored. This mode is used when data transfer between the host and the FPGA VI needs to be done with minimal delay. This method is thus suited for transferring data frequently in small chunks, e.g., for control or simulation of the FPGA VI Front Panel controls and indicators. The following FPGA elements can be used for this type of communication:

- **Block Memory**: This memory can be used to store/retrieve data using the **Memory Method Node** in the location specified by the **Address** input of the node.
- **Look up Table (LUT)**: This is another predefined storage locations available for access on an FPGA.

LUT vs. Block Memory: Block memory does more resource saving than LUT as it does not consume logic resources on an FPGA, while LUT does. However, reading a block memory takes one cycle more than a LUT if both are configured in read mode (using Read method) inside a SCTL. Also, using LUT can avoid block memory wastage if the required memory for data storage is less than the minimum available on the FPGA. For example, if the minimum block memory available on an FPGA is 4Kbytes, and one needs to store 1Kbytes of data, 3Kbytes of memory will be wasted.

Using Memory Items: Both LUT and block memory can be used as either **Target scoped** or **VI-defined**. Target scoped memory stores data that can be accessed from multiple VIs. For example, if a VI has three copies of a subVI configured as non-reentrant, and the subVI uses Target scoped memory, then all the calls to the subVI access this memory. On the other hand, one can create VI-defined memory items that can be accessed from a single VI, or a reentrant subVI. In the latter case, as a VI-defined memory item is included in a reentrant subVI, separate instance of the memory item is maintained, and so different data are maintained in those locations.

DMA Transfer: DMA or Direct Memory Access transfer between host and FPGA VI ensures a burst transfer of large amount of data. This mode needs to be used for applications such as data logging, where one needs to stream large amount of data at a time. Similar to the case of Programmatic Front Panel Communication, DMA transfer can be done using FIFOs as block memory, LUT, or flip flops. While the block memory FIFO does not consume any logic resources, the other two do so. Again, Target–scoped FIFO or VI defined FIFO can be defined. Although, a DMA FIFO can be looked upon as one contiguous logical memory, in reality, there is a separate memory space allocated on both the host computer and the FPGA target. The FPGA VI writes into or reads from the FIFO one element at a time using the **FIFO Method Node**, while the host VI reads from or writes to the FIFO using the **Invoke Method** function.

Difference Between Block Memory FIFO and Block Memory/LUT: Although both the block memory FIFO and block memory LUT denote memory or storage locations, the only difference between them is how the data is accessed. For the block memory RAM, data can be accessed randomly, e.g., a data value may get stored at a location in the block memory by specifying the location with the address input of **Memory Method Node**. If the same location is accessed randomly for three more writes, before read, the previous three values will be overwritten and only the last value will be read. On the other hand, for FIFO or First In First Out, memory cannot be accessed randomly, but there is a register to point to the latest data value.

As a block of data is written into a FIFO using the FIFO Method Node, new elements are added to the end of the FIFO, which can be read out in the same order in which they arrived in the FIFO. Also, when a FIFO is full, new values cannot be written into the FIFO unless they are flushed out (if they have been read), which means that unlike RAM memory/LUT, new values do not replace old values.

It should be noted that all the above memory elements are available in the FPGA fabric, and so they can be used in other FPGA testing modes. In this case, the user must clear the FIFOs if they have been used in the FPGA VI by using **Clear Method** of the FIFO method, noting that FIFOs do not get cleared automatically. Similarly, for the host-FPGA VI communication, it is required to clear DMA FIFOs on the host side by using **Reset Method** in the Invoke Method function. The speed advantage of a FIFO comes from the fact that the host processor can do computation while the FPGA target (or simulated target) transfers data to the host computer memory directly without the host computer interfering.

4.4.4 Testing on PC/FPGA Target with Custom VI via Programmatic Front Panel Communication

Using a custom VI, one can provide stimulus to a test bench and verify results. Thus, custom IO VIs allow creating a repeatable scenario for testing, and changing data for a particular input. For creating a test bench, refer to the LabVIEW 2009 FPGA help.

4.5 FPGA Testing Based on Programming Approach

FPGA test methods can be classified based on the programming approach, as indicated in Figure 4.6.

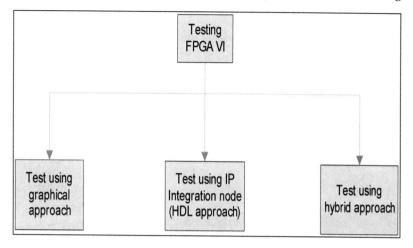

Figure 4.6: Test methods for FPGA VI based on type of programming

a. **Test using graphical approach**: In this method, FPGA/host VIs are created using graphical programming only
b. **Test using IP integration node**: The IP integration node allows HDL designers to rapidly use their code within the LabVIEW FPGA paradigm. This node when placed in a Block Diagram can be used to run and compile existing VHDL or Verilog code
c. **Hybrid approach**: More often than not, it is advantageous to use a combination of graphical and HDL language approach for implementing various designs on an FPGA target.

4.6 Optimization Methods for Saving Space on FPGA

Resources on an FPGA should be used judiciously so that more and more logic can fit onto a single chip. The following methods list the guidelines as how to do so:

a. **Avoid input coercion if not needed**: Numeric functions coerce all the input data to the largest data type, which may not always be desirable. For example, while multiplying an 8-bit and a 16-bit number via the multiply function, the 8-bit input is coerced to 16-bit before multiplication. If this is not required, one could use fixed-point data types for both inputs and use high throughput multiply.

b. **Index array function**: By default, the Index array function (Figure 4.7) uses a 32-bit integer as the default for the input **index**, assuming that the input array is of size 2^{32}-1. However, one could use an 8 or 16-bit representation for the index input if the input array has a size that justifies this representation.

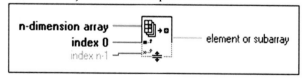

Figure 4.7: Index array function

c. **Rotate 1D array**: This function is shown in Figure 4.8.

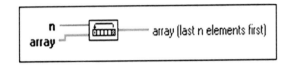

Figure 4.8: Rotate 1D array function

It rotates the input data, the number of places and the direction are decided by **n**. If the input is an array [a b c d] and n=1, the output array becomes [d a b c]. If a control is used for the input **n**, the execution time would be the rotation time plus two clock cycles of overhead to enter and leave the function. Thus, for this example, it would take 3 clock cycles to execute. However, if a constant is wired to the input, the function takes a negligible amount of time to execute and utilizes none of the FPGA resources, giving speed and space savings.

d. **While loop execution**: The iteration terminal of a while loop has a default data type of 32 bits, which is too large. There are two ways to work around this problem. One may use a representation that is less than 32 bits, but supports the maximum loop count needed. Alternatively, one may avoid the iteration terminal using a shift register instead as shown in Figure 4.9, where the while loop executes 10 times.

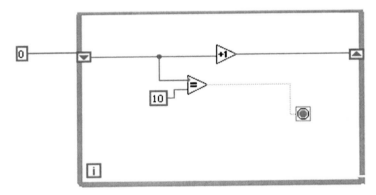

Figure 4.9: Creating while loop that runs for fixed number of times

e. **Fixed-point functions**: Use fixed-point logic wherever possible. Also, many of the high throughput fixed-point functions adapt to the source data type, or might have default data type, which can be of much higher precision than needed. Hence, reducing the fraction length, with a tolerable loss of precision, will lead to improvement in speed and space.

f. **Quotient and Remainder**: Quotient and Remainder function can be used in variety of ways.However, if this function is to be used to implement a divide by two operation, a better option would be to use **Scale by power of two**.Since constants can use less power than controls on an FPGA, use this function by wiring a constant to the **n** input of the function.

For other methods, refer to the LabVIEW 2009 help document.

Lab 4 – Part 1:
LabVIEW FPGA

L4.1 Fixed-Point Adder Example

Let us implement a fixed-point adder in an FPGA VI using simulated FPGA testing and random IO.

L4.1.1 Creating LabVIEW FPGA Project

Start LabVIEW 2009, and select **Targets>>FPGA Project>>GO** as shown in Figure L4.1. The window shown in Figure L4.2 displays options for creating various kinds of LabVIEW projects. It displays options for creating various kinds of LabVIEW projects. In the absence of a target, select **R Series Intelligent DAQ on My Computer >>next** and then Create **new system>>next,** see Figure L4.3. Next, choose an R series FPGA target from the list of the targets supported and select **PCI-7811R>>next** (see Figure L4.4). On the window indicated in Figure L4.5, uncheck the option **Expansion system for C series module** considering that no target board and expansion system is actually used. Select **next** and **finish**. The message **Creating FPGA I/O** appears as shown in Figure L4.6. The project explorer displays the target as **My computer>>FPGA target (PCI-7811R)**. Save the project as **memory_lut_pract.lvproj.**

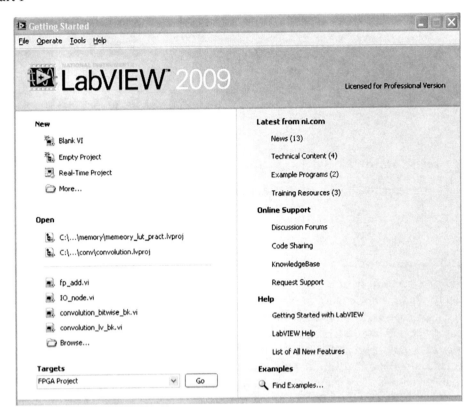

Figure L4.1: Creating a LabVIEW FPGA project

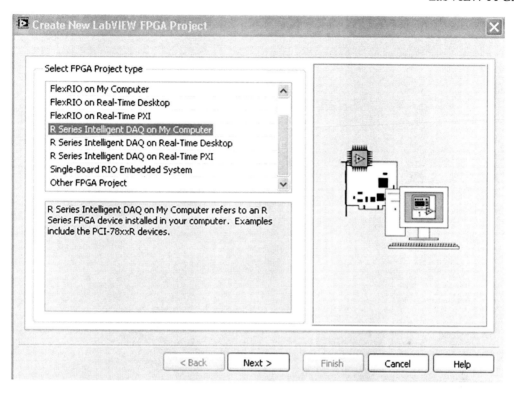

Figure L4.2: Choosing various FPGA project types

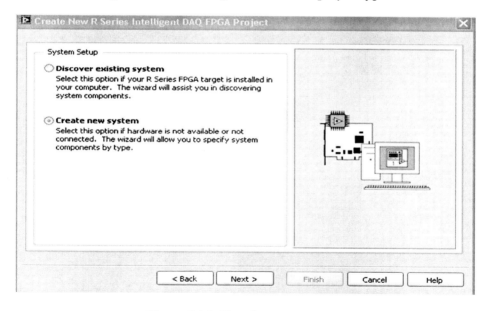

Figure L4.3: Creating new system

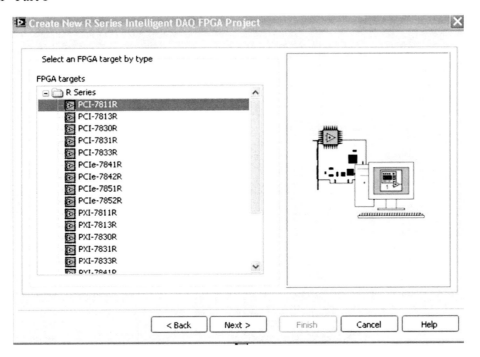

Figure L4.4: Choosing FPGA target type

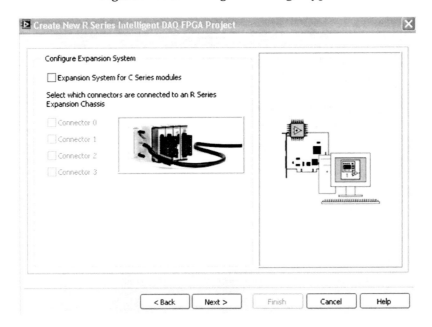

Figure L4.5: Last stage of FPGA project creation

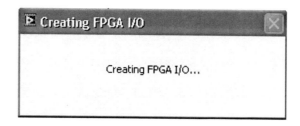

Figure L4.6: Creating FPGA IO

L4.1.2 Create FPGA VI

Right click on **FPGA Target** under the Project Explorer and select **new>>VI**. Save the VI as **fp_add.vi**. View these changes in the project explorer window as shown in Figure L4.7. Choose the type of IO. Right click on the **FPGA target** and select **Properties**. The window shown in Figure L4.8 opens up. Choose **Category>>Debugging>>Execute VI on development computer with Simulated IO** and from the drop down menu, select **Use Random Data for FPGA IO Read>>OK**. The menu displays other options from which the option **Use Custom VI for FPGA IO Read** needs to be selected for creating a custom VI.

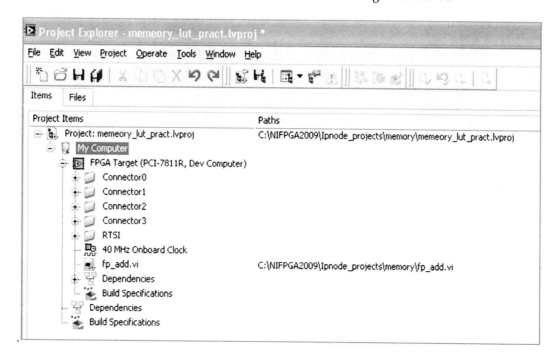

Figure L4.7: Project Explorer for FPGA project

L4.1.3 FPGA VI Block Diagram

Here, two FPGA numbers are read and stored in a LUT memory on the FPGA, then they are added and the result is stored back in the LUT memory. Select **Functions>>memory and**

FIFO>> VI-Defined Memory Configuration. This defines the kind of memory that will be used.

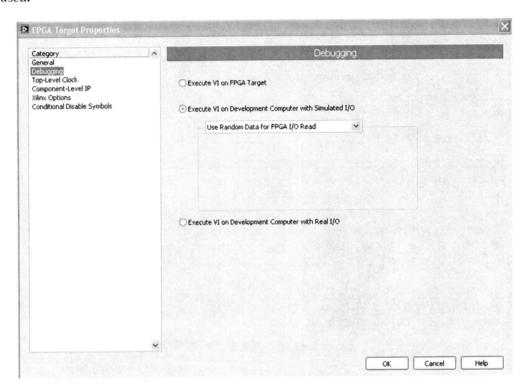

Figure L4.8: Debugging option in FPGA target properties page

The function looks like as displayed in Figure L4.9. Right click and select **Configure** and a window shown in Figure L4.10 appears.

VI-Defined Memory Configuration

Figure L4.9: VI defined memory configuration

Under the category **General,** the following parameters for configuring the memory are available:

- **Data type**: This represents the type of data that the memory will store. All LabVIEW data types are available in this option. For the present example, select **Fxp** for fixed-point data. Define **encoding** as **signed**, **Wordlength** as **8 bits** and **Integer Word Length** as **1 bits**. So, this way two Q8 numbers get stored.
- **Number of elements**: This represents the size of the memory. The default is **1024** memory locations. Change to **2** for storing two fixed-point numbers

- **Implementation:** This represents the type of memory, either **Look Up table** or **Block memory**. The default is **Block memory**. Select **Look Up table**.

With these changes, the diagram shown in Figure L4.11 is obtained

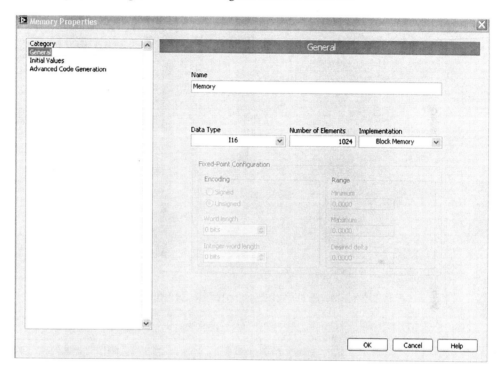

Figure L4.10: General properties page of the memory properties

Next, let us set the initial values for the memory locations that will be written using memory element. Select the category **Initial Values** from memory properties. As seen in Figure L4.12, the default value is 0. Any value can be filled if one wishes to change the default value. Also, note that **Start address** is 0 and **End address** is 1, corresponding to two memory locations of 0 and 1. If the default **Number of elements** is set to 1024, the start and end address will be 0 and 1023. Thus, in general, if a memory of size N is created, the start and end address would be 0 and N-1. These addresses can be changed to any value provided that they match the number of elements created in an earlier step.

Next, let us store two fixed-point values of 0.33 and -0.21 at the location 0 and 1. For this, it is required to place two copies of the function **Memory Method Node** in the Block Diagram. Select **Functions>>memory and FIFO>> Memory Method Node** and place it on the Block Diagram as shown in Figure L4.13.

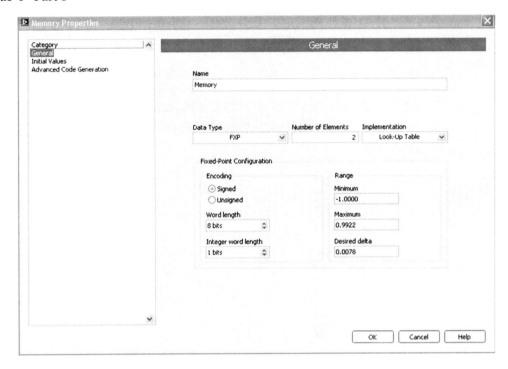

Figure L4.11: Final properties of the memory element

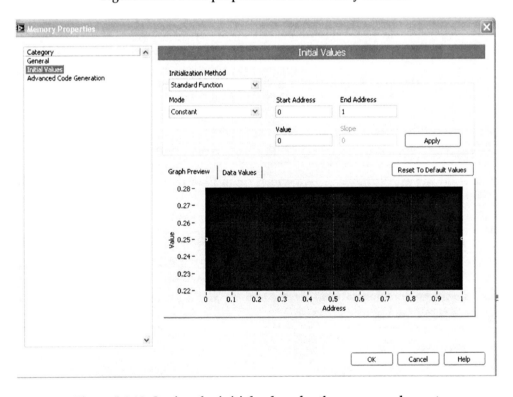

Figure L4.12: Setting the initial values for the memory element

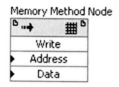

Figure L4.13: Memory Method Node

By default, the node is configured for write operation. It has three terminals:

- **Memory In**: It specifies the FPGA memory. Connect this to the **Memory OUT** terminal of **VI-Defined Memory Configuration.**
- **Data:** It specifies the data to be written to the memory on the FPGA target. Create a **numeric control**, right click on the control, select **Representation** and choose **Fxp** for fixed-point numbers. Connect it to the **Data** terminal. Enter the value of **0.25** and name it **Sample1.**
- **Address:** It specifies the location of the data in memory on the FPGA target. Wire a numeric control to it to specify the address. For the above case, wire address **0**, and name it **Address for sample 1.**

Similarly, create another node **Memory Method Node 2** for writing the second value of -0.125 with the address 1. Now, the structure for writing two fixed-point data values of 0.25 and -0.125 at locations 0 and 1 of the LUT memory is in place. Put this block in a **Sequence** structure and add two frames.

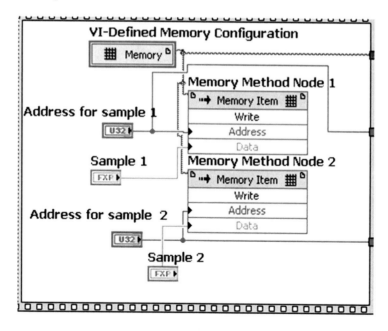

Figure L4.14: Frame 1

In the next frame, place two Memory **Method Node** functions. Right click on each of them and choose **Select Method >> Read** to configure them for read operations. Connect the **Memory In** terminal of each to the **Memory OUT** terminal of the **VI-Defined Memory Configuration** function created in the first frame in order to reference the same memory. Also, connect the address inputs to the address inputs of Memory **Method Node 1** and **Memory Method Node 2,** respectively (Figure L4.14). Now, the two fixed-point values can be read from locations 0 and 1.

Next, let us add these two fixed-point numbers. Place the add function in the Block Diagram by choosing **Functions>>FPGA Math and Analysis>>High throughput Math>>Add**. Right click and choose **configure** to match the **x type** and the **y type** to the data type of the fixed-point inputs, i.e. Q8 numbers. Wire the **x** and **y** terminals to the data terminals of Memory **Method Node 3** and Memory **Method Node 4,** respectively, as shown in Figure L4.15.

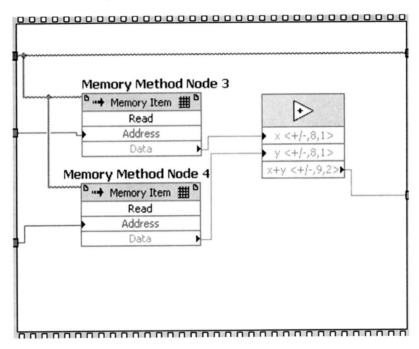

Figure L4.15: Frame 2

Finally, in the last frame, create **Memory 2** (VI-Defined Memory Configuration) in order to store the fixed-point addition result. As done for the first memory, choose two-element LUT memory to store the fixed-point numbers of the format <+-,9,2> because the sum of two Q8 numbers is a number of wordlength 9 and integer length 2. Create **Memory Method Node 5** in the **write** mode, and connect it to Memory 2, so that the addition result is stored at the address 0 of Memory 2. Finally, create a data, address and sum indicator, and place the entire Block Diagram in **Single Cycle Timed Loop (Functions>>Structures>>Timed Structures>>Timed Loop)**. This way, the final Block Diagram looks like the one shown in Figure L4.16.

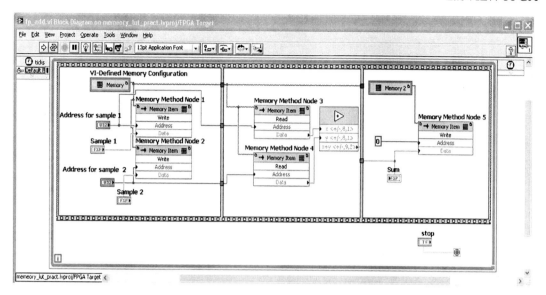

Figure L4.16: Block Diagram of fp_add VI

Run the VI. As seen in Figure L4.17, the two fixed-point numbers 0.25 and -0.125 should get added correctly to yield 0.125.

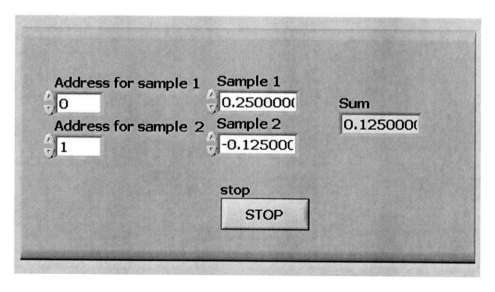

Figure L4.17: Front Panel of fp_add VI

Lab 4 – Part 2:

LabVIEW IP Integration Node

L4.2 IP Integration Node

L4.2.1 And Gate VHDL Code in IP Node

In this section, it is shown how the IP Integration Node can be used to insert existing VHDL codes into the LabVIEW environment. This is better understood by going through an example. This example involves the VHDL implementation of the 'And' operation.

- Create a LabVIEW Project **IP_nd_exmp.lvproj** as an FPGA project running on the PC, and an FPGA VI **andgate.vi** with the option **Execute VI on development computer with simulated IO>>use random data for FPGA I/O Read.**
- In the Block Diagram of the FPGA VI, add the IP node by selecting **Addons>>IP Integration Node** as shown in Figure L4.18.

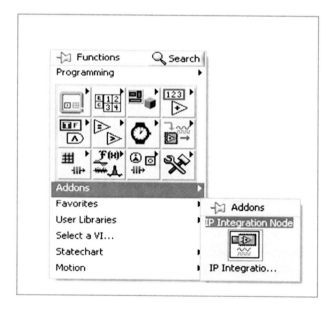

Figure L4.18: Selecting IP node

- Double click on the IP node. The first page that displays the properties of IP node should open up (see Figure L4.19). Here, specify the path to the VHDL file, and then compile it. Choose **IP source>>Add File** and browse to locate the file **ndgt_inside_process.vhd**. Then, compile the file by clicking **Generate**. If the compilation is successful, the message **generated simulation model successfully** will get displayed in the window below Generate. Go to the next window by clicking **Next**.

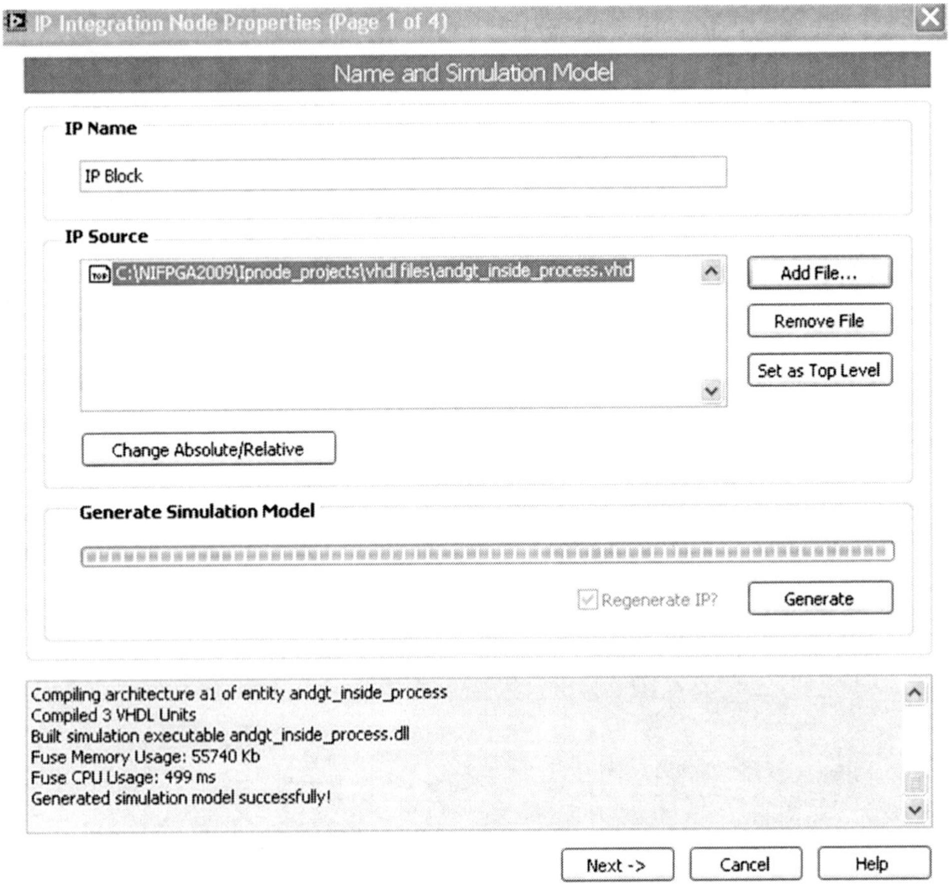

Figure L4.19: First page of IP Node Properties

- The next window displays the clock signals, and other input/output port signals (output signals are grayed). As can be seen in Figure L4.20, there is no clock signal, and three signals vis.a.vis **a, b** and **c** appear under **IP enable Signal(s)**. Click **Next**.

- On the third page, the synchronous and asynchronous signals are specified. Click on the drop down menu **Asynchronous reset signal** and select the reset signal from the signal list, assuming that such signal had been specified in the VHDL file. If no reset is to be specified, select **no asynchronous reset**. Since no reset is used, select this option and go to page 4 (see Figure L4.22).

- On the last fourth page, choose the type of data types compatible with what is specified in the VHDL code. For the VHDL type **Std_logic/Bit**, choose the option **Boolean** for the VHDL type **Std_logic_vector/Bit_ vector**, the option **Boolean array** for the VHDL type **sfixed/ufixed**, the option **FXP** for fixed-point. Then, click on the terminals and select **Terminal Properties>>Data Type**. Finally, choose the encoding. This is shown in Figure L4.22 where all the signals are of type **Boolean**.

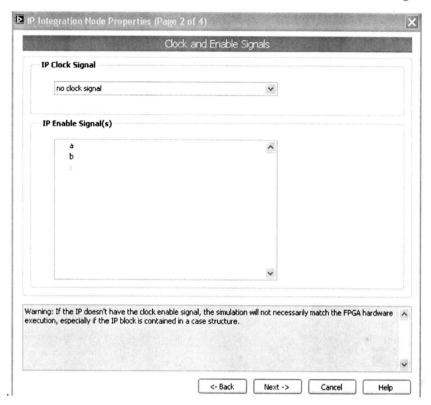

Figure L4.20: Second page of IP Node Properties

- Click **Finish**, and place the IP node within a Single Cycle Timed Loop. Then, create appropriate controls and indicators and run the VI. The Block Diagram is illustrated in Figure L4.23, and the Front Panel in Figure L4.24.

As seen from the Front Panel, the VI implements c= a and b with an asynchronous reset.

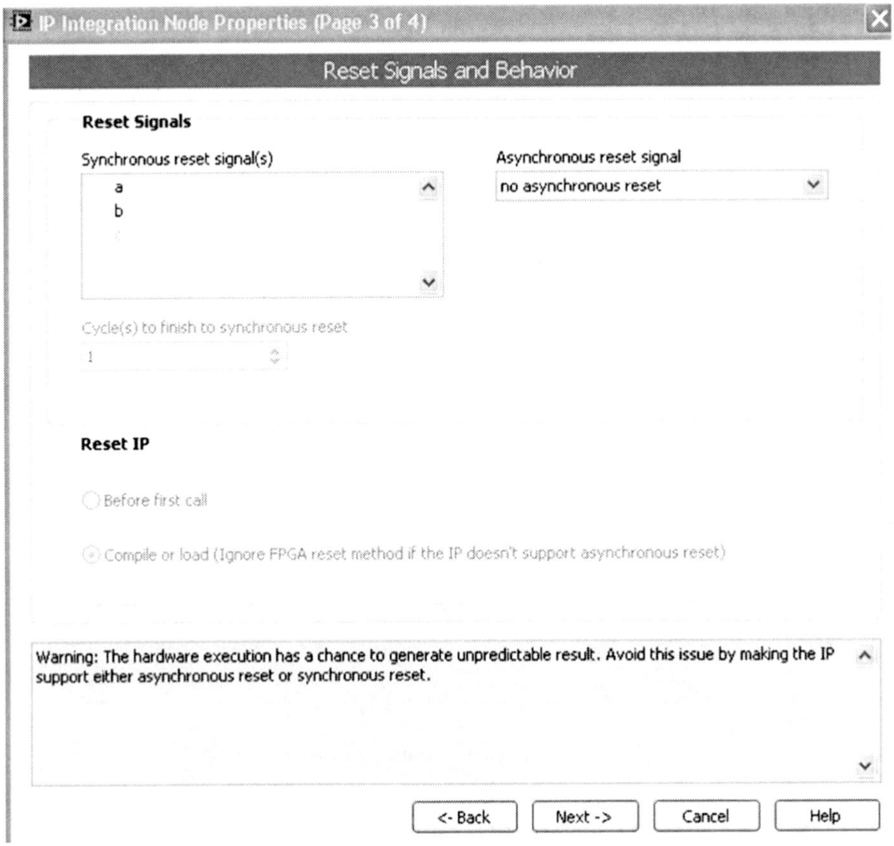

Figure L4.21: Third page of IP Node Properties

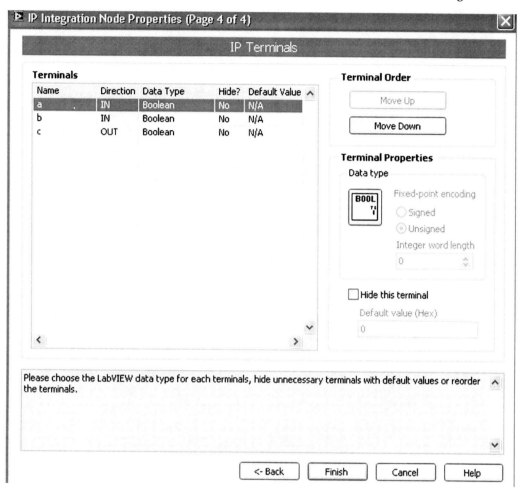

Figure L4.22: Last page of IP Node Properties

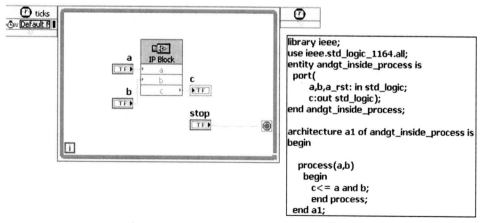

Figure L4.23: Block Diagram of FPGA VI

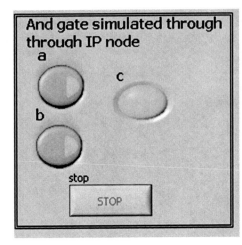 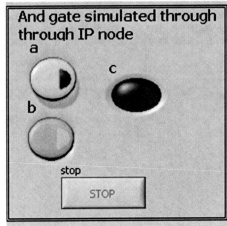

Figure L4.24: Front Panel of FPGA VI

L4.2.2 Key Properties of IP Integration Node

The key properties of IP Node are summarized below:

- The IP node only supports the types **std_logic_vector, std_logic, fixed point, signed** and **unsigned** within the numeric_std or std_arithmetic packages. The type std_logic_vector is displayed as Boolean array, and the type std_logic as Boolean within the LabVIEW IP Node Properties. However, in case of arrays, it is better to write the code using std_logic_vector, and if any other type such as signed is to be used, write the code so that input/output port signals are of std_logic_vector type, and then use type conversions for the signals within the architecture.
- If the VHDL code does not compile, the relevant file(s) have not been added to the VI within the IP node. This requires adding the file again after having corrected the code.
- More than one IP node can be executed within a while/SCTL loop, even if the referenced VHDL file is the same. However, if the referenced VHDL files are different, then one must compile each IP node separately to see the corresponding results.
- If array controls or indicators are used, as in a non-FPGA LabVIEW VI, they need to be represented in the ascending range 0 to array size-1. For example, an array of length 4 needs to be represented by the LabVIEW FPGA as having range 0 to 3 with the elements as x(0) x(1) x(2) x(3). However, the VHDL code within the IP node is compiled assuming the descending range array size-1 down to 0. This means that there are two options for giving input to the IP node and getting results:
  - Write the code with the descending range std_logic_vector (size-1 downto 0) with the arrays on the Front Panel having the ascending range. For example, if it is desired to have the input controls x(3) x(2) x(1) x(0) = [1 1 0 0], the control needs to be done with the ascending range (LabVIEW default) and the input reversed as x(0) x(1) x(2) x(3) = [0 0 1 1]. The result is interpreted in the same manner noting that LSB is to the left and MSB to the right.
  - Write the code with the ascending range std_logic_vector (0 to size -1) with arrays on the Front Panel as having an ascending range (but interpreting these arrays con-

nected to I/O ports in VHDL that are of descending range type). However, this leads to a problem indicated in the next VHDL example.

```vhdl
library ieee;
use ieee.std_logic_1164.all;
use ieee.numeric_std.all;

entity array_dir is
  generic (nx:natural :=4);
  port (x: in std_logic_vector(0 to nx-1);
        y : out std_logic_vector(1 downto 0);
        z: out std_logic_vector(0 to 1)
       );
end array_dir;

architecture a1 of array_dir is
  signal h: std_logic_vector(0 to nx-1);
  signal g: std_logic_vector(1 downto 0);
    begin
      h<=x;
      g(1)<=h(0);
      g(0)<=h(1);
      z<=g;
      y<=g;
  end a1;
```

Figure L4.25: VHDL code illustrating range declaration problem in IP Node

In the program shown in Figure L4.25, $x(n)$ is declared with an ascending range with the values of
$$x(n) = [x(0)\ x(1)\ x(2)\ x(3)] = [1\ 1\ 1\ 0]$$
These values are passed to the signal $h(n)$ having the same range. Thus,
$$h(n) = [x(0)\ x(1)\ x(2)\ x(3)] = [1\ 1\ 1\ 0]$$
$g(n)$ and $y(n)$ are declared with a descending range and the output signal $z(n)$ with an ascending range as indicated below
$$g(n) = [g(1)\ g(0)]$$
$$y(n) = [y(1)\ y(0)]$$
$$z(n) = [z(0)\ z(1)]$$

Next, the leftmost 2 bits of $h(n)$ are extracted and assigned to $g(n)$ as
$$h(n) = [h(0)\ h(1)\ h(2)\ h(3)] = [1\ 1\ 1\ 0]$$

$$g(n) = [g(1)\ g(0)\ ] \quad = [1\ 1]$$
$g(n)$ is then assigned to $z(n)$ and $y(n)$ as follows:

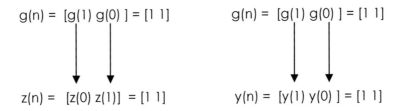

$$g(n) = [g(1)\ g(0)\ ] = [1\ 1] \qquad\qquad g(n) = [g(1)\ g(0)\ ] = [1\ 1]$$

$$z(n) = [z(0)\ z(1)] = [1\ 1] \qquad\qquad y(n) = [y(1)\ y(0)\ ] = [1\ 1]$$

Hence, y(n) declared with the descending range at the end of the simulation should have the values y(n) = [y(1) y(0)] = [1 1].

z(n) declared with the ascending range at the end of the simulation should have the values z(n) = [y(0) y(1)] = [1 1].

Now, let us see how this would work with the LabVIEW IP Node.

1) The control is given as x(n) = [x(0) x(1) x(2) x(3)]= [1 1 1 0] to the IP node.

2) The IP node, however, interprets this with the descending range as
$$x(n) = [x(3)\ x(2)\ x(1)\ x(0)] = [1\ 1\ 1\ 0]$$

2) The VHDL code then works as explained earlier going through the following steps:

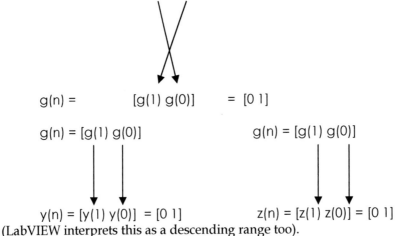

$$x(n) = [x(3)\ x(2)\ x(1)\ x(0)] = [1\ 1\ 1\ 0\ ]$$

h(n) = [h(3) h(2) h(1) h(0)]= [1 1 1 0] interprets it as a descending range.

$$g(n) = \qquad [g(1)\ g(0)] \qquad = [0\ 1]$$

$$g(n) = [g(1)\ g(0)] \qquad\qquad g(n) = [g(1)\ g(0)]$$

$$y(n) = [y(1)\ y(0)] = [0\ 1] \qquad\qquad z(n) = [z(1)\ z(0)] = [0\ 1]$$

(LabVIEW interprets this as a descending range too).

3) LabVIEW indicator shows the result with the ascending range as
$$y(n) = [y(0)\ y(1)] = [1\ 0] \qquad\qquad z(n) = [z(0)\ z(1)] = [1\ 0]$$
instead of [1 1] for both.

As can be seen, the outcome obtained is not what is expected. To address this problem, declare the control array x(n) as an ascending range array, reverse it, and give a descending array as input to the IP node. Read the result as an ascending array without reversing (if declared in VHDL with a descending range). This time, the correct outcome is the result.

1) Create a control $xa(n)$ with an ascending range as
$$xa(n) = [x(0)\ x(1)\ x(2)\ x(3)] = [1\ 1\ 1\ 0]$$
2) Reverse this array, and use it as input to the IP node. That is, the input to the IP node is given by
$$x(n) = [x(0)\ x(1)\ x(2)\ x(3)] = [0\ 1\ 1\ 1]$$
3) The IP node, however, interprets this with the descending range as
$$x(n) = [x(3)\ x(2)\ x(1)\ x(0)] = [0\ 1\ 1\ 1]$$
4) The VHDL code then works as explained earlier going through the following steps:

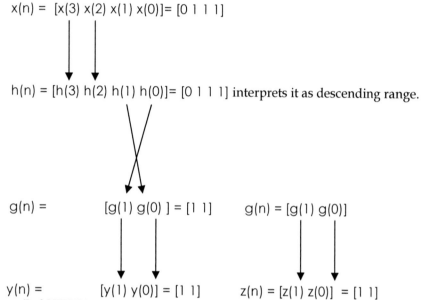

$$x(n) = [x(3)\ x(2)\ x(1)\ x(0)] = [0\ 1\ 1\ 1]$$

$h(n) = [h(3)\ h(2)\ h(1)\ h(0)] = [0\ 1\ 1\ 1]$ interprets it as descending range.

$g(n) =$ $[g(1)\ g(0)\ ] = [1\ 1]$ $g(n) = [g(1)\ g(0)]$

$y(n) =$ $[y(1)\ y(0)] = [1\ 1]$ $z(n) = [z(1)\ z(0)] = [1\ 1]$
(LabVIEW interprets this as a descending range as well.)

4) LabVIEW indicator shows the result with the ascending range as
$$y(n) = [y(0)\ y(1)\ ] = [1\ 1] \qquad z(n) = [z(0)\ z(1)] = [1\ 1]$$

Note that while the result for $y(n)$ is correct, the result for $z(n)$ would have been different if $g(n)$ was $[0\ 1]$, thus instead of $g(n) >= z(0)\ z(1) = [0\ 1]$, LabVIEW would have taken it as $z(1)\ z(0) = [0\ 1]$ and displayed the result as $z(0)\ z(1) = [1\ 0]$. Hence, if the indicator array in the LabVIEW FPGA is mapped to an output port signal in VHDL with an ascending range, reverse the indicator to read the correct result.

This is shown in Figures L4.26 and L4.27 for the corresponding VI **array_dir** together with its Front Panel. The BD for the same is shown in Figure L4.28

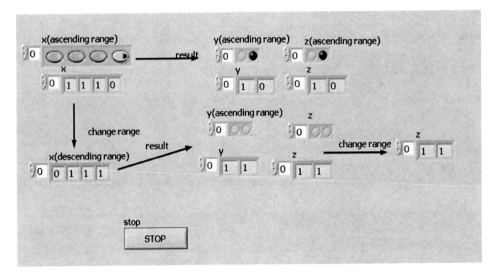

Figure L4.26: Front Panel of array_dir.vi

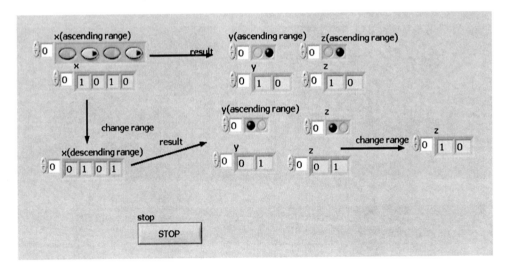

Figure L4.27: Front Panel of array_dir.vi

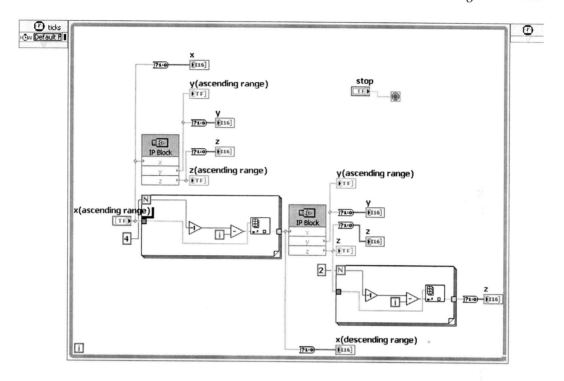

Figure L4.28: Block Diagram of array_dir.vi

L4.3 Lab Experiments

- Refer to the example in section L4.1.1. In this experiment, it is desired to multiply fixed-point numbers in an FPGA VI. Create an FPGA VI using the defined memories M1, M2 and M3 of size 10 elements with the data type being fixed-point Q8, Q8, and Q16 (<+/-,16,2>), respectively. Create two Q8 fixed-point arrays A1 and A2. Write the array elements into the corresponding memory locations M1 and M2. Then, read these memory locations, multiply the corresponding elements using the high throughput multiply function, and write the result into the corresponding locations of memory M3. Display the results using array Mul.

- Implement the full adder system of L3.1.4 in LabVIEW IP Integration Node in LabVIEW FPGA and verify the results with those of Lab 3.

- Implement the counter example of L3.1.5 using the LabVIEW IP Integration Node in LabVIEW FPGA and verify the results with those of Lab 3.

CHAPTER 5
Fixed-Point Data Type

From an arithmetic point of view, there are two ways a signal processing algorithm can be implemented: fixed-point and floating-point data type formats. In this chapter, the issues related to these formats are discussed noting that the implementation on a FPGA processor has to be done in fixed-point format.

In a fixed-point processor, numbers are represented and manipulated in integer format. In a floating-point processor, in addition to integer arithmetic, floating-point arithmetic can be handled. This means that numbers are represented by the combination of a mantissa (or a fractional part) and an exponent part, and the processor possesses the necessary hardware for manipulating both of these parts. As a result, in general, floating-point processors are slower than fixed-point ones.

In a fixed-point processor, one needs to be concerned with the dynamic range of numbers, since a much narrower range of numbers can be represented in integer format as compared to floating-point format. For most applications, such a concern can be virtually ignored when using a floating-point processor. Consequently, fixed-point processors usually demand more coding effort than do their floating-point counterparts.

5.1 Q-format Number Representation

The decimal value of an N-bit 2's-complement number, $B = b_{N-1}b_{N-2}...b_1b_0, b_i \in \{0,1\}$, is given by

$$D(B) = -b_{N-1}2^{N-1} + b_{N-2}2^{N-2} + ... + b_1 2^1 + b_0 2^0 \tag{5.1}$$

The 2's-complement representation allows a processor to perform integer addition and subtraction by using the same hardware. When using the unsigned integer representation, the sign bit is treated as an extra bit. This way only positive numbers can be represented.

There is a limitation of the dynamic range of the foregoing integer representation scheme. For example, in a 16-bit system, it is not possible to represent numbers larger than $2^{15} - 1 = 32767$ and smaller than $-2^{15} = -32768$. To cope with this limitation, numbers are often normalized between -1 and 1. In other words, they are represented as fractions. This normalization is achieved by the programmer moving the implied or imaginary binary point (note that there is no physical memory allocated to this point) as indicated in Figure 5.1. This way, the fractional value is given by

$$F(B) = -b_{N-1}2^0 + b_{N-2}2^{-1} + ... + b_1 2^{-(N-2)} + b_0 2^{-(N-1)} \tag{5.2}$$

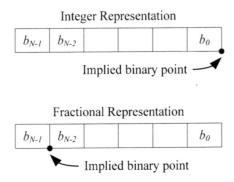

Figure 5.1: Number representations

This representation scheme is referred to as the Q-format or fractional representation. The programmer needs to keep track of the implied binary point when manipulating Q-format numbers. For instance, let us consider two Q15 format numbers. Each number consists of 1 sign bit plus 15 fractional bits. When these numbers are multiplied, a Q30 format number is generated (the product of two fractions is still a fraction), with bit 31 being the sign bit and bit 32 another sign bit (called an extended sign bit). Assuming a 16-bit wide memory, not enough bits are available to store all 32 bits, and only 16 bits can be stored. It makes sense to store the sixteen most significant bits. This requires storing the upper portion of the 32-bit product by doing a 15-bit right shift. In this manner, the product would be stored in Q15 format, see Figure 5.2.

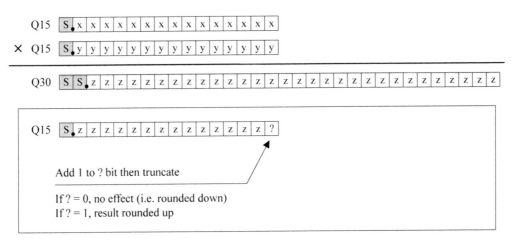

Figure 5.2: Multiplying and storing Q15 numbers

Based on the 2's-complement representation, a dynamic range of $-2^{N-1} \leq D(B) \leq 2^{N-1} - 1$ can be covered, where N denotes the number of bits. For illustration purposes, let us consider a 4-bit system where the most negative number is –8 and the most positive number 7. The decimal representations of the numbers are shown in Figure 5.3. Notice how the numbers change from most positive to most negative with the sign bit. Since only the integer numbers falling within the limits –8 and 7 can be represented, it is easy to see that any multiplication or addition resulting in a number larger than 7 or smaller than –8 will cause overflow. For example, when 6 is multiplied by 2, one gets 12. Hence, the result is greater than the representation limits and will be wrapped around the circle to 1100, which is –4.

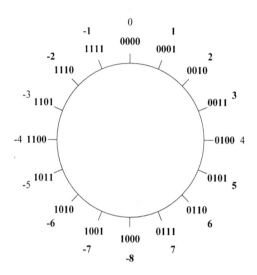

Figure 5.3: Four-bit binary representation

The Q-format representation solves this problem by normalizing the dynamic range between –1 and 1. This way, any resulting multiplication will be within these limits. Using the Q-format representation, the dynamic range is divided into 2^N steps, where $2^{-(N-1)}$ is the size of a step. The most negative number is always -1 and the most positive number is $1-2^{-(N-1)}$.

The following example helps one to see the difference in the two representation schemes. As shown in Figure 5.4, the multiplication of 0110 by 1110 in binary is equivalent to multiplying 6 by –2 in decimal, giving an outcome of –12, a number exceeding the dynamic range of the 4-bit system. Based on the Q3 representation, these numbers correspond to 0.75 and –0.25, respectively. The result is –0.1875, which falls within the fractional range. Notice that the hardware generates the same 1's and 0's, what is different is the interpretation of the bits.

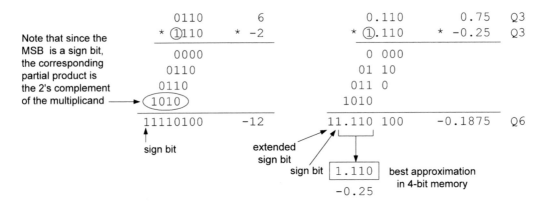

Note that since the
MSB is a sign bit,
the corresponding
partial product is
the 2's complement
of the multiplicand

Figure 5.4: Binary and fractional multiplication

When multiplying Q-N numbers, it should be remembered that the result will consist of 2N fractional bits, one sign bit, and one or more extended sign bits. Based on the data type used, the result has to be shifted accordingly. If two Q15 numbers are multiplied, the result will be 32 bits wide, with the most significant bit being the extended sign bit followed by the sign bit. The imaginary decimal point will be after the 30th bit. So a right shift of 15 is required to store the result in a 16-bit memory location as a Q15 number. It should be realized that some precision is lost, of course, as a result of discarding the smaller fractional bits. Since only 16 bits can be stored, the shifting allows one to retain the higher precision fractional bits. If a 32-bit storage capability is available, a left shift of 1 can be done to remove the extended sign bit and store the result as a Q31 number. The extended sign bit is removed from the result. This is done to provide the maximum number of fractionsl bits (31) that one can get using a 32-bit register.

To further understand a possible precision loss when manipulating Q-format numbers, let us consider another example where two Q12 numbers corresponding to 7.5 and 7.25 are multiplied. As can be seen from Figure 5.5, the resulting product must be left shifted by 4 bits to store all the fractional bits corresponding to Q12 format. However, doing so results in a product value of 6.375, which is different than the correct value of 54.375. The reason for this is that the multiplicands have ranges outside -1 to 1. Thus, if the fractional length of 12 is kept for the multiplicands, it will be done at the expense of an integer wordlength of 4 leading to the wrong result. However, if the product is stored in a lower precision Q-format – say, in Q8 format - then the correct product value can be obtained and get stored since this way a higher integer wordlength (8 in this example) is used at the loss of some of the fractional bits.

Although Q-format solves the problem of overflow during multiplications, addition and subtraction still pose a problem. When adding two Q15 numbers, the sum could exceed the range of the Q15 representation. To solve this problem, the scaling approach, discussed later in this chapter, needs to be employed.

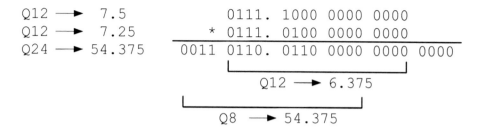

Figure 5.5: Q-format precision loss example

5.1.1 Negative Integer Length

When the dynamic range of data values is much less than 1, it is often advantageous to increase the number of fractional bits at the expense of integer bits while keeping the word-length the same. This way a higher precision is gained. For example, as shown in Figure 5.6, the value 0.31 in Q3 format is represented as ceiling($0.31*2^3$), where ceiling denotes round towards positive infinity or 2=0010 in this case. As a result, this value gets quantized to 0.25 with an error of 0.06. Alternatively, it can get represented in Q(-1,5) format with -1 as integer bit and 5 as fractional bits with the same wordlength of 4. In other words, it can get represented as ceiling($0.31*2^5$)=10 with the binary value of 01010, where the binary point is considered to be to the left of the MSB. This gives a quantized value of 0.3125 which is closer to the actual value.

```
     Q3(1,3)              0  .  0     1     0
0.31 ───►  0.25              0.5  0.25  0.125    (weights)
                          = 0.5*0 +0.25*1+0.125*0=0.25

     Q3(-1,5)             .  0   1    0     1      0
0.31 ───►  0.3125           0.5  0.25 0.125  0.0625 0.03125 (weights)
                         = 0.5*0+0.25*1+0.125*0+ 0.0625*1+0.03125*0= 0.3125
```

Figure 5.6: Negative integer length

5.2 Finite Word Length Effects

Due to the fact that the memory or registers of a processor have a finite number of bits, there could be a noticeable error between desired and actual outcomes on a fixed-point processor. The so called finite word length quantization effect is similar to the input data quantization effect introduced by an A/D converter.

Consider fractional numbers quantized by a $b+1$ bit converter. When these numbers are manipulated and stored in a $M+1$ bit memory, with $M < b$, there is going to be an error (simply because $b - M$ of the least significant fractional bits are discarded or truncated). This finite word length error could alter the behavior of a signal processing algorithm by an unacceptable degree. The range of the magnitude of truncation error ε_t is given by $0 \leq |\varepsilon_t| \leq 2^{-M} - 2^{-b}$. The lowest level of truncation error corresponds to the situation when all the thrown-away bits are zeros and the highest level to the situation when all the thrown-away bits are ones.

This effect has been extensively studied for FIR and IIR filters, for example see [1]. Since the coefficients of such filters are represented by a finite number of bits, the roots of their transfer function polynomials, or the positions of their zeros and poles, shift in the complex plane. The amount of shift in the positions of poles and zeros can be related to the amount of quantization errors in the coefficients. For example, for an Nth-order IIR filter, the sensitivity of the ith pole p_i with respect to the kth coefficient a_k can be derived to be (see [1]),

$$\frac{\partial p_i}{\partial a_k} = \frac{-p_i^{N-k}}{\prod\limits_{\substack{l=1 \\ l \neq i}}^{N}(p_i - p_l)} \tag{5.3}$$

This means that a change in the position of a pole is influenced by the positions of all the other poles. That is the reason the implementation of an N th order IIR filter is normally achieved by having a number of second-order IIR filters in cascade or series in order to decouple this dependency of poles.

Also, note that as a result of coefficient quantization, the actual frequency response $\hat{H}\left(e^{j\theta}\right)$ is different than the desired frequency response $H\left(e^{j\theta}\right)$. For example, for an FIR filter having N coefficients, it can be easily shown that the amount of error in the magnitude of the frequency response, $\left|\Delta H\left(e^{j\theta}\right)\right|$, is bounded by

$$\left|\Delta H\left(e^{j\theta}\right)\right| = \left|H\left(e^{j\theta}\right) - \hat{H}\left(e^{j\theta}\right)\right| \leq N2^{-b} \tag{5.4}$$

In addition to the above effects, coefficient quantization can lead to limit cycles. This means that in the absence of an input, the response of a supposedly stable system (poles inside the unit circle) to a unit sample is oscillatory instead of diminishing in magnitude.

5.3 Floating-Point Number Representation

Due to relatively limited dynamic ranges of fixed-point processors, when using such processors, one should be concerned with the scaling issue, or how big the numbers get in the manipulation of a signal. Scaling is not of concern when using floating-point processors, since the floating-point hardware provides a much wider dynamic range.

In LabVIEW, floating-point numbers conform to the ANSI/IEEE Standard 754-1985. There are three floating-point data representations in LabVIEW: **single precision (SP)**, **double precision (DP)** and **extended-precision (EXT)**. In the single precision format, a value is expressed as

$$-1^s \times 2^{(exp-127)} \times 1.\mathit{frac} \tag{5.5}$$

where s denotes the sign bit (bit 31), exp the exponent bits (bits 23 through 30), and $frac$ the fractional or mantissa bits (bits 0 through 22), see Figure 5.7.

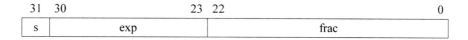

Figure 5.7: Single precision floating-point data representation

Consequently, numbers as big as 3.4×10^{38} and as small as 1.4×10^{-45} can be processed. In the double precision format, more fractional and exponent bits are used as indicated below

$$-1^s \times 2^{(exp-1023)} \times 1.frac \qquad (5.6)$$

where the exponent bits are from bits 20 through 30 and the fractional bits are all the bits of a first word and bits 0 through 19 of a second word, see Figure 5.8. In this manner, numbers as big as 1.7×10^{308} and as small as 4.9×10^{-324} can be handled.

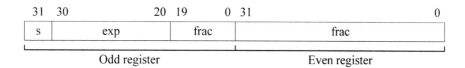

Figure 5.8: Double precision floating-point representation

For floating-point computations, all the steps needed to perform floating-point arithmetic are done by the floating-point hardware. For example, consider adding two floating-point numbers represented by

$$a = a_{frac} \times 2^{a_{exp}}$$
$$b = b_{frac} \times 2^{b_{exp}} \qquad (5.7)$$

The floating-point sum c has the following exponent and fractional parts:

$$
\begin{aligned}
c &= a + b \\
&= \left(a_{frac} + \left(b_{frac} \times 2^{-\left(a_{exp} - b_{exp}\right)} \right) \right) \times 2^{a_{exp}} \quad \text{if } a_{exp} \geq b_{exp} \\
&= \left(\left(a_{frac} \times 2^{-\left(b_{exp} - a_{exp}\right)} \right) + b_{frac} \right) \times 2^{b_{exp}} \quad \text{if } a_{exp} < b_{exp}
\end{aligned} \qquad (5.8)
$$

These parts are computed by the floating-point hardware. It should thus be realized that although it is possible to perform floating-point arithmetic on a fixed-point processor, it takes an excessive and in many cases an unacceptable number of clock cycles to do so.

5.4 Overflow and Scaling

As stated before, fixed-point processors have a much smaller dynamic range than their floating-point counterparts. It is due to this limitation that the Q15 representation of numbers is normally considered. For instance, a 16-bit multiplier can be used to multiply two Q15 numbers and produce a 32-bit product. Then the product can be stored in 32 bits or shifted back to 16 bits for storage or further processing.

When multiplying two Q15 numbers, which are in the range of –1 and 1 as discussed earlier, the product will be in the same range. However, when two Q15 numbers are added, the sum may fall outside this range, leading to an overflow. Overflows can cause major problems by generating erroneous results. When using a fixed-point processor, the range of numbers must be closely examined and if necessary be adjusted to compensate for overflows. The simplest correction method for avoiding overflows is scaling.

The idea of scaling is to scale down the system input before performing any processing and then to scale up the resulting output to the original size. Scaling can be applied to most filtering and transform operations. An easy way to achieve scaling is by shifting. Since a right shift of 1 is equivalent to a division by 2, one can scale the input repeatedly by 0.5 until all overflows disappear. The output can then be rescaled back to the total scaling amount.

As far as FIR and IIR filters are concerned, it is possible to scale coefficients to avoid overflows. Let us consider the output of an FIR filter $y[n] = \sum_{k=0}^{N-1} h[k]x[n-k]$, where h's denote coefficients or unit sample response terms and x's input samples. In case of IIR filters, for a large enough N, the terms of the unit sample response become so small that they can be ignored. Let us suppose that x's are in Q15 format (i.e., $|x[n-k]| \le 1$). Therefore, one can write $|y[n]| \le \sum_{k=0}^{N-1} |h[k]|$. This means that, to insure no output overflow (i.e., $|y[n]| \le 1$), the condition $\sum_{k=0}^{N-1} |h[k]| \le 1$ must be satisfied. This condition can be satisfied by repeatedly scaling (dividing by 2) the coefficients or unit sample response terms.

5.5 Data Types in LabVIEW

The numeric data types in LabVIEW together with their symbols and ranges are listed in Table 5.1.

Terminal symbol	Numeric data type	Bits of storage on disk
SGL	Single-precision, floating-point	32
DBL	Double-precision, floating-point	64
EXT	Extended-precision, floating-point	128
CSG	Complex single-precision, floating-point	64
CDB	Complex double-precision, floating-point	128
CXT	Complex extended-precision, floating-point	256
I8	Byte signed integer	8
I16	Word signed integer	16
I32	Long signed integer	32
U8	Byte unsigned integer	8
U16	Word unsigned integer	16
U32	Long unsigned integer	32
X	128-bit time stamp	<64.64>

Table 5.1: Numeric data types in LabVIEW [3]

Note that, other than the numeric data types shown in Table 5.1, there exist other data types in LabVIEW, such as cluster, waveform, and dynamic data type, see Table 5.2. For more details on all the LabVIEW data types, refer to [2, 3].

Terminal symbol	Data type
	Enumerated type
	Boolean
	String
	Array—Encloses the data type of its elements in square brackets and takes the color of that data type.
	Cluster—Encloses several data types. Cluster data types are brown if all elements in the cluster are numeric or pink if all the elements of the cluster are different types.
	Path
	Dynamic—(Express VIs) Includes data associated with a signal and the attributes that provide information about the signal, such as the name of the signal or the date and time the data was acquired.
	Waveform—Carries the data, start time, and dt of a waveform.
	Digital waveform—Carries start time, delta x, the digital data, and any attributes of a digital waveform.
	Digital—Encloses data associated with digital signals.
	Reference number (refnum)
	Variant—Includes the control or indicator name, the data type information, and the data itself.
	I/O name—Passes resources you configure to I/O VIs to communicate with an instrument or a measurement device.
	Picture—Includes a set of drawing instructions for displaying pictures that can contain lines, circles, text, and other types of graphic shapes.

Table 5.2: Other data types in LabVIEW [3]

Lab 5:
Fixed-Point FPGA Implementation

Fixed-Point arithmetic can be handled in the LabVIEW FPGA Module in two ways, either by using graphical programming or via the IP Node. In this lab, both of these implementation approaches are examined.

L5.1 Fixed-Point Arithmetic in VHDL

There are many VHDL packages that allow fixed-point computation. Here, the package **floatfixlib.fixed_pkg** defined in the library **floatfixlib** is used. This package defines unsigned and signed fixed-point representations as vectors of type std_ulogic as follows:

> type UNRESOLVED_ufixed is array (INTEGER range <>) of STD_ULOGIC;
> type UNRESOLVED_sfixed is array (INTEGER range <>) of STD_ULOGIC;

This way, fixed-point numbers can be declared with a range having the lower index of integer length -1 and the higher index of fractional length. Thus, a x(m,n) fixed-point number, with 'm' representing integer length and 'n' fractional length, can be declared in VHDL as follows:

> signal x: sfixed (m-1 downto –n)   for signed number
> signal x: ufixed (m-1 downto –n)   for unsigned number

For example, the fixed-point representation for 0.875 in Q3 or (1, 3) format is 0.111 and can be declared as

> signal x: sfixed (0 downto –3) = "0111"   (m-1=0, so m=1, n=3)

Similarly, integer 5 can be declared as

> signal x: sfixed (3 downto 0) = "0101"

One may declare numbers in which the radix point is not within the index range of a vector, that is to say when fractional lengths are more than wordlengths or negative integer lengths. It should be noted that fixed-point packages define fixed-point numbers with a descending range.

Let us consider the following examples to get a better understanding of specifying fixed-point numbers in VHDL:

> signal x: sfixed (2 downto -3) = "010100"   (010.100)

This example represents the fixed-point number 2.5 $((0*2^2) + (1*2^1) + (0*2^0) + (1*2^{-1}) + (0*2^{-2}) + (0*2^{-3}))$.

x<= to_sfixed (2.5, 2, -3)
This example indicates another way of declaring the above fixed-point number.

y<= x + 3
This example indicates overloading with integer (2.5+3 assuming y has a desired word-length and fractional length to represent 5.5).

y<= x+3.5
This example indicates overloading with real number. However, although real numbers are supported as floating-point numbers in the same package, floating-point arithmetic is not supported in the LabVIEW FPGA. Consequently, any VHDL code, if written in IP Node, should not use floating-point numbers.

L5.1.1 Overflow and Rounding

The package **floatfixlib.fixed_pkg** defines the following constants:

constant fixed_round_style: fixed_round_style_type := fixed round;
This constant determines the rounding behavior for fixed-point operations, whether the result is to be rounded to the nearest integer or truncated towards zero.

constant fixed_overflow_style: fixed_overflow_style_type := fixed_saturate;
This constant determines the overflow behavior for fixed-point operations, whether the result is to be saturated or wrapped around.

constant fixed_guard_bits: NATURAL := 3;
This constant denotes extra bits that are used in the divide routines.

constant no warning: BOOLEAN := (false)
This constant is used for suppression of warning messages.

A number of conversion and resize functions are also defined in this package. A short list is given in Table L5.1.

	Source	Destination
to_ufixed	integer, real, unsigned, sfixed, std_logic_vector	ufixed
to_sfixed	integer, real, signed, sfixed, std_logic_vector	sfixed
resize	ufixed/sfixed	ufixed/sfixed
to_real	ufixed/sfixed	real (scalar)
to_integer	ufixed/sfixed	integer (scalar)
to_unsigned	ufixed	unsigned
to_signed	sfixed	signed

TableL5.1: Conversion and resize functions

For details on how to use these functions, refer to the IEEE VHDL standard [4]. Now, let us go through the following example code for a better understanding of the fixed-point arithmetic computation.

```vhdl
library ieee;                                      --1
use ieee.std_logic_1164.all;                       --2
use ieee.numeric_std.all;                          --3
library floatfixlib;                               --4
use floatfixlib.fixed_pkg.all;                     --5

entity fp_adder is                                 --6
port (                                             --7
a1, a2, a3: in  sfixed(0 downto -3);               --8
a4,a5:out sfixed (1 downto -3)                     --9
      );                                           --10
end entity;                                        --11

architecture a1 of fp_adder is                     --12

  begin                                            --13

  process(a1,a2)                                   --14
    variable x1,x2: sfixed(0 downto -3);           --15
    variable x3: sfixed(1 downto -3);              --16
    variable x4: std_logic_vector(3 downto 0);     --17
    variable x5: signed(3 downto 0);               --18
    variable x6: sfixed(0 downto -3);              --19
    variable x7,x8,x9: sfixed(0 downto -3);        --20
    variable x10: sfixed(1 downto -3);             --21
     begin                                         --22
     x1:= a1;                                      --23
     x2:= a2;                                      --24
     x3:= x1+x2;                                   --25
     x6:= a3;                                      --26
     x4:= "0001";                                  --27
     x5:= signed(x4);                              --28
     x7:= to_sfixed(x5,0,-3);                      --29
     x9:= to_sfixed(x4,0,-3);                      --30
     x10:= x6+x9;                                  --31
     a4<= x3;                                      --32
     a5<= x10;                                     --33

   end process;                                    --34
end a1;                                            --35
```

Figure L5.1: Fixed-point VHDL code

L5.1.2 Code Explanation

- Line 8 declares the fixed-point fractions a1, a2, a3 of wordlength 4 and the fractional length 3 (Q3 format) as

$$a1 = 0.110 \qquad (0.75)$$
$$a2 = 1.110 \qquad (-0.25)$$
$$a3 = 0100 \qquad (0.5)$$

- Line 9 declares the sum variables a4, a5 as fixed-point fractions with the wordlength 1 more than the operands, that is the wordlength of 5.
- Line 23 and 24 assign the inputs signals a1, a2 to the variables x1, x2, and line 25 sums them up as a fixed-point addition and assigns the sum to the variable x3, that is

$$x1 = 0.110 \ (0.75)$$
$$+ \ x2 = 1.110 \ (-0.25)$$
$$\overline{x3 = 10.100 \ (-1.5) \quad \text{(incorrect)}}$$

However, the above fixed-point math package performs the arithmetic in full precision. Hence, x1 and x2 are cast to the full precision of wordlength and then added as indicated below

$$x1 = 00.110 \ (0.75)$$
$$+ \ x2 = 11.110 \ (-0.25)$$
$$\overline{x3 = 11.100 \ (0.5) \quad \text{(correct)}}$$

The correct result can be observed in the ModelSim simulation stated below:

- Line 27 declares a std_logic_vector x4 with a descending range and assigns it the bits "0001".
- Line 28 converts it into a signed integer x5, thus x5 appearing as "0001" is interpreted as integer 1.
- Line 29 uses the conversion **to_sfixed** to convert from signed integer x5 to signed fixed-point fraction x7. The variable x7 has a wordlength of 4, a fractional length of 3, so it can represent numbers in fixed-point format if they fall within the range -1 to 0.875. Since x5 has a value of 1, the result saturates to the positive maximum of 0.875 or 0.111. The overflow mode can be changed from saturation to wrap by changing the overflow mode parameter using the to_sfixed conversion.
- Alternatively, the conversion can be done directly from std_logic_vector to signed fixed fraction, as done in line 30. In other words, line 30 maps the corresponding bits of x4 into x9, and interprets x9 as a fixed-point number. This is illustrated in Figure L5.2.

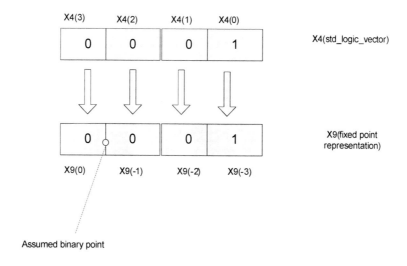

Figure L5.2: Std_logic to fixed-point conversion

- Line 31 adds the fixed-point numbers x9 and x6 (assigned signal value a3 = 0.110 or 0.25 at line 26) and stores the result in the fixed-point variable x10 of wordlength 5,

$$
\begin{array}{r}
x6 = 0.\,1\,0\,0\ (0.5) \\
+\ x9 = 0.\,0\,0\,1\ (0.125) \\
\hline
x10 = 0\ 0.1\,0\,1\ (0.625)\quad\text{(correct)}
\end{array}
$$

This can be verified from the simulation result shown in Figure L5.3.

/fp_adder/a1	0110	0110
/fp_adder/a2	1110	1110
/fp_adder/a3	0100	0100
/fp_adder/a4	00100	00100
/fp_adder/a5	00101	00101
/fp_adder/line__19/x4	0001	0001
/fp_adder/line__19/x5	0001	0001
/fp_adder/line__19/x6	0100	0100
/fp_adder/line__19/x7	0111	0111
/fp_adder/line__19/x9	0001	0001
/fp_adder/line__19/x10	00101	00101

Figure L5.3: Simulation result in ModelSim for fixed-point adder

L5.2 Fixed-Point Arithmetic in LabVIEW IP Node

For using fixed-point computation in IP Node, instead of the package **floatfixlib.fixed_pkg,** the package **ieee_proposed.fixed_pkg** needs to be used. Thus, the code for the above program is similar except for lines 4 and 5 which need to be replaced by the following lines:

> library ieee_proposed;
> use ieee_proposed.fixed_pkg.all;

It should be noted that LabVIEW defines fixed-point numbers in the format

> Fxp<+/-, m+n, m,> for signed and Fxp<+, m+n,m,> for unsigned number

where m+n= wordlength, m= integer length, and n= fraction length. For example, the signed fixed-point number 2.5 with the wordlength of 6 and integer length 3 is represented as

> Fxp<+/-, 6, 3,>

Figure L5.4 illustrates this fixed-point representation in the above VHDL code.

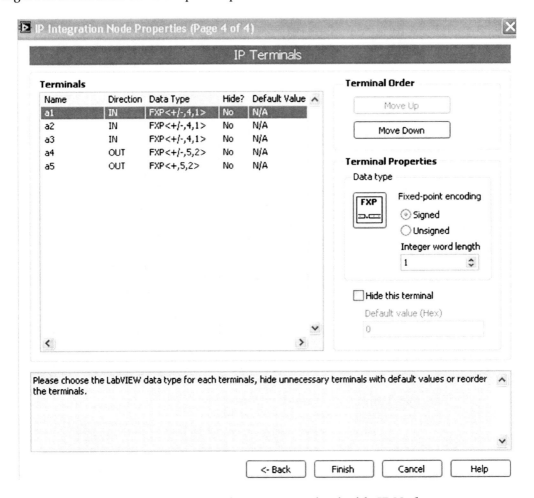

Figure L5.4: Fixed-point representation inside IP Node

It is important to be aware of different representations of fixed-point numbers in VHDL and VHDL used within IP Node/LabVIEW FPGA.

The Front Panel and Block Diagram shown in Figures L5.5 and L5.6, respectively, illustrate the outcomes when the file **fp_adder.vhd** is added to the project fixed_point_lvproj, and compiled on the PC using random I/O.

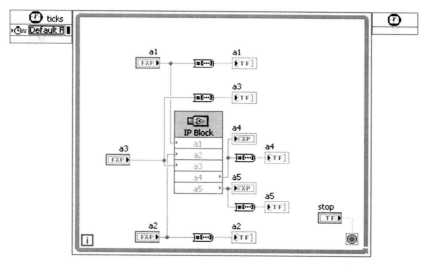

Figure L5.5: Block Diagram of fixed-point adder

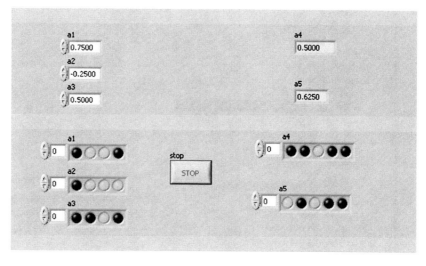

Figure L5.6: Front Panel of fixed-point adder

L5.3 Fixed-Point Arithmetic in LabVIEW FPGA

Suppose it is required to add two real numbers 0.33 and 0.24 using the Q7 representation. The integer representation for 0.33 in Q7 format is ceiling($0.33*2^7$) = 42 = 00101010 (in 2's complement) and in fixed-point is 0.0101010 or 0.328125. Of course, precision is lost while gaining faster integer arithmetic. Similarly, the Q7 representation for the real number 0.24 is ceiling($0.24*2^7$) = 31 = 00011111 (in 2's complement) and in fixed-point is 0.0011111 or 0.2421875.

Now, let us show an example using the Host-FPGA VI Communication. The following steps need to be taken:

- Create an FPGA project **fixed_point.lvproj** (PCI-7811R on development computer), add the FPGA target, and set its properties to **simulated I/O and random I/O read.** Create the host VI **real_to_fp_Host VI**. Similarly, create the FPGA VI **adder_lv_fpga.vi.** The project explorer appears as shown in Figure L5.7.
- In the FPGA VI, open the Block Diagram, create a numeric constant, right click on it and choose **representation>>FXP** to add a fixed-point control. By default, the object created has a fixed-point configuration <+- 16, 16>. Right click on the control and choose category **data type** as shown in Figure L5.8. Change the wordlength to **8** bits, and integer wordlength to **1** bit for the Q7 representation and type **signed**, as shown in Figure L5.9. Name the control **a1_fp**. Copy this control into the Block Diagram, and create a second fixed-point control with the name **a2_fp.**

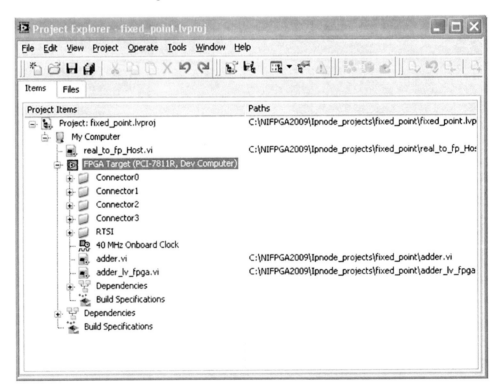

Figure L5.7: Host-FPGA VI fixed-point addition project

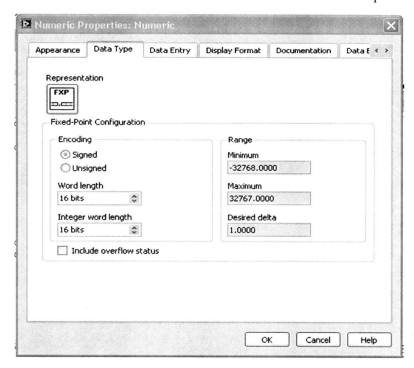

Figure L5.8: Default configuration for fixed-point control

Add the fixed-point numbers a1_fp and a2_fp by using the function **High Throughput Add.** This function can be located via **functions>>FPGA Math & Analysis>> High Throughput math>>Add.** Set the properties of this function to conform to Q7 operands, and the execution mode to **inside single cycle timed loop.** Wire the **a1_fp** and **a2_fp** controls to the **x** and **y** input terminals of the **Add** function, respectively. As the inputs are wired, the **Add** function again adapts to the source and shows the Q7 configuration as shown in Figure L5.10. Also, create the sum output **sum_fp** by right clicking on the **x+y** output terminal of the **Add** function. The **Add** output has the format <+- 9, 2> as desired.

- Finally, place the Block Diagram inside a Single Cycle Timed Loop and save the VI. The corresponding Block Diagram and Front Panel are shown in Figure L5.11 and Figure L5.12, respectively. Next, let us add the elements to the host VI. Open the Block Diagram of the host VI. Create two controls of data type Extended Precision, name them **a1_real** and **a2_real**. The range of numbers representable in Q7 format is -1 to 0.9922 for signed values. Therefore, to check whether the input falls in this range, use the **In range and Coerce** function by selecting **(Functions>>Programming>>Comparison>>In range and Coerce).** Connect a1_real to the **x** input of the function, and wire **1** and **-1** as constants for the **maximum** and **minimum** inputs. To represent this real value in Q7 format, wire the **coerced(x)** terminal of the **In range and Coerce** function to the **To Fixed point** function. Change the wordlength to 8 bits, and the integer wordlength to 1 bit for the Q7 representation.

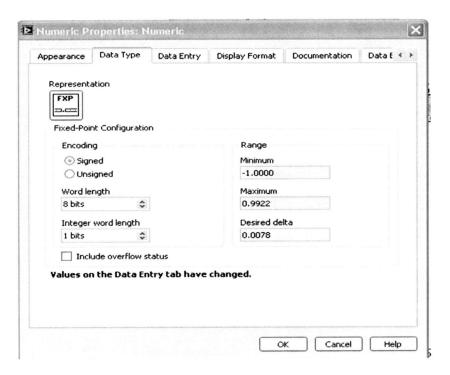

Figure L5.9: Fixed-point configuration for Q7 format

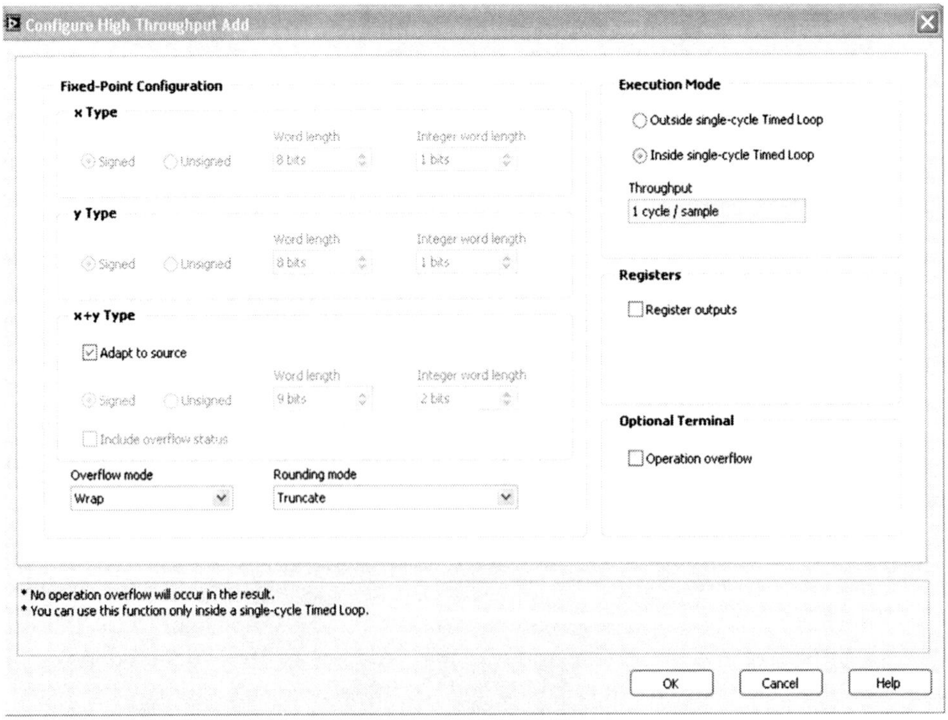

Figure L5.10: Final configuration for High Throughput Add function

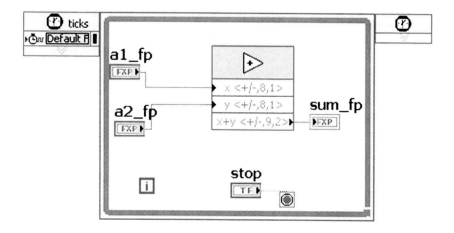

Figure L5.11: Block Diagram of adder_lv_fpga.vi

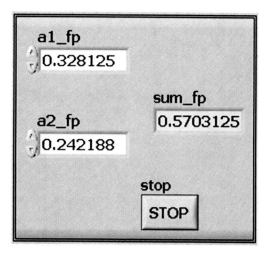

Figure L5.12: Front Panel of adder_lv_fpga.vi

- Similarly, wire the output of the **a2_real** control to the **In range and Coerce** function, and wire the **coerced(x)** output terminal to convert it into a fixed-point number. Create the corresponding fixed-point indicators **a1_fp** and **a2_fp**, and then wire them to the respective fixed-point outputs. The VI appears as shown in Figure L5.13.

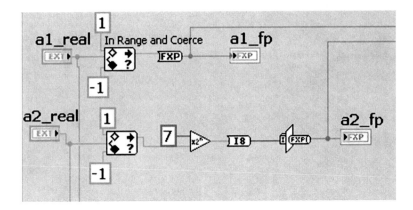

Figure L5.13: Block Diagram of a portion of the host VI

- Next, add the real values **a1_real** and **a2_real** by using the **Add** function, and create the indicator **float_sum** to indicate floating-point or real sum.
- Then, create a call to the FPGA VI. Open a reference to the FPGA VI from the host VI. Place the function **Open FPGA VI reference** on the Block Diagram. Select the VI **adder_lv_fpga.vi** as the FPGA VI, for which this host VI is configured and press ok (see Figure L5.14).

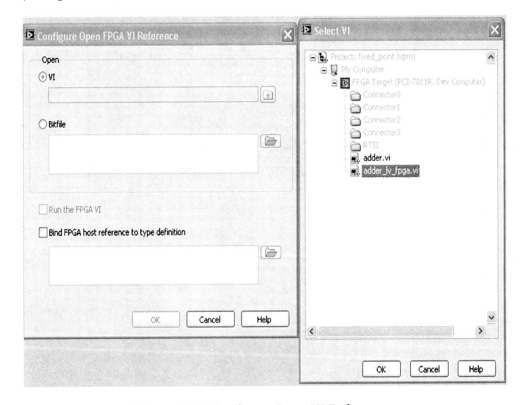

Figure L5.14: Configure Open VI Reference

- Place the function **Read Write Control** on the Block Diagram. This function reads a value from or writes a value to a control or indicator in the FPGA VI on the FPGA target. It is desired to write the two fixed-point numbers a1_fp and a2_fp on the host VI into the corresponding controls a1_fp and a2_fp on the FPGA VI. The FPGA VI computes the fixed-point addition and stores the result in the indicator sum_fp, which is then read into the host VI. Wire the **FPGA VI Reference Out** terminal of the **Open FPGA VI Reference** function to the **FPGA VI Reference In** terminal of the **Read Write Control** function.

- Right click on the function and choose **Controls>>a1_fp** to create reference to the **a1_fp** control of the FPGA VI. By right clicking on the function and selecting **Controls**, all the controls and indicators on the FPGA VI are displayed. The read/write property of this control is **write**. This property can be changed by right clicking on the function and choosing **Change To Read**. Now, resize the function to two more elements so that the control **a2_fp** and indicator **sum_fp** on the FPGA VI appear automatically. Finally, wire the wires from the **a1_fp** and **a2_fp** indicators on the Block Diagram of the host VI to the corresponding terminals of the **Read write** function, and create an indicator for the **sum_fp** terminal renaming it **sum_fp (read back from FPGA VI**.

- **Close the reference to the FPGA VI:** Close the reference to the FPGA VI by placing the function **Close FPGA VI Reference.** Place a while loop over the structure so that **Open FPGA VI Reference** and **Close FPGA VI Reference** are outside the loop (reference is to be opened or closed only once).

  The completed Block Diagram is shown in Figure L5.15. Some other elements are displayed in this figure, which are created to show saturation and wrap mode in fixed-point arithmetic.

- Create four integer controls on the Block Diagram, and name them as **a, b, c, d** as shown in Figure L5.15. Convert them to equivalent fixed-point numbers using the function **Integer to Fixed-Point Cast (functions>>numeric>>fixed point>>to fixed point cast).** This function interprets the bits from an integer input to give equivalent bits in terms of a fixed-point number. As in the function **To Fixed Point**, the numbers are represented in Q3 format by changing the property of the function **Integer to Fixed-Point Cast** via right clicking on it and choosing **Properties.**

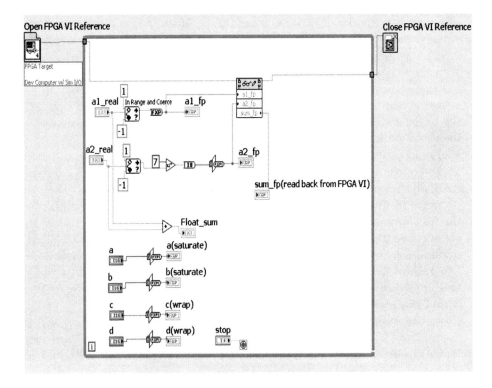

Figure L5.15: Block Diagram of Real_to_fp_host.VI

As seen, the default **Rounding mode** is **Truncate** and **Overflow mode** is **wrap**. Their effects are discussed next.

L5.3.1 Effects of Overflow Modes

These modes affect FPGA resource usage in the following ways:

- **Saturate** — Requires FPGA resources to determine whether the input value is within the range of the desired output type.
- **Wrap** — Requires fewer FPGA resources than the saturate mode, and is thus the default mode in many functions.

L5.3.2 Effects of Rounding Modes

These affect FPGA resource usage in the following ways:

- **Truncate** — This truncates the bits and therefore does not require any FPGA resources, but also produces the largest mean error of all the rounding modes. This mode is the default mode for integer operations.
- **Round-Half-Up** — This adds to the least significant bit of the output data type.

- **Round-Half-Even**— This requires the most FPGA resources and results in the longest combinatorial path of the three rounding modes, but is the least error producing mode, and is therefore the default rounding mode for the fixed-point data type.

Now, as shown in Figure L5.15, connect the controls **a, b, c, d** to each of the **Integer to Fixed-Point Cast** function, and create the corresponding indicators **a(saturate), b(saturate), c(saturate), d(saturate)**, respectively. Set the overflow mode of the cast function that are connected to the **a** and **b** controls to **saturate** and **c** and **d** controls to **wrap**. This completes the host VI as shown in Figure L5.16.

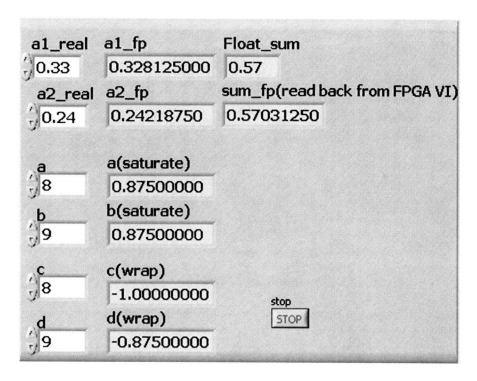

Figure L5.16: Front Panel of Real_to_fp_host.VI

Now, the program is ready to be run. Enter the values 0.33 and 0.24 for the controls **a1_real** and **a2_real** in the host VI, respectively. Also, enter the integer value 8 for controls **a** and **c** and 9 for controls **b** and **d**. Run the host VI.

When the host VI is run, it converts the real values 0.33 and 0.24 to the fixed-point numbers a1_fp and a2_fp, passes them to the FPGA VI, which computes the high throughput addition, and then the sum sum_fp is read back by the host VI. As indicated in Figure L5.16, the read back value is 0.57031250 instead of the real value 0.57. This loss of precision is inevitable at the expense of high speed integer arithmetic inside the FPGA VI. Figure L5.17 illustrates the Host and FPGA VI running simultaneously.

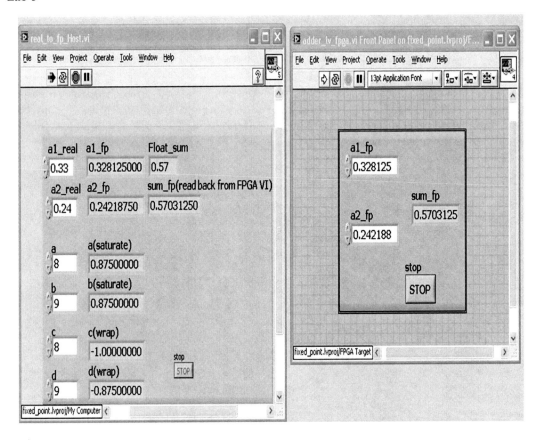

Figure L5.17: Communication between host and FPGA VI

Also, consider the effect of overflow. The fixed-point number 1.0 is represented as a signed number with a wordlength of 5 and a fractional length of 3 (01.000). The 2's complement number that corresponds to it is 01000 or integer 8. If integer 8 is represented in Q3 format, it will lead to overflow as the range this Q3 number can represent is -1 to 0.875. If the overflow mode is set to saturate, the result of this representation will be 0.875, and for the number 9 too. However, if the overflow mode is set to wrap, the result wraps around generating -1 (lower range) for integer 8, and -1+0.125 or -0.875 for integer 9. This is illustrated in the snapshot appearing in Figure L5.18.

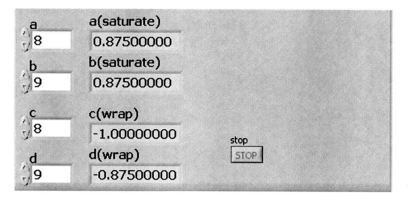

Figure L5.18: Illustrating wrap and saturation mode in fixed-point

L5.4 Bibliography

[1] J. Proakis and D. Manolakis, *Digital Signal Processing: Principles, Algorithms, and Applications*, Prentice-Hall, 1996.

[2] National Instruments, *LabVIEW Data Storage, Application Note 154*, Part Number 342012C-01, 2004.

[3] National Instruments, *LabVIEW User Manual*, Part Number 320999E-01, 2003.

[4] Language Reference Model, IEEE Std 1076, 1987.

L5.5 Lab Experiments

- Create a host and an FPGA VI similar to Figures L5.11 and L5.15. In the host VI, do not create the controls a, b, c, d and the associated graphics. Instead, in the FPGA VI, create seven more high throughput add functions, and add the inputs a1_fp and a2_fp using these functions. All these add functions are similar except for their different overflow and rounding modes. For two overflow and four rounding modes, there are eight combinations of overflow and rounding modes. Read back the results in the host VI. With diffferent inputs, compare the results of these additions.

- Build a VI (host VI) to compute the equivalent decimal magnitude of 32-bit integers using the following formats (i) Q-25, (ii) Q-23 and (iii) Q-20. For example, in the case of Q-25 format, the 7 MSB bits of a 32-bit integer should correspond to the integer part of the number and the remaining 25 bits to the fractional part of the number. For negative integers, first generate the 2's complement bits, then use these bits to compute the equivalent decimal magnitude, followed by negation to obtain the final result.

- Create four fixed point controls with the formats <+/- ,16,1>,<+/- ,8,1>,<+/-,8,-4>, <+/-,8,-8>. Enter the value 0.00345 in these. What do you observe? Does increasing the integer length on the negative side provide correct results?

CHAPTER 6
Analog to Digital Conversion

The process of analog-to-digital signal conversion involves converting a continuous time and amplitude signal into discrete time and amplitude values. Sampling and quantization constitute the steps needed to achieve analog-to-digital signal conversion. To minimize any loss of information that may occur as a result of this conversion, it is important to understand the underlying principles behind sampling and quantization.

6.1 Sampling and Aliasing

Sampling is the process of generating discrete time samples from an analog signal. First, it is helpful to mention the relationship between analog and digital frequencies. Let us consider an analog sinusoidal signal $x(t) = A\cos(\omega t + \phi)$. Sampling this signal at $t = nT_s$, with the sampling time interval of T_s, generates the discrete time signal

$$x[n] = A\cos(\omega nT_s + \phi) = A\cos(\theta n + \phi), \quad n = 0, 1, 2, \ldots, \qquad (6.1)$$

where $\theta = \omega T_s = \dfrac{2\pi f}{f_s}$ denotes digital frequency with units being radians (as compared to analog frequency ω with units being radians/sec).

The difference between analog and digital frequencies becomes more evident by observing that the same discrete time signal is obtained from different continuous time signals if the product ωT_s remains the same. An example is shown in Figure 6.1. Likewise, different discrete time signals are obtained from the same analog or continuous time signal when the sampling frequency is changed. An example is shown in Figure 6.2. In other words, both the frequency of an analog signal f and the sampling frequency f_s define the frequency of the corresponding digital signal θ.

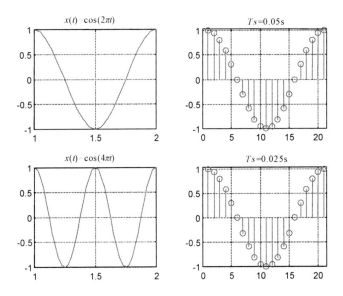

Figure 6.1: Sampling of two different analog signals leading to the same digital signal

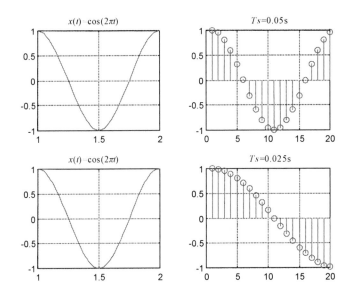

Figure 6.2: Sampling of the same analog signal leading to two different digital signals

It helps to understand the constraints associated with the above sampling process by examining signals in the frequency domain. The Fourier transform pairs in the analog and digital domains are given by

Fourier transform pair for analog signals

$$\begin{cases} X(j\omega) = \int_{-\infty}^{\infty} x(t)e^{-j\omega t}dt \\ x(t) = \dfrac{1}{2\pi}\int_{-\infty}^{\infty} X(j\omega)e^{j\omega t}d\omega \end{cases}$$

(6.2)

Fourier transform pair for discrete signals

$$\begin{cases} X(e^{j\theta}) = \displaystyle\sum_{n=-\infty}^{\infty} x[n]e^{-jn\theta}, \ \theta = \omega T_s \\ x[n] = \dfrac{1}{2\pi}\int_{-\pi}^{\pi} X(e^{j\theta})e^{jn\theta}d\theta \end{cases}$$

(6.3)

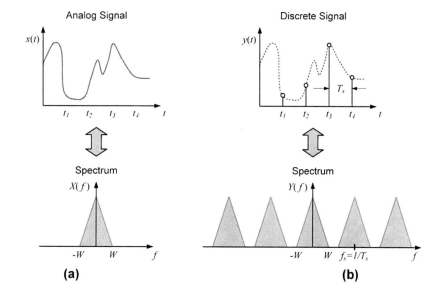

Figure 6.3: (a) Fourier transform of a continuous-time signal, and (b) its discrete time version

As illustrated in Figure 6.3, when an analog signal with a maximum bandwidth of W (or a maximum frequency of f_{max}) is sampled at a rate of $T_s = \dfrac{1}{f_s}$, its corresponding frequency response is repeated every 2π radians, or f_s. In other words, the Fourier transform in the digital domain becomes a periodic version of the Fourier transform in the analog domain. That is why, for discrete signals, the frequency range $[0, f_s/2]$ is of interest.

In order to avoid any aliasing or distortion of the frequency content of the discrete signal, and hence to be able to recover or reconstruct the frequency content of the original analog signal, we must have $f_s \geq 2f_{max}$. This is known as the Nyquist rate; that is, the sampling frequency should be at least twice the highest frequency in the signal. Normally, before any digital manipulation, a front-end antialiasing lowpass analog filter is used to limit the highest frequency of the analog signal.

The aliasing problem can be further illustrated by considering an undersampled sinusoid as depicted in Figure 6.4. In this figure, a 1 kHz sinusoid is sampled at f_s=0.8 kHz, which is less than the Nyquist rate of 2 kHz. The dashed-line signal is a 200 Hz sinusoid passing through the same sample points. Thus, at the sampling frequency of 0.8 kHz, the output of an A/D converter would be the same if either of the 1 kHz or 200 Hz sinusoids was the input signal. On the other hand, oversampling a signal provides a richer description than that of the signal sampled at the Nyquist rate.

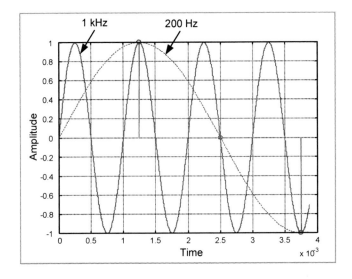

Figure 6.4: Ambiguity caused by aliasing

6.2 Quantization

An A/D converter has a finite number of bits (or resolution). As a result, continuous amplitude values get represented or approximated by discrete amplitude levels. The process of converting continuous into discrete amplitude levels is called quantization. This approximation leads to errors called quantization noise. The input/output characteristic of a 3-bit A/D converter is shown in Figure 6.5 to see how analog voltage values are approximated by discrete voltage levels.

A quantization interval depends on the number of quantization or resolution levels, as illustrated in Figure 6.6. Clearly the amount of quantization noise generated by an A/D converter depends on the size of the quantization interval. More quantization bits translate into a narrower quantization interval and hence into a lower amount of quantization noise.

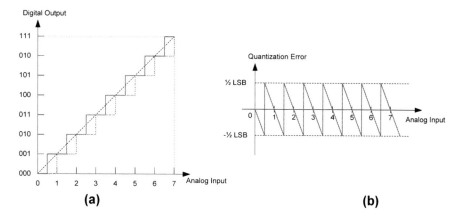

Figure 6.5: Characteristic of a 3-bit A/D converter:
(a) input/output transfer function, and (b) additive quantization noise

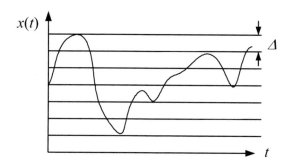

Figure 6.6: Quantization levels

In Figure 6.6, the spacing Δ between two consecutive quantization levels corresponds to one least significant bit (LSB). Usually, it is assumed that quantization noise is signal independent and is uniformly distributed over –0.5 LSB and 0.5 LSB. Figure 6.7 shows the quantization noise of an analog signal quantized by a 3-bit A/D converter.

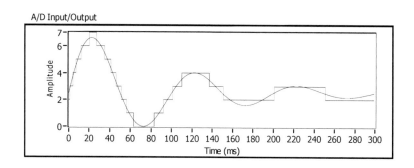

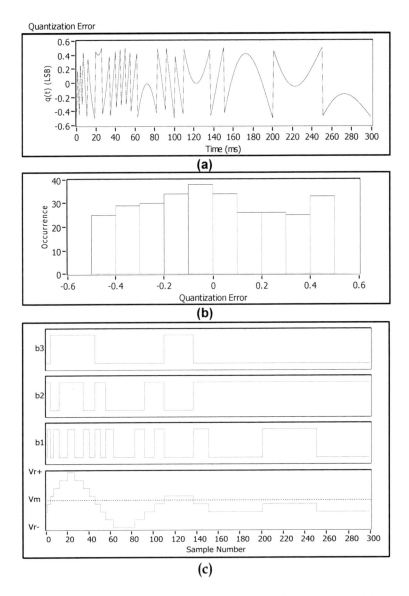

Figure 6.7: Quantization of an analog signal by a 3-bit A/D converter: (a) output signal and quantization error, (b) histogram of quantization error, and (c) bit stream

6.3 Signal Reconstruction

So far, the forward process of sampling is discussed. It is also important to understand the inverse process of signal reconstruction from samples. According to the Nyquist theorem, an analog signal v_a can be reconstructed from its samples by using the following equation:

$$v_a(t) = \sum_{k=-\infty}^{\infty} v_a\left[kT_s\right]\left[\text{sinc}\left(\frac{t - kT_s}{T_s}\right)\right] \qquad (6.4)$$

From this equation, it can be seen that the reconstruction is based on the summations of shifted sinc functions. Figure 6.8 illustrates the reconstruction of a sinewave from its samples.

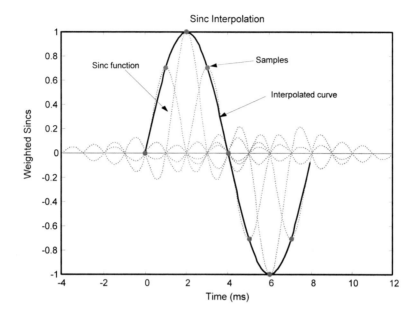

Figure 6.8: Reconstruction of an analog sinewave based on its samples,

f = 125 Hz, and f_s = 1 kHz

It is very difficult to generate sinc functions by electronic circuitry. That is why, in practice, a pulse approximation of a sinc function is used. Figure 6.9 shows a sinc function approximated by a pulse, which is easy to realize in electronic circuitry. In fact, the well-known sample and hold circuit performs this approximation [2].

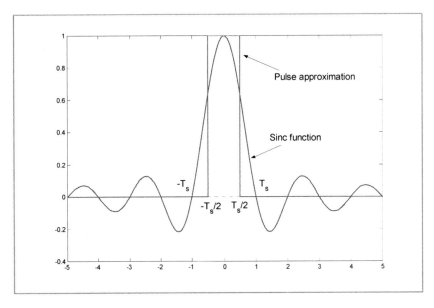

Figure 6.9: Approximation of a sinc function by a pulse

Lab 6:
Sampling in LabVIEW FPGA

L6.1 Sampling and Aliasing in LabVIEW FPGA

Create a LabVIEW FPGA project **sampling_and_reconstruction.prj** to run in simulation mode. Select the NI FlexRIO target. Add the host VI to the project by choosing **My Computer>>New>VI** and naming it as **Alisaing_host.vi**. Similarly, add the FPGA VI to the project by choosing **FPGA target>>New>VI** and naming it as **aliasing_trgt_sim.vi**. Figure L6.1 shows the project explorer window with these entries. In the first part of this lab, a discrete signal is generated by sampling a sinusoidal signal. The sampled signal is generated both on the host VI and on the FPGA VI. When the normalized frequency f / f_s of the discrete signal becomes greater than 0.5, or greater than the Nyquist frequency, the aliasing effect becomes evident.

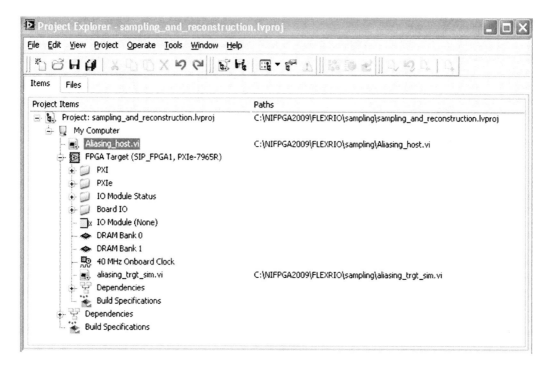

Figure L6.1: Project explorer window for the project sampling_and_reconstruction.prj

L6.1.1 Creating Host VI

Figure L6.2 displays the Block Diagram of the host VI. Create the discrete time sampled signal on the host as shown in Figure L6.3. Create a **While Loop** with **Stop Control** wired to the loop Condition terminal, refer to Figure L6.2. Place a sequence diagram (**Programming>>Structures>>Flat Sequence structure**) on the Block Diagram with three frames. Create the function **Sine.Wave.vi (Signal Processing>>Signal generation>>Sine.wave.vi)** in the first frame. This function creates a discrete time sine waveform based on the inputs **samples, phase In, amplitude frequency.** Wire the constants with the values **10** and **90** to the terminals **samples** and **phase in**, respectively. This would create a cosine wave with 10 samples. Keep the amplitude terminal unwired so that by default, amplitude **1** is used. Create two controls **F(hz)** and **Fs** corresponding to the signal frequency and sampling frequency in Hz.

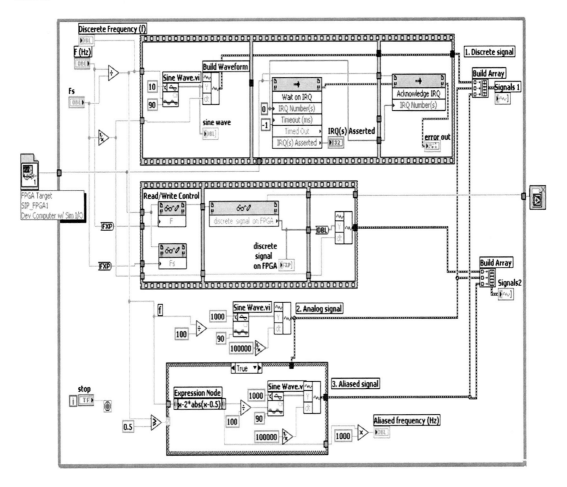

Figure L6.2: BD of the host VI

Create the discrete time frequency f=F/Fs and wire it to the frequency input of the function. To set the range of sampling frequency between 0 to 1000 Hz, right click on **Fs>>properties>>data entry>>Uncheck use Default limits** and set **Minimum** and **Maximum** to **0** and **1000,** respectively, as shown in Figure L6.4. Do the same for the signal frequency F.

Set F to **250** and Fs to **1000**. This way, a sampled signal with 10 samples for period of 10 msec is created. Wire the **Sine wave** output of this function to the **Y** input of the function **Build Waveform (Programming>>Waveform>>Build waveform)** to create sampled signal of the type waveform. Wire the **dt** terminal of this function, which represents the sampling time, to 1/Fs by wiring the **1/x** function **(Programming>>Numeric>>1/x)** to Fs. Create a **Build array** function **(Programming>>Array and Cluster>>Build Array)** outside the sequence diagram, see Figure L6.2. Stretch it to have two more inputs and wire the discrete time cosine waveform signal to the first input.

Figure L6.3 displays the resulting BD. More elements appear in this figure that are explained later after going through the FPGA VI.

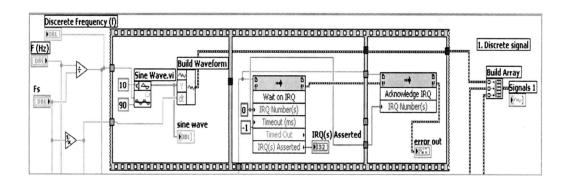

Figure L6.3: Creating discrete signal on the host

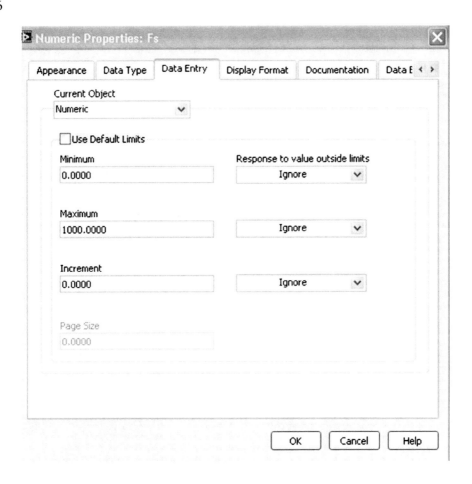

Figure L6.4: Setting the limits of sampling frequency control

L6.1.2 Creating FPGA VI

In the FPGA VI, it is desired to recreate the discrete time cosine signal, as done in Figure L6.3. In other words, it is desired to realize the following

$$x[n] = A\cos\left(\omega n T_s + \phi\right) = A\cos\left(\theta n + \phi\right), n = 0,1,2,\dots, \qquad (6.5)$$

where

f = signal frequency in Hz
Fs = sampling frequency in Hz
F = discrete time frequency in cycles/sample
N = number of samples

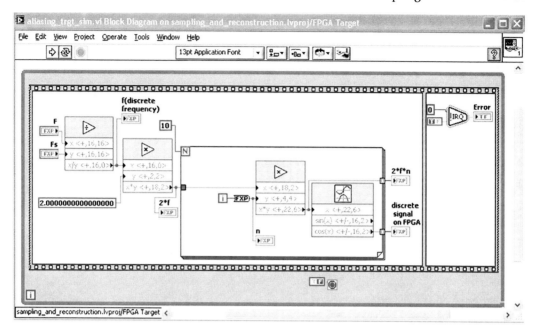

Figure L6.5: BD of the FPGA VI

Refer to Figure L6.5. Create a While Loop with a false constant wired to it and place a sequence diagram within it consisting of two frames. Create two fixed-point controls **F** and **Fs**, representing signal and sampling frequencies. This can be done by choosing **Programming>>Numeric>>numeric Constant>>Right click on the numeric constant>>Configure>>Data type>>Representation>>fixed point>>Uncheck 'adapt to entered data'**, followed by choosing appropriate fixed-point settings, and then changing the constant to control by right clicking on the constant and choosing **Change to control**. As F and Fs hold integer values, set the fixed-point representation to <+16,16> or unsigned 16 wordlength and 0 fractional length.

Create an indicator **f (discrete frequency)**, representing digital frequency f by dividing F by Fs, using the **High Throughput Divide** function **(Programming>>FPGA Math & Analysis>>High throughput Math>>Divide)**. As explained in Chapter 4, these functions are optimized for high throughput results. Right click on the **Divide** function and choose **Properties** to open its properties page to set its fixed-point configuration. As seen in Figure L6.6, the **x** type and **y** type automatically adapt to the fixed-point configuration of the signals connected to these terminals, which is <+16,16>. Note that as F and Fs have a maximum value of 1000, a minimum of 10 bits are needed to represent them in 2's complement representation; <+10,10> would be sufficient too. Since the result is going to be in the range 0 to 1, no integer bit and all fractional bits are used, thus giving the fractional fixed-point configuration of <+16,0>. Next, create the fixed-point value 2*f by using a **High Throughput Multiply** function with the output of the **divide** function wired to its **x** terminal, and 2 for its fixed-point configuration <+2,2> (need 2 bits to represent unsigned 2). 2*f has a fixed-point configuration of <+18,2>, that is 16 fractional bits and 2 integer bits (need 16 bits for fractional result and 2 bits for integer result).

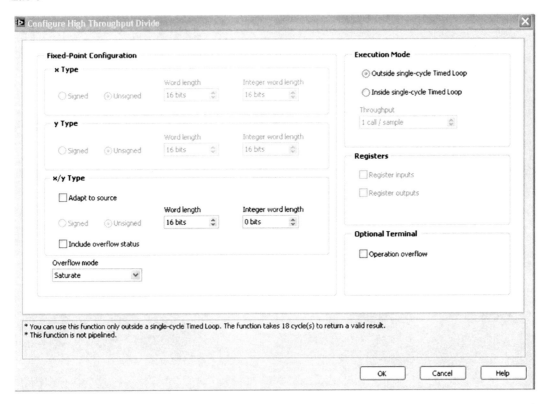

Figure L6.6: Property page of High Throughput Divide function

Create a For Loop with constant 10 wired to it so that one can generate 10 samples of the cosine signal. Now, to generate Equation (6.4) having created 2*f, it is needed to multiply 2*f with pi*n or pi*index of the For Loop and then taking cosine of the result. Hence, wire the **index** terminal of the For Loop (after converting to fixed-point <+4,4> shown by the indicator **n**) to the **y** terminal of the **High Throughput Multiply** function with the **x** terminal wired to 2*f. Now, place **High throughput Sine and Cosine** on the BD, and wire the **x*y** terminal of the **Multiply** function to it. Note that pi is not wired (for a value to appear in radians, it needs to be divided by pi). Right click on the output **cos(x),** choose **Create indicator** and name it **discrete signal on FPGA.** This creates the discrete time signal on the FPGA with the fixed-point configuration of <+16,2> for its elements.

To synchronize the FPGA VI with the host VI, an **Interrupt VI (Programming>>Synchronization>>Interrupt VI)** is used. This ensures that while continuously creating a discrete time signal on the host in real-time, the FPGA VI runs in synchronism with the host VI. This function has two inputs: **IRQ Number** and **Wait until Cleared**. IRQ Number specifies the number of the logical interrupt that is to be asserted. The default is 0. Wire a constant 0 to it. Wait until showing **Cleared** which specifies whether the FPGA VI waits for the host VI to acknowledge the logical interrupt or asserts the interrupt and continues. A **True** constant is wired to it so that the FPGA VI waits until the host VI acknowledges it. Finally, wire an **error** indicator to the output of this function to capture any errors. This finishes the design of the FPGA VI. Its FP is shown in Figure L6.7.

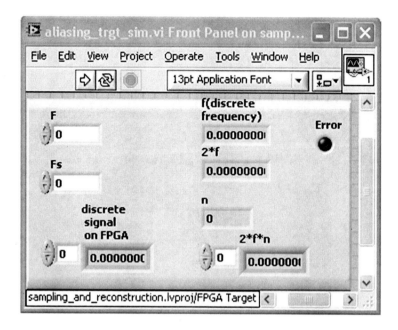

Figure L6.7: FP of FPGA VI

L6.1.3 Completing Host VI

In order to write the sampling frequency **Fs** and signal frequency values **F** from the host to the corresponding controls on the FPGA VI, one needs to open a reference to the FPGA VI from the host, refer to Figure L6.2. This is done by placing the function **Open FPGA VI Reference (FPGA VI Interface>> Open FPGA VI Reference)** on the BD.

In order to pass **F** and **Fs** values to the FPGA VI, they are first needed to be converted to fixed-point types using the function **To Fixed-point (Programming>>Numeric>>Conversion>>To Fixed-point)**, refer to Figure L6.8. This function is shown in Figure L6.9. Right click on the function and select properties. Then, change the fixed-point type to <+16,16> for both **F** and **Fs** to match the corresponding settings on the FPGA VI, see Figure L6.6.

Then, create a sequence diagram with three frames. Place the function **Read Write Control (FPGA VI Interface>> Read Write Control)** in the first frame. By default, the function appears as shown in Figure L6.10. Wire its input **FPGA VI Reference in** to the **Open FPGA VI Reference** function output. Now, clicking on this function will display all the controls and indicators used in the FPGA VI. Since it is desired to write to F, select F, right click and choose the method **change to write.** Wire the output of the **To Fixed-point** function, the one wired to the control F, to the F input of this Read/Write Control. Similarly, repeat this procedure to write Fs into the FPGA VI, as shown in Figure L6.8.

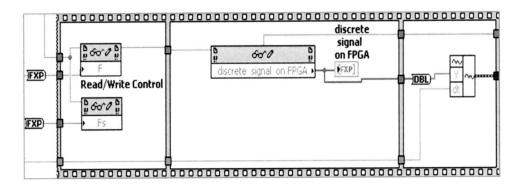

Figure L6.8: Using Read/Write Control on Host VI

Figure L6.9: To Fixed-point function

Figure L6.10: Read/Write Control function

To read back the discrete signal from the FPGA VI to the host, create a Read/Write Control in a manner similar to the previous step, and choose the corresponding indicator **discrete signal on FPGA** with the mode set to **change to read**. Create a corresponding fixed-point indicator on the BD.

In the third frame, convert the signal values to floating-point by using the function **To Double Precision Float (Programming>>Numeric>>Conversion>> Programming>>Numeric>>Conversion)**. Convert these values to waveform by using the function **Build Waveform**. Wire the double signal to its input **Y**, and **1/FS** value (refer to Figure L6.3) to the input **dt**. Finally, wire the FPGA reference to the **Close FPGA VI Reference** function to close the reference to the FPGA VI. As shown in Figure L6.5 where an interrupt structure is created on the FPGA VI, in order to acknowledge the interrupt from the FPGA VI, create the structure as shown in Figure L6.11 (also refer to Figure L6.3).

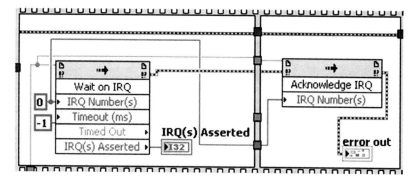

Figure L6.11: Handling interrupts on host VI

Place the function **Invoke Method (FPGA VI Interface>> Invoke Method)** on the BD. Wire its input **FPGA VI Reference in** to the **Open FPA VI Reference** function output. Right click on the function and choose **Wait on IRQ**. The function resizes to display two inputs **IRQ number** and **Timeout**. **IRQ Number(s)** specifies the number of logical interrupts or an array of logical interrupts for which this function waits. Wire a constant **0** to it in order to wait for the interrupt number **0** from the FPGA VI (refer to Figure L6.5). **Timeout (ms)** specifies the number of milliseconds this function waits before timing out. The default is 5000 milliseconds. Wire a constant with value –1 so that the function waits indefinitely. **IRQ(s) Asserted** returns the asserted interrupts. For a single interrupt, a value of –1 indicates that the interrupt was not received. Create an indicator for this value.

Place the function **Invoke Method** in the next frame, and choose its method to be **Acknowledge IRQ**. Wire the FPGA **VI Reference In** and **Error In** inputs of this function to the **FPGA VI Reference Out** and **Error Out** of the function **Wait on IRQ** to acknowledge the interrupt. Wire the **IRQ number** input to constant **0** as in the previous step. Create an error indicator to display any error that might occur.

Next, let us create the portion of the host VI that generates the analog signal and the aliased signal if the signal frequency exceeds the Nyquist Frequency, refer to Figure L6.12. Create the analog signal in a manner analogous to the discrete time signal, as done in Figure 6.2. The only difference is that for the analog signal generation, the value wired to the **f** terminal of the **Sine Wave.vi** is divided by 100 because it is sampled 100 times faster than the discrete signal, and the input **dt** of **Build Waveform** is wired to 100*1000=100000.

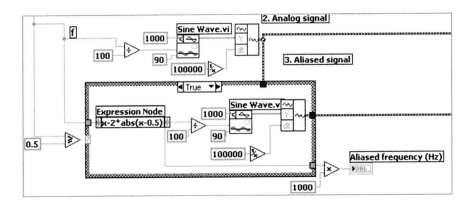

Figure L6.12: Creating analog and aliased signal on the host VI

A **Case Structure (Programming>>Structure >>Case Structure)** is then used to handle the sampling cases with aliasing and without aliasing. If the discrete time frequency is greater than 0.5, corresponding to the **True case**, the third **Sine Wave.vi** generates an aliased signal. All the inputs except for the aliased signal frequency are the same. Note that an **Expression Node (Functions >>Programming >>Numeric >>Expression Node)** is used to obtain the aliased frequency. An Expression Node is usually used to calculate an expression of a single variable. Many built-in functions, e.g. abs (absolute), can be used in an Expression Node to evaluate an equation. More details on the use of Expression Node can be found in [1]. For the **False** case, i.e. sampling without aliasing, there is no need to generate an aliased signal. Thus, the analog signal is connected to the output of the case structure so that the same signal is drawn on the waveform graph and the frequency of the aliased signal is set to 0. This is illustrated in Figure L6.13. It should be remembered that when using a case structure, it is necessary to wire all the outputs for each case.

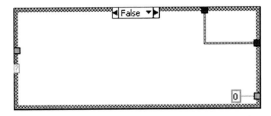

FigureL6.13: False case

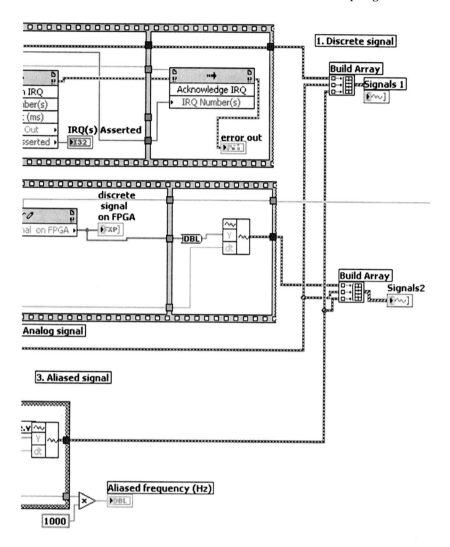

Figure L6.14: Displaying the signals using graphs

Finally, display the discrete signal generated on the host, the analog signal, and the aliased signal (if generated) using the build array and waveform graph **Signals1**. Similarly, display the discrete signal generated on the FPGA, the analog signal, and the aliased signal (if generated) using the build array and waveform graph **Signals2**. Right click on **Signals1** and choose **properties.** Set the properties as shown in Figures L6.15, L6.16 and L6.17.

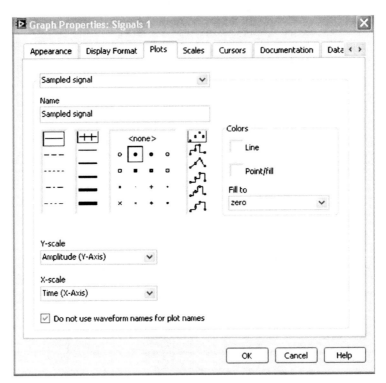

Figure L6.15: Setting properties for the sampled signal in graph

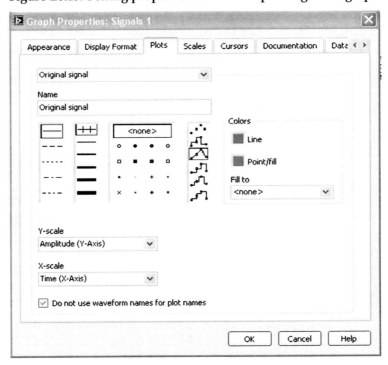

Figure L6.16: Setting properties for the original signal (analog) in graph

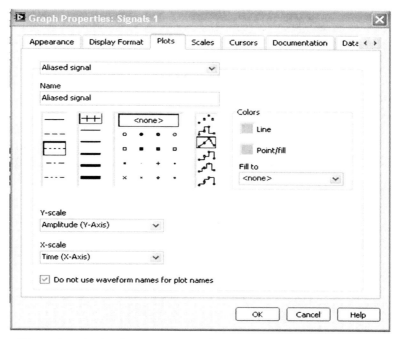

Figure L6.17: Setting properties for the aliased signal in graph

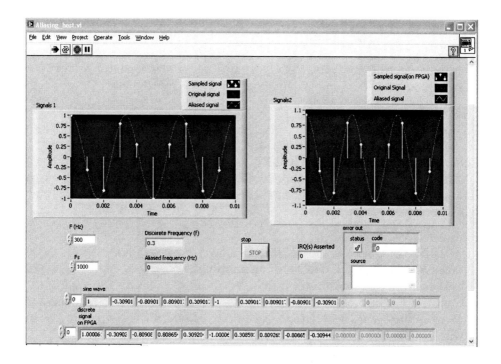

Figure L6.18: Host VI in run mode

Run the host VI. Figure L6.18 shows a snapshot with F=300 Hz and Fs=1000Hz. The sampled and analog signals are clearly discernible with the sampled signal on the FPGA VI in synchronism with the FPGA VI. Note the values of the sampled signal in fixed-point format match closely with the floating-point values.Similarly, Figure L6.19 shows the aliasing effect when F=700Hz and Fs=1000Hz for f=0.7>0.5. As seen in this figure, the 700 Hz tone is aliased to the 300 Hz tone, indicated by the dotted line.

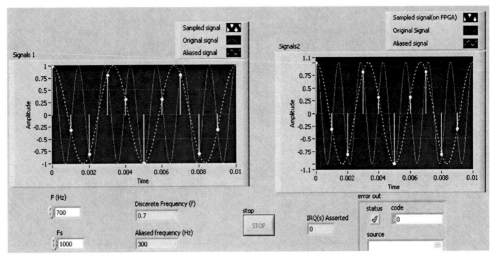

Figure L6.19: Aliasing effect

L6.2 Quantization in LabVIEW FPGA

Let us now build an A/D converter VI to illustrate the quantization effect. The VI is build on the host or without any FPGA implementation. An analog signal given by

$$y(t) = 5.2\exp(-10t)\sin(20\pi t) + 2.5 \tag{6.6}$$

is considered for this purpose. Note that the maximum and minimum values of the signal fall in the range 0 to 7, which can be represented by 3 bits. On the Front Panel, the quantization error, the histogram of the quantization error, as well as the quantized output are displayed as indicated in Figure L6.20. The following steps are to be followed:

- To build the converter BD (**Quantization.vi**), refer to Figure L6.21, the **Formula Waveform VI** (**Functions>>Programming>>Waveform>>Analog Waveform >>Waveform Generation>>Formula Waveform**) is used. The inputs to this VI comprise a string constant specifying the formula, amplitude, frequency, and sampling information. The values of the output waveform, **Y** component, are extracted with the **Get Waveform Components** function.

- To exhibit the quantization process, the double precision signal is converted into an unsigned integer signal by using the **To Unsigned Byte Integer** function (**Functions >>Programming>>Numeric>>Conversion>>To Unsigned Byte Integer**). The resolution of quantization is assumed to be 3 bits, noting that the amplitude of the signal remains between 0 and 7. Values of the analog waveform are replaced by quantized values forming a discretized waveform. This is done by wiring the quantized values to a **Build Waveform** function while the other properties are kept the same as the analog waveform.

- Now the difference between the input and quantized output values can be found by using the **Subtract** function. This difference represents the quantization error. Also, the histogram of the quantization error is obtained by using **Create Histogram Express VI (Functions >>Express >>Signal Analysis >>Create Histogram)**. Placing this VI brings up a configuration dialog as shown in Figure L6.22. The maximum and minimum quantization errors are 0.5 and -0.5, respectively. Hence, the number of bins is set to 10 in order to divide the errors between -0.5 and 0.5 into 10 levels. In addition, for the **Amplitude Representation** option, choose **Sample count** to generate the histogram. A Waveform Graph can be created by right-clicking on the **Histogram** node of the **Create Histogram Express VI** and choosing **Create >>Graph Indicator**.

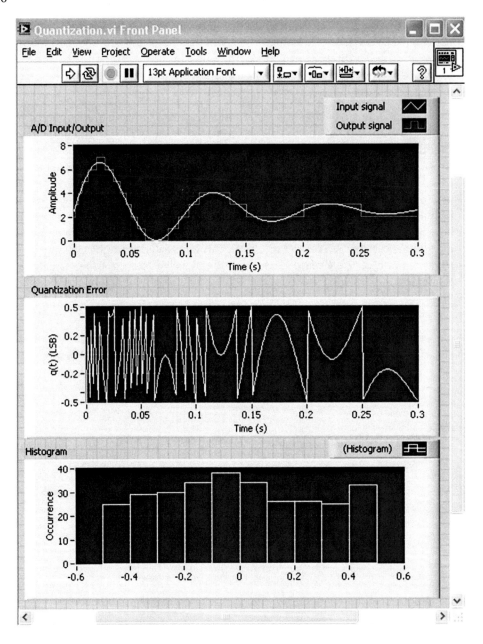

Figure L6.20: Quantization of an analog signal by a 3-bit A/D converter: (top to bottom) output signal, quantization error, and histogram of quantization error

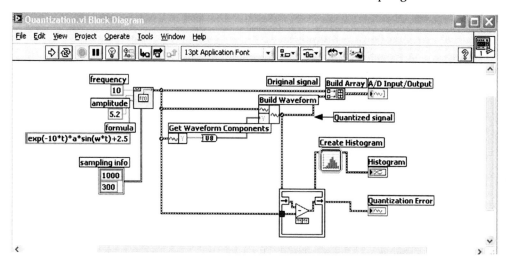

Figure L6.21: Quantization of an analog signal by a 3-bit A/D converter

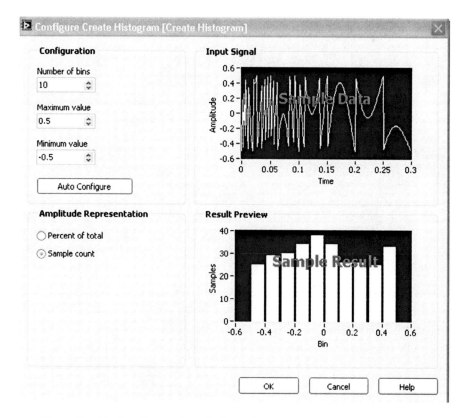

FigureL6.22: Configuration dialog of Create Histogram Express VI

- Return to the FP and change the property of the graph for a more understandable display of the discrete signal. Add the plot legend to the waveform graph and resize it to display the two signals. Rename the analog signal as **Input Signal** and the discrete signal as **Output Signal**. To display the discrete signal, bring up the properties dialog box by right-clicking and choosing **Properties** from the shortcut menu. Click the **Plots** tab and choose the signal plot **Output Signal**. Then, choose **stepwise horizontal**, indicated by ⌐⌐, from the **Plot Interpolation** option as the interpolation method. The VI is complete now.

Next, let us build a VI (**Quantization Bit stream.vi**) which can analyze the quantized discrete waveform into a bitstream resembling a logic analyzer. For a 3-bit A/D converter, the bitstream can be represented by $b_3 b_2 b_1$ in binary format. The discrete waveform and its bit decomposition are shown in Figure L6.23.

- The same analog signal used in the previous example is considered here. The analog signal is generated by a **Formula Waveform VI**, and quantized by using a **To Unsigned Byte Integer** function.

- Locate a **For Loop** to repeat the quantization as many as the number of samples. This number is obtained by using the **Array Size** function (**Functions>>Programming >>Array>> Array Size**). Wire this number to the **Count** terminal of the For Loop. Wiring the input array to the For Loop places a **Loop Tunnel** on the loop border. Note that auto indexing is enabled by default when inputting an array into a For Loop. With auto indexing enabled, each element of the input array is passed into the loop one at a time per loop iteration.

- In order to obtain a binary bitstream, each value passed into the For Loop is converted into a Boolean array via a **Number To Boolean Array** function (**Functions >>Programming>>Boolean>>Number To Boolean array**). The elements of the **Boolean array** represent the decomposed bits of the 8-bit integer. The value of a specific bit can be accessed by passing the **Boolean array** into an **Index Array** function (**Functions>>Programming>>Array>>Index Array**) and specifying the bit location with a **Numeric Constant**. Since the values stored in the array are Boolean, i.e. False or True, they are converted into 0 and 1, respectively, using the **Boolean To (0,1)** function (**Functions>>Programming>>Boolean>>Boolean to (0,1)**). Data from each bit location are wired out of the For Loop. Note that an array output is created with the auto-indexing being enabled.

As configured in the previous example, the **stepwise horizontal** interpolation method is used for the waveform graph of the discrete signal. The completed VI is shown in Figure L6.24.

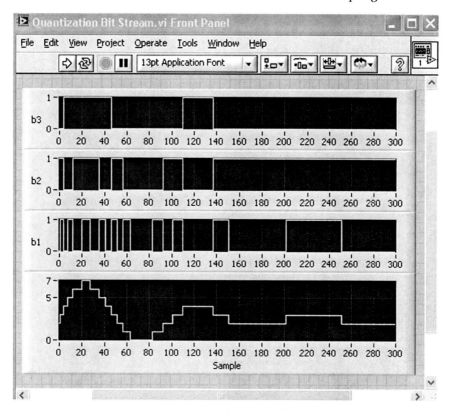

FigureL6.23: Bitstream of 3-bit quantization

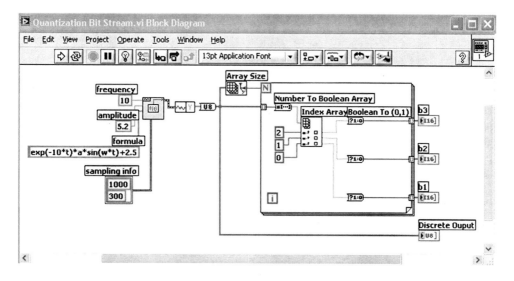

Figure L6.24: BD of logic analyzer

L6.3 Signal Reconstruction in LabVIEW FPGA

As the final part of this lab, a signal reconstruction VI is built. The VI is built as a host VI named **Reconstruction.vi,** added to the same LabVIEW project **sampling_and_recsonstruction.lvproj** with the FPGA VI as **add_Zeroes.vi,** refer to Figure L6.25.

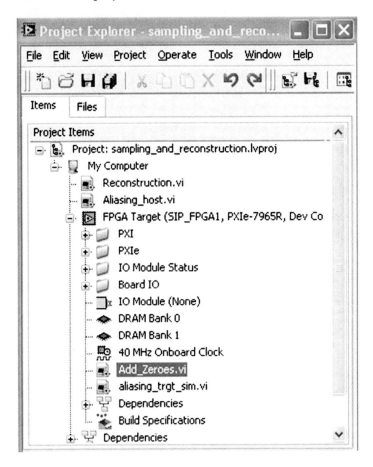

Figure L6.25: Modification in sampling_and_recsonstruction.lvproj

Let us examine the FP shown in Figure L6.26 exhibiting a sampled signal and its reconstructed version. The reconstruction kernel is also shown in this FP.

The sampled signal is shown via bars in the waveform graph **Analog signal**. In order to reconstruct an analog signal from the sampled signal, a convolution operation with a sinc function is carried out as specified by Equation (6.5).

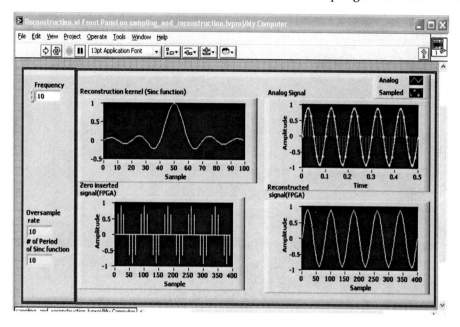

Figure L6.26: FP of a reconstructed sine wave from its samples

Let us build the VI. As shown in Figure L6.27, it is assumed that a unity amplitude sinusoid of 10 Hz is sampled at 80 Hz. To display the reconstructed analog signal, the sampling frequency and the number of samples are set to 100 times those of the discrete signal. The two waveforms are merged and displayed in the waveform graph **Analog signal**.

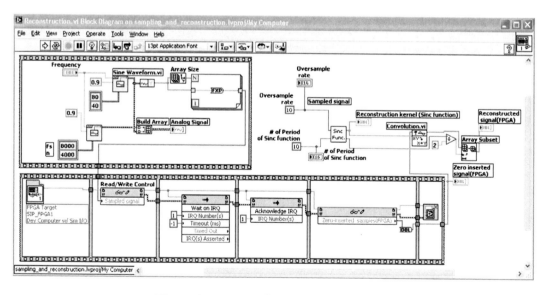

Figure L6.27: BD of signal reconstruction

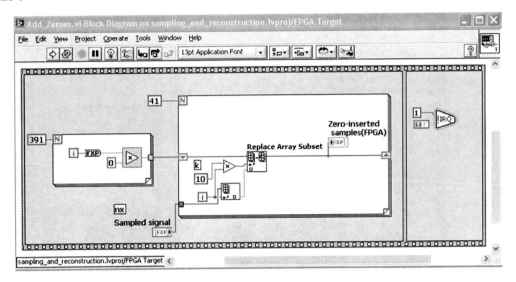

Figure L6.28: BD of FPGA VI

Figure L6.28 displays the BD of the **FPGA VI Add_zeroes.vi**. It inserts zeroes between consecutive samples to simulate oversampling. With 40 samples of input sampled data, oversampled 10 times (this must be constant as arrays in the FPGA VI must be of fixed size), the resultant signal has 391 samples due to 10 samples inserted between any two samples. The output of the VI comprises the array of zero-inserted samples, **zero-inserted samples (FPGA)**. Similar to section 6.2, interrupts are used for synchronization between the host and the FPGA VI.The sub VI **Sinc Func.vi** generates samples of a sinc function based on the number of samples, delay, and sampling interval parameters. Its BD is shown in Figure L6.29.

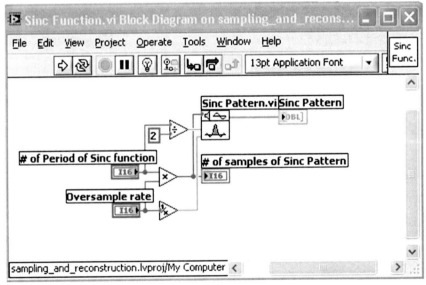

Figure L6.29: Sinc function subVI

Finally, let us return to the BD shown in Figure L6.27. The zero-inserted signal and sinc signal are convolved using a **Convolution VI** function **(Functions>>Signal Processing >>Signal Operation>>Convolution)**. Note that the length of the convolved array obtained from the Convolution VI is one less than the sum of the samples in the two signals.

L6.4 Bibliography

[1] National Instruments, *LabVIEW User Manual*, Part Number 320999E-01, 2003.
[2] B. Razavi, *Principles of Data Conversion System Design*, IEEE Press, 1995.

L6.5 Lab Experiments

- Build a VI to generate the signal exprerssed by Equation (6.7) with the frequency F Hz and amplitude A based on a sampling frequency Fs= 4000 Hz with the number of samples being 200. Set the frequency range from 1 Hz to 1000 Hz and the amplitude range from 0 to 25. As shown in the L6.1 example, create the signal in an FPGA VI, passing F and Fs as controls and reading back the discrete signal on the host VI (use a high throughput square root function for implementing the equation). Use the Q16 format for F, Fs and the constant 3.7. Use interrupts for the synchronization. Generate the quantized bit stream and display it together with the quantization error. Compare the results obtained with and without using the FPGA VI for signal generation.

$$x(t) = \sqrt{A}\sin\left(2\pi Ft\right) + \sqrt{3.7} \qquad (6.7)$$

- Create a signal reconstruction VI similar to L6.3, except in place of Sinc function.vi on the host, create this subVI on the FPGA VI with the corresponding inputs and outputs represented in the Q16 fixed-point format (use integer format wherever necessary). Compare the result with that in L6.3.
- Build a FPGA VI to generate the signal given by Equation (6.8) with the frequencies f_1 Hz and f_2 Hz and the amplitude A with the number of samples being 300. Compute the sampling frequency as (4 * *max* (f_1, f_2)). Set the frequency ranges from 1 Hz to 1 kHz and the amplitude range from 0 to 40. Generate the quantized bit stream (similar to Figure L6.21) and display it together with the quantization error on the host VI.

$$x(t) = A\sin\left(2\pi f_1 t\right) + A\cos\left(2\pi f_2 t\right) \qquad (6.8)$$

CHAPTER 7

Convolution

Consider a discrete time input signal $x(n)$ that is passed through a linear shift invariant filter with the unit sample response $h(n)$ generating a discrete time output signal $y(n)$. The output signal can be obtained by the convolution of the input signal with the filter, that is

$$y(n) = \sum_{k=-\infty}^{\infty} x(k).h(n-k) \tag{7.1}$$

The notation $y(n) = x(n) * h(n)$ is often used to indicate the above convolution summation. For the causal case normally occurring in practice, the above convolution sum can be written as

$$y(n) = \sum_{k=0}^{n} x(k).h(n-k) \tag{7.2}$$

If the sequence $x(n)$ and $h(n)$ are finite length signals of length nx and nh, respectively, the convolution outcome $y(n)$ will be of length $nx+nh-1$. The output at a particular time instant n_0 can be found by evaluating this sum

$$y(n_0) = \sum_{k=0}^{n} x(k).h(n_0-k) \tag{7.3}$$

The following implementation steps are needed to compute $y(n_0)$:

1) **Folding**: Fold or mirror $h(k)$ about $k=0$ to get $h(-k)$.
2) **Shifting**: Shift the mirrored signal $h(-k)$ n_0 samples to the right ($n_0>0$) or to the left ($n_0<0$) to obtain $h(n_0-k)$. If $h(k)$ is causal, $h(-k)$ will be anticausal, and if $x(k)$ is causal, then $h(-k)$ needs to be shifted to the right ($n_0>0$) to cover $x(k)$ in order to obtain a non-zero product in the next step.
3) **Product**: Obtain the product sequence $p_{n0}(k) = x(k).h(n_0-k)$, by multiplying sample by sample the sequence $x(k)$ and $h(n_0-k)$.

4) **Sum**: Obtain $y(n_0)$ as sum of samples in the sequence $p_{n0}(k)$, i.e. $y(n_0) = \sum_{k} p_{n0}(k)$

For a better understanding, let us consider the following example:
$x(n) = \{1, 2, 3\}$ and $h(n) = \{-1, 1\}$

Chapter 7

where *nx* denotes length of signal *x* = 3, *nh* length of signal *h* = 2, *ny* length of signal *h*= *nx+nh-1* = 4.

Let us compute the output graphically as well as numerically based on the above steps.
1) Folding: *h(k)* is mirrored to get *h(-k)* as shown in Figure 7.1,

h(-k)= {1, -1}

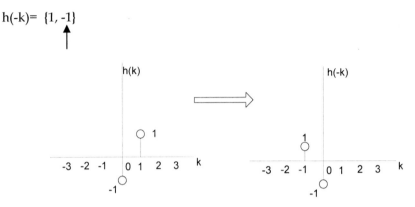

Figure 7.1: Folding

2) The product *p(0)* is computed by multiplying *x(k)* with *h(-k)*,

$p(0)$= {1, 2, 3} . {1, -1} = {1*1, 2*-1, 3*0} = {1, -2, 0}

Figure 7.2 illustrates this operation.

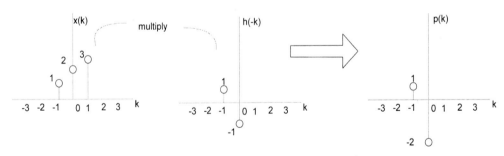

Figure 7.2: Convolution product at k

3) $y(0)$ is the sum of the elements of $p(0) = \sum_{k=-1}^{1} p(k) = 1 + -2 + 0 = -1$

4) To get $y(1) = \sum_{k} x(k).h(1-k)$, *h(1-k)* needs to be computed. *h(1-k)* is obtained from *h(-k)* by shifting *h(-k)* one sample to the right as shown in Figure 7.3, *h(1-k)* ={0, 1, -1}

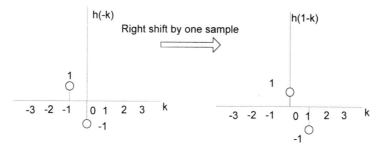

Figure 7.3: Shifting

Then, the product $p(1)$ is formed by $x(k).h(1-k)$ as shown in Figure 7.4 resulting in $y(1) = 0+2+-3 = -1$.

$$p(1)= \{1, 2, 3\} . \{0, 1, -1\} = \{1*0, 2*1, 3*-1\} = \{0, 2, -3\}$$

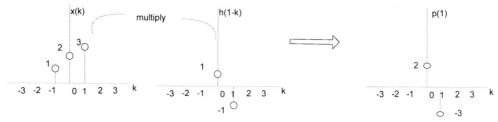

Figure 7.4: Convolution product at 1

5) Similarly, for $y(2)$, the product $x(k).h(2-k)$ is formed, where $h(2-k)$ is $h(k)$ shifted 2 samples to the right, and then the product elements are summed as shown numerically below

$$p(2) = \{1, 2, 3, 0\} . \{0, 0, 1, -1\} = \{1*0, 2*0, 3*1, 0*-1\} = \{0, 0, 3, 0\}$$

resulting in $y(2) = 0+0+3+0 =3$.

Beyond y(3), y becomes zeroes as $h(n-k)$ for $n>2$ aligns zeroes with $x(k)$ making the products $p(k)$ zeroes. On the other hand, one gets non-zero outputs for $y(-1)$ by summing the product elements $x(k) h(n-k)$ for $n=-1$. Thus, $p(-1)$ will be

$$p(-1)= \{0, 1, 2, 3\} . \{1, -1, 0, 0\} = \{0*1, 1*-1, 2*0, 3*0\} = \{0, -1, 0, 0\}$$

and $y(-1) = 0+ -1+0+0= -1$.

This is illustrated in Figure 7.5.

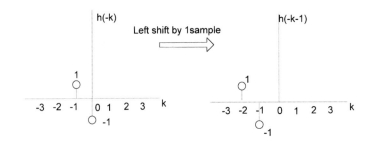

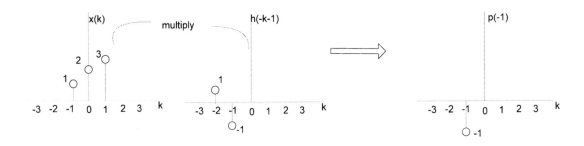

Figure 7.5: Convolution product at -1

Likewise, for $y(-2)$, the following is obtained

$$p(-2) = \{0, 0, 1, 2, 3\} \cdot \{1, -1, 0, 0, 0\} = \{0*1, 0*-1, 1*0, 2*0, 3*0\} = \{0, 0, 0, 0, 0\}$$

and $y(-2)=0$, and so on.

Thus, $\{1,2,3\} * \{-1,1\}$ gives $\{\ldots 0, -1,-1,-1,3,0\ldots\ldots\}$ and the length of $y(n)$ is equal to the number of non-zero elements which is 4 or length of $x(n)$ + length of $h(n)$ - 1.

Lab 7:

FPGA Implementation of Convolution

L7.1 Convolution Algorithm

To perform the FPGA implementation of the convolution algorithm using the IP Integration Node, the algorithm needs to be amenable to the VHDL paradigm. This is best explained through an example.

Let $x(n) = \{3, 2, 1\}$ and $h(n) = \{4, 5\}$

where $nx=3$, $nh=2$, and the length of $y(n) = 3+2-1= 4$.

a) Create $hm(n)$ as mirrored $h(n)$

$hm(n) = \{5, 4\}$, i.e. the array $hm(n)$ is the reversed version of $x(n)$.

b) Align $x(n)$ and $hm(n)$,

$x(n) = \{\quad 3, 2, 1\}$
$hm(n) = \{5, 4\}$

Then, fill $hm(n)$ with length of $x(n)$-1 (2 here) zeroes to the right

$x(n) = \{\quad 3, 2, 1\}$
$hm\_z(n) = \{5, 4, 0 , 0\}$

c) Add length of $h(n)$-1 (1 here) zeroes to the left of $x(n)$
$x\_z(n) = \{\ 0, 3, 2, 1\}$
$hm\_z(n) = \{5, 4, 0 , 0\}$

Now, both $x\_z(n)$ and $hm\_z(n)$ have the equal length of $nx+nh-1$.

d) Take products of $x\_z(n)$ with $hm\_z(n)$ and sum the elements. Then, keep shifting $hm\_z(n)$ to right till the leftmost element of $hm\_z(n)$ reaches the rightmost element of $x\_z(n)$ beyond which zero products are obtained. This process is shown below

$x\_z(n) = \{0, 3, 2, 1\} => \{0, 12, 0, 0\}$ sum $= 12$

201

$hm\_z(n) = \{5, 4, 0, 0\}$

$x\_z(n) = \{0, 3, 2, 1, 0\} \Rightarrow \{0, 15, 8, 0, 0\}$ sum = 23
$hm\_z(n) = \{0, 5, 4, 0, 0\}$

$x\_z(n) = \{0, 3, 2, 1, 0, 0\} \Rightarrow \{0, 0, 10, 4, 0, 0\}$ sum = 14
$hm\_z(n) = \{0, 0, 5, 4, 0, 0\}$

$x\_z(n) = \{0, 3, 2, 1, 0, 0, 0\} \Rightarrow \{0, 0, 0, 5, 0, 0, 0\}$ sum = 5
$hm\_z(n) = \{0, 0, 0, 5, 4, 0, 0\}$

Beyond this point, one gets zeroes.

From the above example, the following points are noted:

- Number of shifts needed is equal to the number of samples (5) needed for $hm\_z(n)$ and $x\_z(n)$ to get aligned with the rightmost element (1) in $x\_z(n)$, see Figure L7.1.

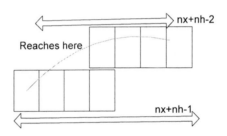

Figure L7.1: Number of shifts

That is the number of shifts is $nx+nh-2$; for $nx=3$, $nh=2$, it becomes 3+2-2 = 3 shifts.

- For VHDL implementation, alternatively, the signal $hm\_z\_sh(n)$ can be created to have $nx+nh-2$ zeroes added to the left as indicated above.
- Then, to avoid unnecessary zero by zero multiplications, slices of $hm\_z\_sh(n)$ having the same elements can be considered to align the non-zero elements of $x\_z(n)$ and $hm\_z(n)$ for taking the sum of the products. For example,

$hm\_z\_sh = \{0, 0, 0, 5, 4, 0, 0\}$ and the first product is obtained to be

$x\_z(n) = \{0, 3, 2, 1\} \Rightarrow \{0, 12, 0, 0\}$ sum = 12
$hm\_z\_sh(3:6) = \{5, 4, 0, 0\}$

Thus, the resulting convolution output becomes {12, 13, 14, and 15}.

L7.2 VHDL Implementation

The above algorithm is implemented in VHDL as indicated in Figure L7.2. It is assumed that the package int.array is declared in the work library which contains the definition for creating integer arrays.

```vhdl
use ieee.std_logic_1164.all;
use ieee.std_logic_arith.all;
use work.int_array.all;--***** using package from separate file****

entity conv_wocomp_wofunc is

  generic (nx:natural :=3;
           nh:natural :=8);
  port (x:in integer_vector(nx-1 downto 0); -- length nx
       h:in integer_vector(nh-1 downto 0);  -- length nh
       y: out integer_vector((nx+nh-1)-1 downto 0));

end conv_wocomp_wofunc;

architecture a1 of conv_wocomp_wofunc is
        signal x_z: integer_vector((nx+nh-1)-1 downto 0);-- length nx+nh-1
        signal  hm_z_sh: integer_vector(((2*(nx+nh))-3)-1 downto 0);
                                -- nx+nh-2(shifts)+nx+nh-1(orig length)

  begin

  P1:  process (h,x)
        variable hm :integer_vector(nh-1 downto 0);
        variable hm_z :integer_vector((nx+nh-1)-1 downto 0);-- length nx+nh-1
  begin

        for i in nh-1 downto 0 loop                      -- mirror h
          hm(i):= h((h'length-1)-i);
        end loop;

-- continued on next page
```

```
        for i in (hm_z'length-1) downto 0 loop                    -- add zeros to right
          hm_z(i):= 0;
        end loop;

        for i in (hm_z'length-1)downto (hm_z'length-hm'length) loop    --copy hm
          hm_z(i):=hm(i-(x'length-1));
        end loop;

        for i in (hm_z_sh'length-1) downto 0 loop -- add left zeros(right shift v_hm_z)
          hm_z_sh(i)<= 0;
        end loop;

        for i in (hm_z'length-1) downto 0 loop
          hm_z_sh(i)<=hm_z(i);                        -- variable to signal assignment
        end loop;

     end process;

P2: process (x)
      begin

        for i in (x_z'length-1) downto 0 loop             -- add left zeros to x
          x_z(i)<= 0;
        end loop;

        for i in (x'length-1)downto 0 loop
          x_z(i)<=x(i);
        end loop;

  end process;

-- continued on next page
```

```
P3: process(x_z,hm_z_sh)
   variable sum:integer :=0;
   variable p:integer_vector((nx+nh-1)-1 downto 0);
   variable hm_z_sh_sl :integer_vector((nx+nh-1)-1 downto 0);
                           --sliced from hm_z_sh to match final length
      begin
      for i in (p'length-1) downto 0 loop
        p(i):= 0;
      end loop;

      for k in (y'length -1) downto 0 loop
        hm_z_sh_sl := hm_z_sh((2*(y'length-1)-k) downto (y'length-1 -k));
                     -- slice hm_z_sh, note:signal to variable assignment

        for i in (hm_z_sh_sl'length -1) downto 0 loop
          p(i):= x_z(i)* hm_z_sh_sl(i);
                                    -- take product
        end loop;

        for j in (p'length -1)downto 0 loop
          sum := sum+ p(j);              --take sum
        end loop;

        y(k)<= sum;
        sum:=0 ;
      end loop;

   end process;
end a1;
```

Figure L7.2: Convolution in VHDL

L7.3 LabVIEW FPGA Implementation

The above algorithm can also be implemented inside the LabVIEW FPGA using the graphical programming approach. The advantage gained would be in speed since the entire Block Diagram gets executed inside a single cycle timed loop (SCTL) taking only 1 clock cycle to execute.

The VI **convolution_lv_bk.vi** shows how this can be achieved. This VI uses two subVIs: **add_arr_ele.vi** (see Figure L7.6 and L7.7) and **mirror.vi** (see Figure L7.5), which add the elements in an array and mirror the array, respectively. Both of these subVIs must be added to the FPGA target under the same project.

The following points are to be noted:

a. All the array sizes must be constant since the array functions inside the FPGA VI use fixed size arrays. The diagrams shown in Figures L7.3 and L7.4 are for implementing the convolution of two signals of length 4 elements and of length 3 elements.

b. A large number of indicators are shown in the BD. This is for explanation purposes so that the variables used in the VHDL code can be understood in the Block Diagram by establishing a one to one correspondence. However, in actual implementation, avoid using many indicators in order to minimize the FPGA resources.

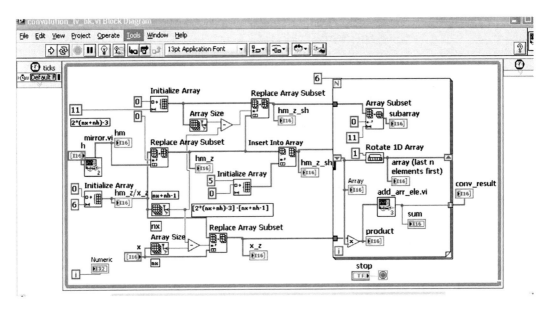

Figure L7.3: Block Diagram of convolution_lv_bk

L7.4 Bit-wise VHDL Implementation of Convolution in IP Node

The above code assumes that integer arrays are declared in a separate package and can be compiled within the calling program. However, if the same code is to be implemented inside an IP Node, it does not compile since the IP Node assumes the input and output port signals to be of the type std_logic_vector or signed/unsigned array. Therefore, in order to implement convolution using std_logic_vector port signals, the following approach is needed.

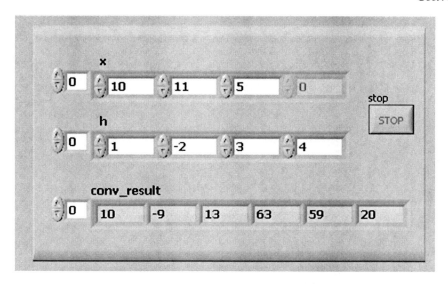

Figure L7.4: Front Panel of convolution_lv_bk

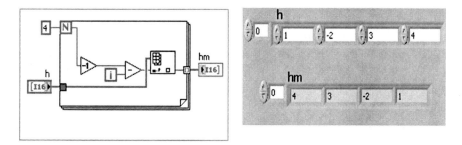

Figure L7.5: Block Diagram and Front Panel of mirror.vi

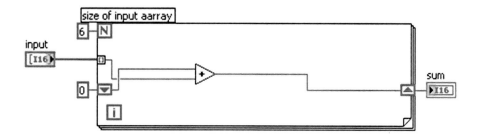

Figure L7.6: Block Diagram of add_arr_ele.vi

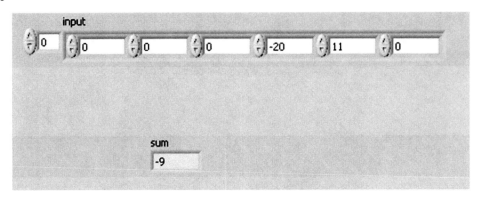

Figure L7.7: Front Panel of add_arr_ele.vi

Consider $x(n) = \{3, 2\}$ and $h(n) = \{4, 5, 6\}$

Now, $x(n) * h(n) = x(n) * [h(0).\delta(n) + h(1). \delta(n-1) + h(2). \delta(n-2)]$
$$= h(0).x(n) + h(1.x(n-1) + h(2).x(n-2)$$
$$= 4.x(n) + 5.x(n-1) + 6.x(n-2)$$

where

$$4.x(n) = \{12, 8\}$$
$$5.x(n-1) = \{0, 15, 10\}$$
$$6.x(n-2) = \{0, 0, 18, 12\}$$

Hence,

$y(n) =$	12	8	0	0
+	0	15	10	0
+	0	0	18	12
	12	23	28	12

The above shift and add algorithm can get implemented in VHDL. However, it should be noted that if $x(n)$ is of length nx and there are 12 bits per integer, then $x(n)$ will be of length $nx*12$. For example, a binary representation of $x(n)$ would be

$$x(n) = [3\ 2] = [000000000011\ \ 000000000010]$$
$$= [0000\ 0000\ 0011\ 0000\ 0000\ 0010]$$
$$= [0\ 0\ 3\ 0\ 0\ 2]\ \text{in hexadecimal format}$$

By taking the above into account, the VHDL convolution code is written as shown in Figure L7.8

```
library ieee;
use ieee.std_logic_1164.all;
use  ieee.numeric_std.all;

entity convolution_bit_by_bit is

generic (nx:natural :=2;  -- x=[ 3 2]
        nh:natural :=3; -- h=[ 4 5 6];
        nb:natural:=8); -- no of bits/integer

port (x:in std_logic_vector(0 to (nx*nb)-1);
 -- 8 bits/integer,where integer  no is between 0-9 , nx=2, 0 to 15
     h:in std_logic_vector(0 to (nh*nb)-1);
     y:out std_logic_vector(0 to ((nx+nh-1)*nb)-1));

end convolution_bit_by_bit;
architecture a1 of convolution_bit_by_bit is
 begin
   process(x,h)
   variable a ,b,p: std_logic_vector(0 to nb-1);
   variable a1,b1,p1_slic: signed(0 to nb-1);
   variable p1: signed(0 to (2*nb)-1);
   variable p2: std_logic_vector(0 to x'length-1);
   variable p3: std_logic_vector(0 to (nh*p2'length)-1);
   variable p3_us: signed(0 to p3'length-1);
   variable p3_us_slic: signed(0 to x'length-1);
   variable y1,s1_us,p3_us_app: signed(0 to ((nx+nh-1)*nb)-1);
   variable c: signed(0 to ((nh-1)*nb)-1);
 begin
     p2:= (others=> '0');
     for i in 0 to nh-1 loop
         b:=h((nb*i)to ((nb*i)+(nb-1))) ;     -- take each digit of h
         for j in 0 to nx-1 loop
           a:= x((nb*j)to ((nb*j)+(nb-1))) ; -- take each digit of x
           a1:= signed(a);
           b1:= signed(b);
           p1:= a1 * b1;
                             -- multiply the two, result will be 2*nb bits
           p1_slic:=p1(nb to (2*nb)-1);
           p:=std_logic_vector(p1_slic);
           p2((nb*j)to ((nb*j)+(nb-1))):=p;
                             -- form product sequence for each digit of h
         end loop;
         p3( (nx*nb*i) to ((nx*nb*i)+(p2'length-1))):= p2;
     end loop;
--Continued on next page
```

209

```
s1_us:= (others=> '0');
        c:= (others=>'0');
        p3_us:= signed(p3);

        for i in 0 to nh-1 loop
                p3_us_slic := p3_us((nx*nb*i) to ((nx*nb*i)+(p2'length-1)));
                p3_us_app:= c&p3_us_slic;
                s1_us:=s1_us + shift_left(p3_us_app,(nh-1-i)*nb);

        end loop;
        y1:= s1_us;
        y<= std_logic_vector(y1);

  end process;
end a1;
```

Figure L7.8: VHDL code for convolution using bitwise operations

The code can be verified from the ModelSim simulation indicated in Figures L7.9 and L7.10
using the following example
1) conv {9, 8, 7, 6} and {9, 9} = {81, 153, 135, 117, 54}
The hex equivalent of the convolution outcome (8 bits for each digit) becomes
[51 99 87 75 36]
or in 12 bits,
[051 099 087 075 036]

Figure L7.9: ModelSim simulation outcome 1

2) conv {-3, 2} and {1, 4, 5} = {-3, -10, -7, 10} with the
binary and hex equivalents of
x= [1111 1111 1101 0000 0000 0010] or x = [ffd 002]
h= [0000 0000 0001 0000 0000 0100 0000 0000 0101]; or h=[001 004 005];
The convolution outcome in binary and hex is
        y = [1111 1111 1101 1111 1111 0110 1111 1111 1001 0000 0000 1010]

y = [ffd ff6 ff9 00a]

This is shown in Figure L7.10.

Figure L7.10: ModelSim simulation outcome 2

L7.4.1 LabVIEW FPGA Implementation

The above VHDL code can also get implemented inside the IP Node within the LabVIEW FPGA. However, it should be noted that the above code has been written using an ascending range while the IP Node supports only a descending range. This requires that the inputs are provided in a reverse order, and the outputs are interpreted in a reverse order as well. Thus, for the above example,

x = [003 002] = [000000000011 000000000010] should be specified with the reverse bits, i.e.

x = [010000000000 110000000000]

h = [004 005 006] = [000000000100 000000000101 000000000110] should be specified as

h = [011000000000 10100000000 001000000000]

Likewise, the convolution outcome

y = [00C 017 01C 00C] (hex) = [000000001100 000000010111 000000011100 000000001100] should be specified as

y = [001100000000 001110000000 111010000000 001100000000]

The Front Panel and Block Diagram of the implementations appear in Figures L7.11 and L7.12, respectively.

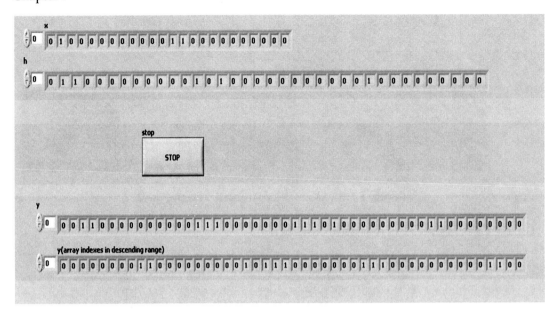

Figure L7.11: Front Panel of convolution_bitwise_bk.vi

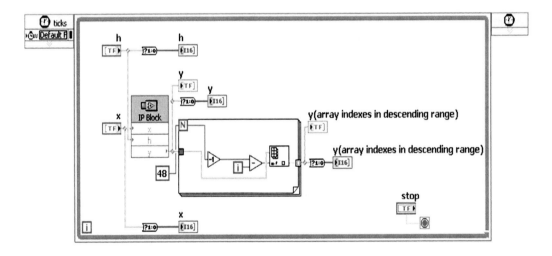

Figure L7.12: Block Diagram of convolution_bitwise_bk.vi

L7.5 Lab Experiments

- Digital correlation is similar to digital convolution. Correlation can be implemented by using the convolution algorithm and taking the mirror image of the kernel signal. Use this concept to write a bitwise VHDL code for implementing correlation similar to L7.4.

- Implement the above program in LabVIEW IP Integration Node and verify the results with the ModelSim simulation.

- Implement the correlation operation using the standard definition of correlation (see Equation (7.4)) in the LabVIEW FPGA using the fixed-point representation and high throughput functions (numbers would be real, thus use the representation <+,-16,1>).

$$y(n) = \sum_{k=-\infty}^{\infty} x(k).h(k+n) \tag{7.4}$$

CHAPTER 8

Digital Filtering

Filtering is a fundamental concept in digital signal processing. Here, it is assumed that the reader is already familiar with the theory of digital filtering and the design of FIR (Finite Impulse Response) and IIR (Infinite Impulse Response) filters.

In this chapter, the structure of digital filters is briefly mentioned followed by a discussion on the LabVIEW Digital Filter Design (DFD) toolkit. This toolkit provides various tools for the design, analysis, and simulation of digital filters.

8.1 Digital Filtering

8.1.1 Difference Equations

As a difference equation, an FIR filter is expressed as

$$y[n] = \sum_{k=0}^{N} b_k x[n-k] \tag{8.1}$$

where b's denote the filter coefficients and N the number of zeros or filter order. As described by this equation, an FIR filter operates on a current input $x[n]$ and a number of previous inputs $x[n-k]$ to generate a current output $y[n]$.

The equi-ripple method, also known as the Remez algorithm, is often used to produce an optimal FIR filter [1]. Figure 8.1 shows the filter responses using the design methods consisting of equi-ripple, Kaiser window and Dolph-Chebyshev window. Among these methods, the equi-ripple method generates a response whose deviation from the desired response is evenly distributed across the passband and stopband [2].

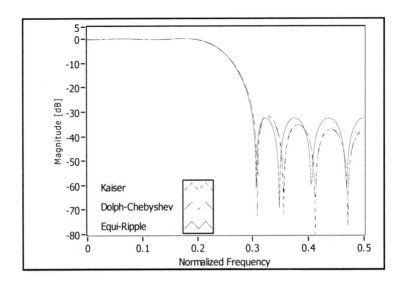

Figure 8.1: Responses of different FIR filter design methods

The difference equation of an IIR filter is given by

$$y[n] = \sum_{k=0}^{N} b_k x[n-k] - \sum_{k=1}^{M} a_k y[n-k] \qquad (8.2)$$

where b's and a's denote the filter coefficients and N and M the number of zeroes and poles, respectively. As indicated by Equation (8.2), an IIR filter uses a number of previous outputs $y[n-k]$ as well as a current and a number of previous inputs to generate a current output $y[n]$.

Several methods are widely used to design IIR filters. They include Butterworth, Chebyshev, Inverse Chebyshev, and Elliptic methods. In Figure 8.2, the magnitude response of an IIR filter designed by these methods having the same order are shown for comparison purposes. For example, the elliptic method generates a relatively narrower transition band and more ripples in passband and stopband while the Butterworth method generates a wider transition band response [2]. Table 8.1 summarizes some of the key characteristics of these design methods.

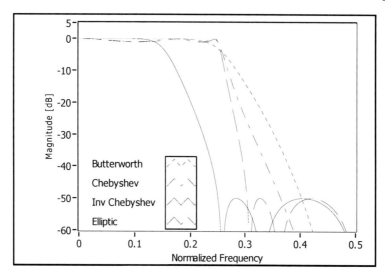

Figure 8.2: Responses of different IIR filter design methods

IIR filter	Ripple in pass-band?	Ripple in stop-band?	Transition bandwidth	Needed order for given filter specifications
Butterworh	No	No	Widest	Highest
Chebyshev	Yes	No	Narrower	Lower
Inverse Chebyshev	No	Yes	Narrower	Lower
Elliptic	Yes	Yes	Narrowest	Lowest

Table 8.1: Comparison of different IIR filter design methods [1]

8.1.2 Fixed-Point Implementation

When any filter including an FIR or IIR is to be realized on a fixed-point processor including FPGAs, one needs to address possible overflows due to data/coefficient quantizations. In what follows, the scaling approach is examined for addressing overflows.

8.1.2.1 Scaling

The idea of scaling is to scale down the input before performing any processing and then to scale up the resulting output to the original scale. Scaling can be applied to most filtering and transform operations. An easy way to achieve scaling is by shifting. Since a right shift of 1 is equivalent to a division by 2, the input can be repeatedly scaled by 0.5 until all overflows disappear. The output can then be rescaled back to the total scaling amount. However, it should be realized that scaling cannot be done indefinitely as the signal power and thus SNR (signal to noise ratio) gets reduced as a result of scaling, which in turn degrades the filtering performance [3].

To handle overflows, the following norms are often used to define the scaling amount:

- **L1 norm:** This norm is given by

$$\sum_{m=-\infty}^{\infty} |h(m)| \tag{8.3}$$

where $h(m)$ refers to the impulse response of the filter. This is the most conservative scaling norm which may lead to excessive scaling of the input, especially for sinusoids, by reducing SNR to the extent that the signal would get buried in quantization noise.

- **L2 norm:** This is norm is given by

$$\left[\sum_{m=-\infty}^{\infty} |h(m)|^2 \right]^{1/2} \tag{8.4}$$

- **L∞ norm:** This norm is expressed in terms of frequency response samples, which does the scaling in between those of L1 and L2 norms. It is given by

$$\max_{w} |H(w)| \tag{8.5}$$

where $H(w)$ denotes the frequency response of the filter.

In general, as compared to IIR filters, FIR filters require less precision and are computationally more stable. The stability of an IIR filter depends on whether its poles are located inside the unit circle in the complex plane. Consequently, whenever an IIR filter is implemented on a fixed-point processor, its stability can be affected. In [1], a summary of the differences between the attributes of FIR and IIR filters is listed.

Attribute	FIR filter	IIR filter
Stability	Always stable	Conditionally stable
Fixed-point implementation	Easier to implement	More involved
Computational complexity	More operations	Fewer operations
Datapath precision	Lower precision required	More precision required

Table 8.2: Some FIR filter attributes versus IIR filter attributes [1]

Let us now discuss the stability and structure of IIR filters. The transfer function of an IIR filter is expressed as

$$H(z) = \frac{b_0 + b_1 z^{-1} + \ldots + b_N z^{-N}}{1 + a_1 z^{-1} + \ldots + a_M z^{-M}} \tag{8.6}$$

It is well known that as far as stability is concerned, the direct-form implementation is sensitive to coefficient quantization errors. Noting that the second-order cascade form produces a more robust response to quantization noise [2], the above transfer function can be rewritten as

$$H(z) = \prod_{k=1}^{N_s} \frac{b_{0k} + b_{1k}z^{-1} + b_{2k}z^{-2}}{1 + a_{1k}z^{-1} + a_{2k}z^{-2}} \tag{8.7}$$

where $N_s = \lfloor N/2 \rfloor$, $\lfloor \cdot \rfloor$ represents the largest integer less than or equal to the inside value. This serial or cascaded structure is illustrated in Figure 8.3.

Figure 8.3: Cascaded filter stages

It is worth mentioning that each second-order filter is often considered to be of direct-form II, see Figure 8.4, in order to have a more memory efficient implementation.

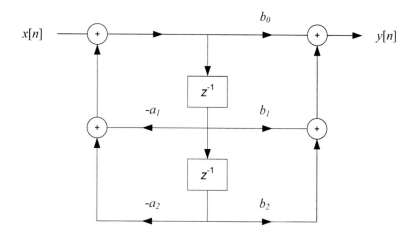

Figure 8.4: Second order direct-form II

8.2 LabVIEW Digital Filter Design Toolkit

There exist various software tools for designing digital filters. Here, the LabVIEW Digital Filter Design (DFD) toolkit is used. Any other filter design tool may be used to obtain the coefficients of a desired digital filter. The DFD toolkit provides various tools to design, analyze, and simulate floating-point and fixed-point implementations of digital filters [1].

L8.3 Bibliography

[1] National Instruments, *Digital Filter Design Toolkit User Manual*, Part Number 371353A-01, 2005.

[2] J. Proakis and D. Manolakis, *Digital Signal Processing: Principles, Algorithms, and Applications*, Prentice-Hall, 1995.

[3] W. Padgett and D. Anderson, *Fixed-Point Signal Processing*, Morgan and Claypool, 2009.

Lab 8:

FPGA Implementation of Digital Filtering

This lab covers the FPGA implementation of digital filtering using the LabVIEW FPGA in simulation mode. FIR filtering is covered in this lab. IIR filtering is covered in Chapter 11, where an actual FPGA hardware implementation is discussed. The following steps need to be taken to reach the implementation:

- Create a composite signal
- Filter it via a floating-point FIR filter
- Observe the spectrum of the filtered signal for correctness
- Convert the signal and the filter coefficients to fixed-point representation
- Save the filter coefficients in a file
- Use the Xilinx Core Generator to generate the IP for fixed-point filtering. Then, use the IP within an FPGA VI.
- Read the fixed-point filtered values on a host VI, convert it to an equivalent floating-point representation, and observe the spectrum again.

L8.1 Designing Floating-Point Filter

To start with, create a new LabVIEW FPGA project **FIR_filter.lvproj,** configure it to run in **simulation mode** with **random IO.** Create a host VI and an FPGA VI as **Fir_hst_Xilinx_coregen.vi** and **FIR_FPGA_Xilinx_coregen.vi,** respectively. This is shown in Figure L8.1.

The main host VI uses three subVIs **sig_gen.vi, filt_dsgn.vi** and **wrt_coeff.vi.** Now, let us go though the above mentioned steps one at a time.

L8.1.1 Creating Composite Signal

Figure L8.2 illustrates the BD of the subVI sig_gen.vi which creates the composite signal. Create four numeric controls **Fs, T, F1, F2** corresponding to **sampling frequency, time duration** of the signal, **signal frequency1** and **signal frequency2** on the Front Panel of the VI. Set the frequencies to **F1= 2000Hz** and **F2=7000Hz** with the sampling frequency of **Fs=32000Hz.** Set the time duration to **T=0.02 sec** (see Figure L8.3) so that the number of samples is T*Fs= **640.** In order to create sine wave signals with these frequencies, create the control **sine waveform.vi** in the Block Diagram as shown in Figure L8.2.

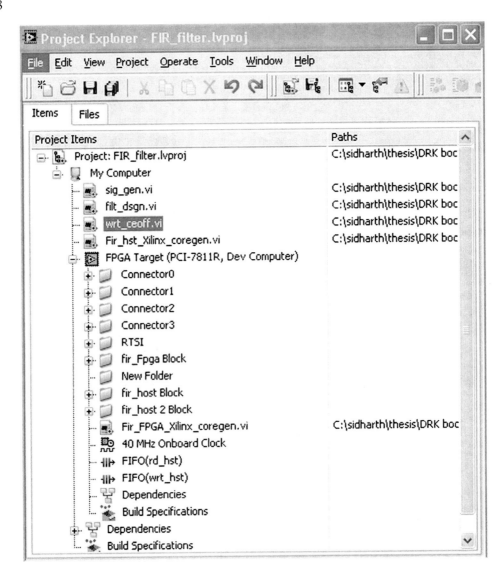

Figure L8.1: Project FIR_filter.lvproj

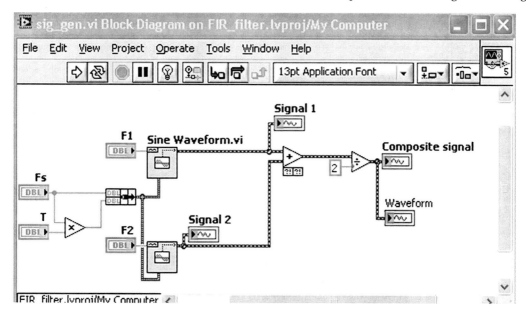

Figure L8.2: Composite signal generation

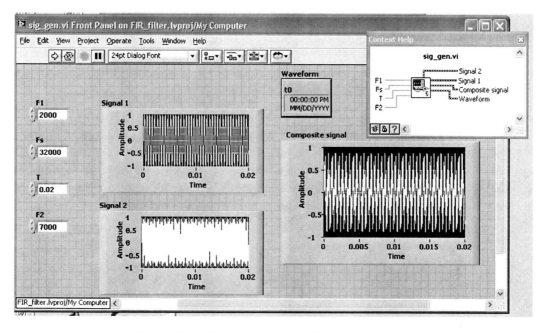

Figure L8.3: Signal generation FP controls

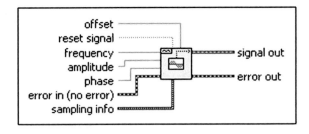

Figure L8.4: Sine waveform.vi

The sine waveform function can be located via **Functions>>signal processing>>waveform generation>>sine waveform.vi.** As shown in Figure L8.4, it has a number of input parameters. To configure its frequency, wire **F1** to the **frequency** input of this function. The amplitude and phase inputs are to be left unwired so that the signal has the default amplitude of 1 with a phase of '0' which corresponds to a sine wave. The function accepts the sampling information as a cluster input with the first element representing the sampling frequency and the second element as the number of samples. To create these inputs, first create the number of samples by wiring **F1** and **T** controls to the **Multiply** function, and then wire **F1** and the output of the multiplier to the **Bundle** function **(Programming>>Cluster class and variant>>Bundle)**. Connect the cluster output of this function to the **Sampling Info** control of the **sine waveform** function. Create the Graph indicator **Signal 1**, and connect it to the **signal out** terminal of the sine function. Similarly, create another sine wave signal of frequency F2, unit amplitude, sampled at the same rate, as indicated by the graph indicator **Signal 2**. Next, add the two signals using the **add** function, and divide them by 2 to get the composite signal while ensuring that the final amplitude is less than 1. Wire this output to the graph indicator **Composite signal.** Figure L8.3 displays these graphs.

Figures L8.5(a) and L8.5(b) display the first frame of the sequence diagram within the BD of the host VI. As seen in the upper portion of this VI, sig_gen.vi is used as a subVI. Wire the output of the waveform indicator corresponding to the composite signal to the input of the function **Spectral measurements (Functions>>Express>>Signal analysis>>Spectral measurements)** as shown in Figure L8.5(a). By default, the FFT magnitude of the input gets displayed. Wire the **FFT-peak** output to the graph indicator **FFT of input signal.** As seen in Figure L8.6, the FFT shows 2 peaks at the respective F1 and F2 frequencies.

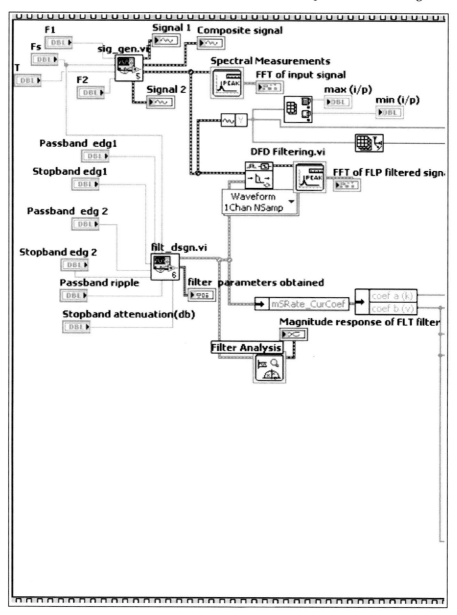

Figure L8.5(a): Creating FFT of the composite signal

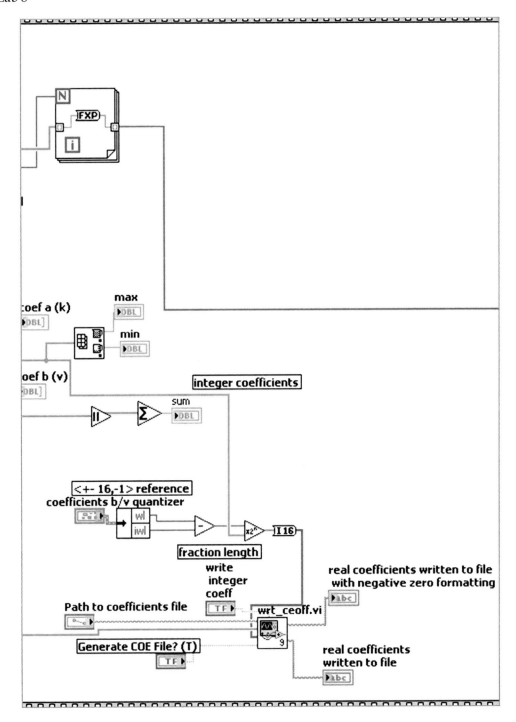

Figure L8.5(b): Creating FFT of the composite signal

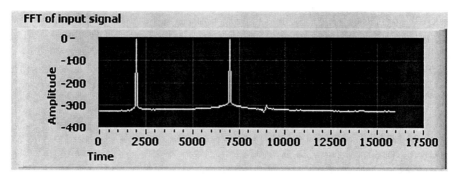

Figure L8.6: FFT of the composite signal

L8.1.2 Quantizing Input Signal

To determine the quantization for the input signal, the maximum and minimum values of the signal are determined by wiring the **composite signal** cluster to the function **get wave-form components**, followed by wiring its output to the array function **array Max & Min** (see Figure L8.7 as expanded version of Figure L8.5 within the host VI). The max input and min input indicators are wired to the max and min outputs of the array function. As seen in the Front Panel, these values are 0.877 and -0.877. Therefore, to represent them with the mini-mum quantization error, the Q16.15 format is chosen, that is <+- 16,1> with the wordlength of 16, the fraction length of 15, and the integer word length of 1 in order to represent the range -1 to 1 with the precision 3.0518E-5. To do the conversion, use the function **To fixed point** as shown within the for loop since this function only accepts scalar data, and thus the output is an array of fixed-point numbers representing the input signal.

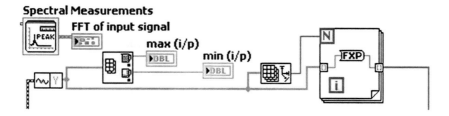

Figure L8.7: Quantizing the input

L8.1.3 Floating-Point Filter Configuration

Let us now create an FIR filter using the Kaiser window. Figure L8.8 displays the FP and BD for the subVI flt_dsgn.vi. The controls on the BD correspond to a bandpass filter with the parameters: Fp1= passband frequency1 = 6500hz, Fp2 = passband frequency2 = 7500 Hz, Fs1 = stopband frequency1 = 5000 Hz, Fs2 = stopband frequency2 = 9000 Hz, stopband attenua-tion of 44dB, and passband ripple of 0.05. Based on these parameters, the filter order and parameter β corresponding to the Kaiser window need to be determined. To do so, place the function **DFD Kaiser Order Estimation (Signal processing>>Digital filter design>>Filter design>>Advance filter design>>DFD Kaiser Order Estimation)** on the Front Panel of the host VI. This function is shown in Figure L8.9. It requires various inputs including **filter**

type with a ring control having values 0, 1, 2, 3 corresponding to lowpass, highpass, bandpass and bandstop filters. As outlined in this figure, right click on this input and select **create constant**, then choose **Bandpass** to specify a bandpass filter. If a different type of filter is desired, one could create a control and choose the appropriate filter type. The option **order option** specifies the filter order. The default is **MinEven** giving the minimum even order for the filter to meet the specifications. The sampling frequency **fs [Hz]** is expressed in Hz. This input needs to be wired to the **Fs** control created earlier. The input **ripple specs** specify the ripple level in the passband and stopband of the filter. It is a cluster of three elements: **passband** specifies the ripple level in the passband (the default is 0.1), **stopband** specifies the ripple level in the stopband (the default is 60), and **dB/linear?** specifies whether this VI accepts the ripple specification in decibel scale (True) or linear scale (False). Wire a **false** constant to the input **dB/linear?** and specify a linear scale for the ripple parameters. Wire the control Passband ripple to the input **passband** of this function, and set the value to 0.05. Now, the stopband attenuation of 44 dB is provided in dB scale by the control stopband attenuation (db). Since this must be converted to the linear scale before being wired to the function, one needs to use the conversion function $10^{(-\text{attenuation in db}/20)}$, as shown in Figure L8.8.

The function **DFD Kaiser Order Estimation** has two outputs: **estimated order** and **beta** which refer to the order and parameter β. These are clustered into a two element cluster using the bundle function, and then wired to the control **filter parameters obtained** (see Figure L8.8). As seen in this figure, the filter order is found to be 54, and so the filter has 55 (nh) coefficients with β of 3.86. The other outputs of the function are high cutoff freq and low cutoff freq which refer to the high cutoff frequency and low cutoff frequency (in Hz) of the filter where the magnitude response falls to one half the maximum. Connect these outputs to the corresponding inputs of the **DFD Kaiser Design** function (**Signal processing>>Digital filter design>>Filter design>>Advance filter design>>DFD Kaiser Design**). The other inputs of this function are **order** and **beta** (β), which are also wired to the corresponding outputs of the DFD Kaiser Order Estimation function, and **filter type** and **fs**, which are wired to the same controls as before.

Next, to get the filter coefficients, the **filter out** signal from **DFD Kaiser Design** is used. Refer to the host VI BD of Figure L8.5, or the expanded vesion in Figure L8.10 to see how this can be done. The **Filter out** signal is a cluster of several elements. To get the current coefficients, place the function **unbundle by name** on the Front Panel, and wire the **filter out** signal to this function. Resize it to display various options as displayed in Figure L8.11.

Since the current coefficients are of interest, resize back to display the option **mSRate_CurCoef** as shown in Figure L8.10. Again, this is a cluster of two elements having two arrays for coefficients **a** and **b.** Wire this output to another **unbundle by name** function, and wire the indicators **coef a(k)** and **coef b(k)** to the output of this function. Next, as was done for the input signal, create two indicators **max** and **min** to indicate the maximum and minimum values of the coefficients. Since this is an FIR filter, only **b** coefficients are needed and these are used in the rest of the BD. Also, to display the magnitude response of the filter, **filter out** of the **DFD Kaiser Design** function is wired to the **Filter in input** of the **filter analysis** function (**Signal processing>>Digital filter design>>filter analysis>>filter analy-**

sis), and the output magnitude of this function is wired to a graph indicator **Magnitude response of FLT filter** (see Figure L8.5 and Figure L8.12).

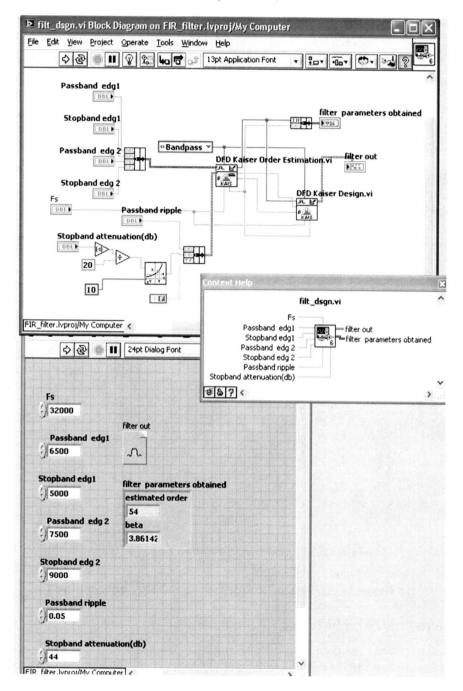

Figure L8.8: SubVI of filter design using Kaiser window

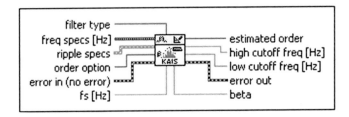

Figure L8.9: DFD Kaiser order estimation function

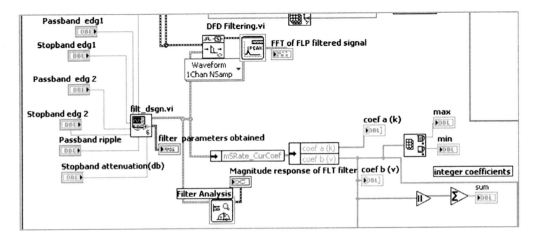

Figure L8.10: Designing Kaiser window filter

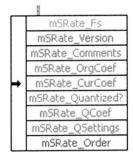

Figure L8.11: Options of filter out signal after unbundling

L8.1.4 Floating-Point Filtering

After having created the floating-point FIR filter based on the Kaiser window, let us now create the portion of the BD that implements the floating-point filtering. To do so, place the function **DFD Filtering (Signal processing>>Digital filter design>>Processing>>DFD Filtering)** on the BD of the host VI (see Figure L8.5). The function is shown in Figure L8.13.

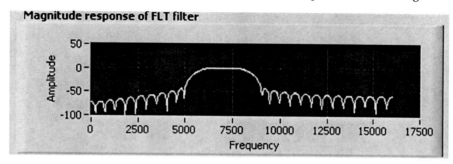

Figure L8.12: Magnitude response of floating-point filter

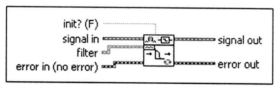

Figure L8.13: DFD Filtering function

This function is a polymorphic function, and thus to wire the inputs as waveforms, choose **single channel>>multiple samples>>waveform,** as shown in Figure L8.14.

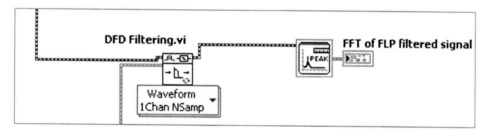

Figure L8.14: Choosing the polymorphic instance waveform in the DFD Filtering function

Wire the **signal in** input of the function to the composite signal wire, and the **filter in** input to the **filter out** wire of **DFD Kaiser Design** in order to connect the input signal and the filtered signal. The filtered signal is available at the **signal out** terminal of the function. As done for the input signal, display the FFT of this filtered signal. The FFT outcome is illustrated in Figure L8.15. It can be seen that, as expected, the filter filters out F1=2000 Hz and shows a peak near the frequency F2=7000 Hz. Also, note the window effect.

L8.2 Designing Fixed-Point Filter

L8.2.1 Coefficient Quantization

From the BD of the host VI, note that the maximum and minimum values for the coefficients are **0.15625** and **-0.137373**. These coefficients are represented using the Q16.15 format, but it is seen that the range of the coefficients is less than 0.25 (absolute), while Q16.15 has a range of -0.99996 to 1. Thus, it is better to reduce the number of bits while keeping the precision the same. This is achieved by using the Q14.15 format (wordlength of 14, fraction length of 15, integer word length of -1) as this format has a range of -0.25 to 0.25 and the same precision as Q16.15. However, to avoid overflow/saturation errors, consider the Q16.17 format (wordlength of 16, fraction length of 17, integer word length of -1) which has the same range of -0.25 to 0.25 but having a better precision (3.8147E-6). Thus, create a cluster of 6 elements (Figure L8.16) in order to provide the quantization parameters. An easy way to do so is to create an indicator out of the **mSRate_Qsetting** output. Then, delete the cluster and change the indicator to the control. In the control, choose **coefficients b/v** in the source corresponding to the FIR coefficients. Set the control **wl** to 16, and **iwl** to -1, and **signed** to checked corresponding to the word length of 16, integer wordlength of -1 for the Q16.17 format. The other two controls are for defining the overflow and saturation mode. Leave these options as default (Figure L8.17).

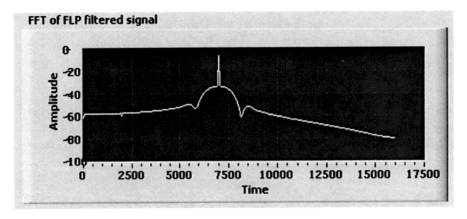

Figure L8.15: FFT of floating-point filtered signal

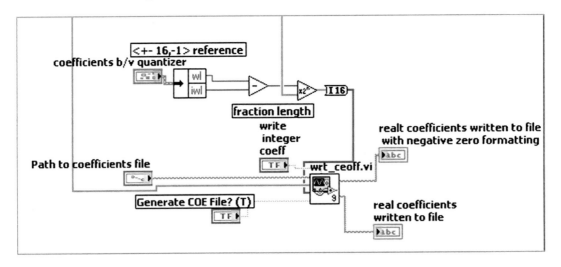

Figure L8.16: Converting real coefficients to equivalent Q16 integer representation

As seen in Figure L8.16, the **unbundle by name** function is used to get the wordlength and integer wordlength from the cluster, and then the difference of the two gives the fraction length, which is used to convert the real coefficients to the Q16 format.

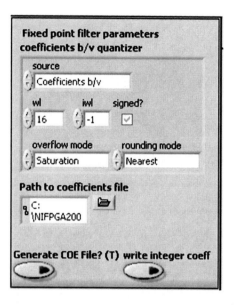

Figure L8.17: Filter coefficients quantization and file saving

Next, it is required to save the coefficients in the **Xilinx Core Generator** [1] file format. The Xilinx Core Generator allows one to store the coefficients as integers or floating-point/real numbers. Let us create the BD of the subVI **wrt_ceoff.vi** for storing the numbers in both formats. However, once the coefficients are stored and the IP is generated for use in the FPGA VI, one may desire to replace the coefficients. For this purpose, first create a case structure and wire the Boolean control **generate COE file** to it (Figure L8.18). Then, create the subsequent BD in the **true case**, and leave the **false case** empty. To provide the option for writing integer or real coefficients, create the Boolean control **Write integer coeff** and wire it to the inner case structure. In the inner **true case**, first create a structure to save integer coefficients.

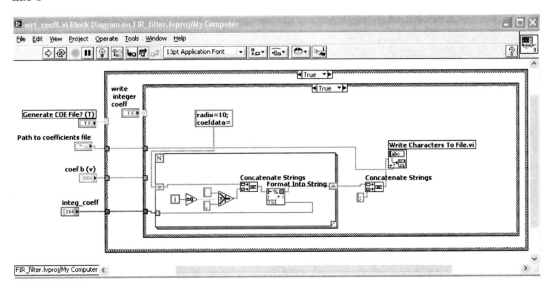

Figure L8.18: BD of the subVI for writing integer coefficients

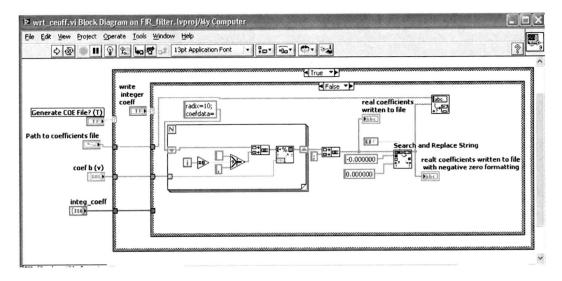

Figure L8.19: BD of the subVI for writing real coefficients

The format for the coefficients file is given in the appendix at the end of the chapter. For integer coefficients, and radix 10, the coefficient format is

radix=coefficient_radix;
coefdata= h(0), h(1),.... h(N-1);

Thus, create a For Loop and wire its index terminal to the control **Equal to 0?** (see Figure L8.18). Wire its output to the **s** input of the control **select**. Wire an **empty** string constant to terminal **t** and a string constant with the value **comma** to the **f** terminal of this control. Create a string constant with the following value

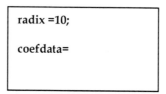

```
radix =10;

coefdata=
```

and connect it to one of the inputs of the control **concatenate strings (programming>>String>>concatenate strings).** Connect the output of the **select** output to the other terminal of **concatenate strings.** Thus, when the loop index is 0, this creates the first line of the coefficients file and the second line upto '='. For other indices of the file, to append the coefficients with the already created string, separated by commas, first convert the coefficients to a string format, and then append them with the already created string using the control **Format into String (programming>>String>>Format into String).** Wire the output of **concatenate strings** to the **initial string** input and the coefficients array to the **input1** of the function. Wire the **resulting string** output to the right side of the For Loop, and create a **shift register** with the initial terminal connected to the string constant. The shift register ensures that coefficients are written in the specified format. Finally, wire a string constant **semicolon** to one of the inputs of the second **concatenate strings control,** and the output of the shift register to the other input of the control. This ensures that the file is terminated. Finally, to write the coefficients in the specified format into a text file, create a control **Write Characters to file (Programming>>File I/O>>Write Characters to file)** and place it on the BD. Note that LabVIEW 2009 does not support this function, so use the control **write to text file** instead. Create a file path control **Path to coefficients file** and wire it to the **file path** input of this function. Create a new text file **fir.txt**, and change the extension to **.coe** so that a blank coefficient file **fir.coe** is created in the folder where it is desired to place the file. Using the **file path** control, browse to this folder and select **fir.coe.** The other input of the **Write Characters to file** function is **character string** which accepts the string data to be written to the file. So wire the coefficient data from the output of the second **concatenate strings** to the **character string input** of this control. Hence, this way the coefficient data is ready to be written into the file.

Next, choose the **false** case for the inner case structure to create the BD for saving real coefficients into the **fir.coe** file. Figure L8.19 shows this structure. As seen from this figure, the real coefficients array created is wired to the For Loop in a similar manner as was done for the integer coefficients. This way, real coefficients are written to the file as shown by the BD of the indicator **real coefficients written to file**. A sample result of the FP of this indicator is shown in Figure L8.20.

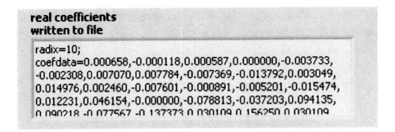

Figure L8.20: Real coefficients

Now, as seen for integer coefficients, one cannot write integer coefficients directly since the Xilinx Core Generator cannot recognize -0.000 or -0 as 0. If the IP is compiled using the FIR Compiler, it will give the error

Customizing IP...
ERROR:sim - Invalid decimal value detected in COE file: -0.000000.

Thus, it is needed to replace all -0.000000 with 0.000000. For this purpose, use **search and replace string (programming>>String>>search and replace string)**. This function is shown in Figure L8.21.

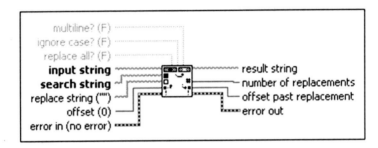

Figure L8.21: Search and replace string function

The input **input string** is the complete string that the function searches. Wire the indicator **real coefficients written to file** to this input. The input **search string** is the string to search and get replaced. Wire a string constant with value -0.000000 to it. The input **replace string** replaces the string wired to **search string** with the string wired to this input. Wire the string constant with value 0.000000 to it. Since there are multiple instances of negative zero, wire a **true constant** to the control **replace all**. This way the resultant string has only positive zeroes, as shown in Figure L8.22.

Finally, wire the string to the **character string** input of the function **Write Characters to file**, and wire the **Path to coefficients file** control to the **file path** input of this function.

Now, before creating the FPGA VI and the part of the host VI which calls and communicates with the FPGA VI, one needs to create the IP core for the filter using the FIR Compiler (version4) within the Xilinx Core Generator. First, save the coefficients file by running the host VI and saving the coefficients in the file **fir.coe**.

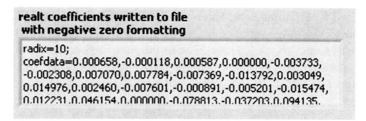

```
realt coefficients written to file
with negative zero formatting
radix=10;
coefdata=0.000658,-0.000118,0.000587,0.000000,-0.003733,
-0.002308,0.007070,0.007784,-0.007369,-0.013792,0.003049,
0.014976,0.002460,-0.007601,-0.000891,-0.005201,-0.015474,
0.012231,0.046154,0.000000,-0.078813,-0.037203,0.094135,
```

Figure L8.22: Real coefficients with negative zeroes removed

L8.2.2 IP Generation via FIR Compiler v4.0

As outlined in the appendix, create a project **coregen_fir.cgp**. Open **FIR compiler v4.0** [2]. The first page of the compiler is displayed in Figure L8.23.

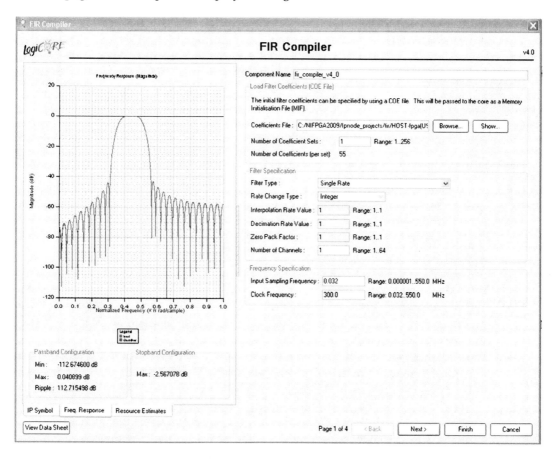

Figure L8.23: FIR Compiler

The options appearing in this figure are:
- **Component Name**: The user-defined filter component instance name - leave the name as is.
- **Coefficients File**: This is the file for filter coefficients; browse and choose **fir.coe.**

- **Number of Coefficient Sets**: The number of filter coefficient sets to be implemented; as there is only one set of coefficients, set the value to **1.**
- **Number of Coefficients (per set)**: The number of filter coefficients per filter set; this value is automatically derived from the COE file contents and the number specified in **number of coefficient sets**. As can be seen, its value is 55 due to the filter being 54th order.
- **Filter Type**: Six filter types are supported: Single-rate FIR, Interpolated FIR, Interpolating FIR, Decimating FIR, Receiver and Transmitter Polyphase filter bank. Select **Single rate.**
- **Number of Channels:** It refers to number of channels processed by the filter. As there is only one channel here, set this to **1.**
- **Input Sampling Frequency:** This refers to the sampling frequency (in MHz) for the system. The upper limit is set based on the clock frequency and filter parameters. For the sampling frequency of 32 kHz, set this to 0.032.
- **Clock Frequency:** This refers to the system clock frequency in MHz. Leave this field at default **300MHz.**

The remaining options are not needed for the design at hand. Click **next** to move to the second screen (see Figure L8.24) that allows various implementation options which are explained below:

- **Filter Archtecture:** Three filter architectures are supported: **Systolic Multiply Accumulate; Transpose Multiply Accumulate, and Distributed Arithmetic.** Choose **Systolic Multiply Accumulate.**
- **Use Reloadable Coefficients**: When the Reloadable option is selected, a coefficient reload interface is provided on the core. For more information on this control, refer to the data sheet.
- **Coefficient Structure:** Five coefficient structures are supported: Non-symmetric; Symmetric; Negative Symmetric; Hilbert transform and Half-band. However, to choose the best structure, one needs to use the option **inferred** where the structure is inferred automatically from the coefficient file.
- **Coefficient Type:** This can be signed or unsigned. When the signed option is selected, the conventional two's complement representation is assumed. Since signed coefficients are loaded from the file, this field displays only **signed**.
- **Quantization:** Specifies the quantization method used. The options available are **Integer Coefficients, Quantize Only,** or **Maximize Dynamic Range.** The Integer Coefficients option appears in the drop down menu where the COE file is specified using only integer values. The Quantize Only option is used when real coefficients are saved in the COE file. Since this is the desired type here, select Quantize Only in the field. The Maximize Dynamic Range option maximizes the dynamic range of the filter by scaling all the coefficients such that the maximum coefficient becomes equal to the maximum possible number in the specified bit width.

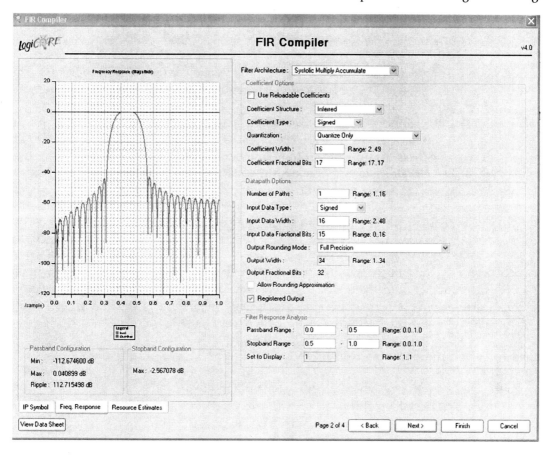

Figure L8.24: Implementation option screen

- **Coefficient Width:** This field is used to set the wordlength of the coefficients.
- **Coefficient Fractional Bits:** This is an indicator field that shows the number of fractional bits that are used for quantizing coefficient values. Its value is equal to the Coefficient Width value minus the required integer wordlength. The integer wordlength value is automatically determined from the integer bit width needed to represent the maximum coefficient. As was seen in step 5, the Q16.17 representation provides a good representation, although Q14.15 can also be used. Choosing the value 16 for Coefficient Width automatically selects 17 for this field, as shown in Figure L8.24. If the value 14 is used for Coefficient Width, this field updates to 15. One can view the effect of these quantization changes in the **frequency response(magnitude)** display, which shows the magnitude response of the fixed-point and floating-point filter versus normalized frequency (for quantize only option).

Next, the datapath options include:

- **Number of Paths:** Place the value 1 as there is only one datapath.
- **Input Data Type:** It specifies whether the input data is **signed** or **unsigned**.

- **Input Data Width:** This field is used to specify the wordlength of the input data. As seen in step 2, the Q16.15 representation is used for the input data, so enter the value **16** in this field.
- **Input Data Fractional Bits:** It specifies the fraction length of the input data. This field together with the Coefficient Fractional Bits determine the **Output Fractional Bits** value.
- **Output Rounding Mode:** This field specifies the rounding mode to be used for the filtered output samples. **Full precision** ensures that output fractional bits are the sum of coefficient fractional bits and input data fractional bits. Other modes are also available. **Full precision** (i.e., the output fractional bits of 17+15=32) is used here.
- **Output Width:** When using full precision, this field is disabled and indicates the word length of the filtered output samples taking into account the bit growth. However, in any other Rounding Mode, one can specify the desired output data wordlength using this field. One can see that for the present design, this field indicates a wordlength of 34 for the output data samples. This includes 32 fractional and 2 integer bits. Thus, the output data has the format Q34.32 or <+- 34, 2>. The remaining fields are not used in the present design, and their discussion is excluded. The interested reader can refer to the data sheet for information on the remaining fields. Click **Next** and move to the third screen, as shown in Figure L8.25.

The available options in this screen are:

- **SCLR:** Specifies whether the core will have a reset pin.
- **DATA_VALID:** Specifies whether the core will have a DATA_VALID pin.
- **CE:** Specifies whether the core will have a clock enable pin.
- **ND:** Specifies whether the core will have a new data pin.
  For the meaning of these pins, refer to the appendix. Since these handshake signals are not used here, leave these fields empty.
- **Optimization Goal:** Specifies whether the design is optimized to run at maximum possible speed (**Speed** option) or minimum area (**Area** option). The default option is **Area** option. At this stage, since the filter is being simulated, choose the **Speed** option, as indicated in Figure L8.25.
- Leave the remaining options at the default settings and click **next.** The last screen (see Figure L8.26) displays the design summary. By clicking **Finish**, the compilation starts and an IP gets generated.

Once compiled, the FIR Compiler generates a number of files in the same folder in which the COE file is present. These files include the VHDL and XCO files (an output file generated by the CORE generator that contains the core parameters and an XCO file which is used as an input file when the core is either recustomized or regenerated). For the present design, **fir_compiler_v4_0.vhd** and **fir_compiler_v4_0.xco** are generated along with the other files.

Next step is to use the IP integration node to call these files and implement the FPGA VI.

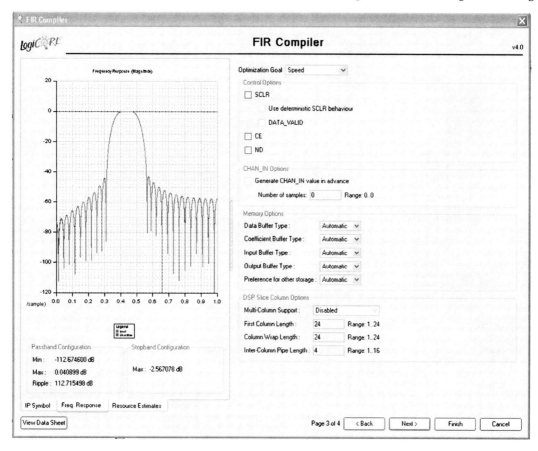

Figure L8.25: Detailed implementation option screen

L8.2.3 Building FPGA VI

Here the DMA (Direct Memory Access) feature is used to communicate between the FPGA VI and the host VI. To read the input fixed-point data in the FPGA VI from the FPGA input memory as part of the IP generator, create a FIFO (**Right click on FPGA target>>New>>FIFO).** To change its name, right click on **FIFO>>rename,** enter the name **FIFO (rd_hst).** Now, double click on it to edit its properties. The **general** page is configured as shown in Figure L8.27.

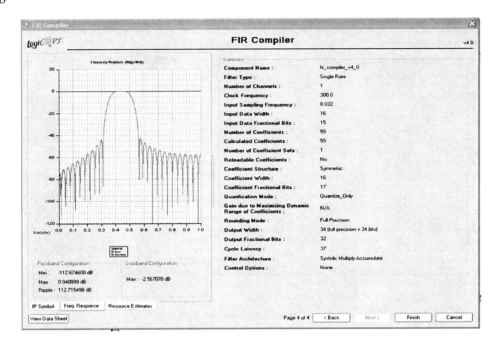

Figure L8.26: Summary screen

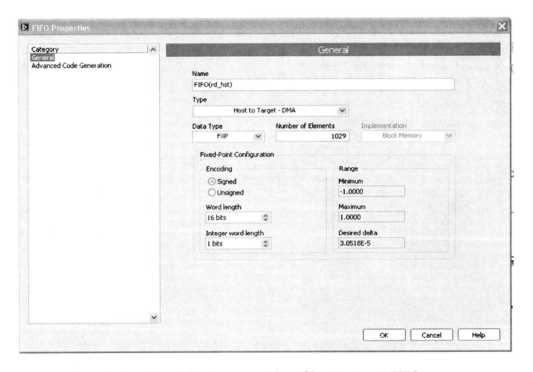

Figure L8.27: Setting properties of host to target FIFO

Its fields are explained below:

- **Name**: It displays the name of the FIFO. Although the FIFO was renamed earlier from the project explorer, the FIFO can also be renamed here.
- **Type**: It specifies three types of FIFO's: **Target scoped**, **Host to target-DMA**, **Target to Host-DMA**. Since this FIFO is to be used to read data from the host VI, choose Host to target-DMA.
- **Number of Elements**: This specifies the size or depth of the FIFO. The value is automatically coerced to the value that allows efficient implementation. For example, for a 640-point input signal, when the value 640 is used in this field, it automatically changes to 1029.
- **Data type**: It specifies the data type of the data that the FIFO will be using. If fixed-point data type is used, then the rest of the controls are enabled; else they remain disabled.
- **Encoding**: This specifies whether the data that the FIFO will hold can be **signed** or **unsigned.**
- **Wordlength and Integer Wordlength:** Specify the wordlength and integer wordlength of data. Since the composite signal is in the Q16.15 format, choose the data type as **Fxp**, encoding as signed, **Wordlength** as 16, and **Fraction length** as 1.

Next, after pressing the **ok** button, create a target to host DMA FIFO **FIFO (wrt_hst)** that will be used to store the filtered signal as generated by the IP core. To store the fixed-point data in the Q34.32 format, set the parameters accordingly, as shown in Figure L8.28.

At this point, create the IP Integration Node and integrate the FIR IP generated through the Xilinx Core Generator or FIR Compiler into the IP generator. The steps involved include:

- Create an IP Integration Node on the Block Diagram of the FPGA VI.
- Double click to open the first page of the IP Node properties. Browse and select the XCO file **fir_compiler_v4_0.xco** as the top level file. Generate the simulation model, and move to the next pages.
- In the last page or screen, various input and output data terminals and their data types are shown. As explained in the fixed-point chapter, set the data types to match the data type that Xilinx Core has generated, as illustrated in Figure L8.29. Click finish to perform the compilation. Next, place the IP Node within a SCTL.

Then, drag the FIFO **FIFO (rd_hst)** from the project explorer and place it on the BD of the FPGA VI. Similarly, drag the FIFO **FIFO (wrt_hst)** from the project explorer and place it on the BD (see Figure L8.30). Enclose everything in a While Loop.

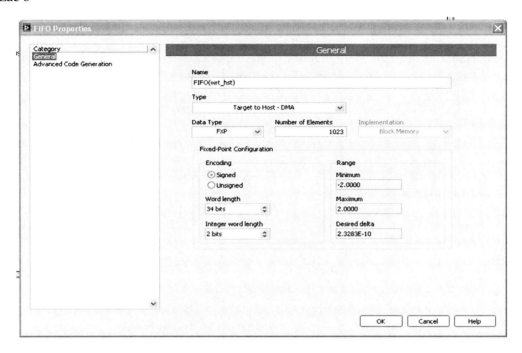

Figure L8.28: Setting properties of target to host FIFO

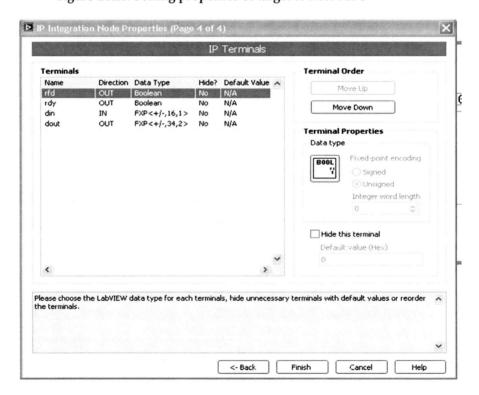

Figure L8.29: Page 4 of the IP node properties

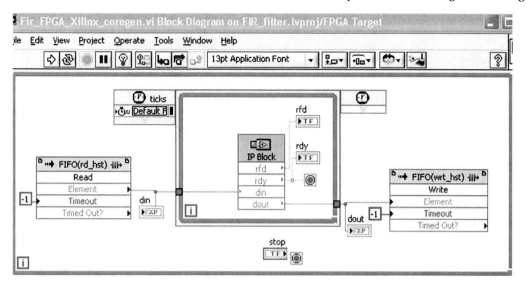

Figure L8.30: BD of the FPGA VI

Wire the **element** output of the **FIFO (rd_hst)** to the **din** input of the IP node, and create the indicator **din** to show each element of the input composite signal as it is read within the While Loop. Similarly, wire the **dout** output of the IP Node that carries the filtered signal, element by element, to the corresponding indicator **dout**, and the **element** input of the FIFO **FIFO (wrt_hst).** Wire the value of -1 for the timeout of both the read and write FIFOs. For read FIFO, **Timeout** specifies the number of clock cycles the function waits for data to become available in the FIFO if the FIFO is empty. A value of –1 prevents the function from timing out, so the function completes execution only when data is available for reading. A value of 0 indicates that the function does not wait. One must wire a 0 if the FIFO is used within a SCTL. For write FIFO, **Timeout** specifies the number of clock cycles the function waits for space to become available in the FIFO if the FIFO is full. A value of –1 prevents the function from timing out. A value of 0 indicates that the function does not wait. Again, one must wire a 0 if the FIFO is used within a SCTL.

Create a numeric indicator **rfd** and connect it to the **rfd** output of the IP node. Similarly, create a numeric indicator **rdy** and connect it to the **rdy** output of the IP node. Also, wire the **rdy** output to the **stop** input of the SCTL so that when filtering is complete, the SCTL is stopped. Create a Stop button for the While Loop. This completes the FPGA VI with its Front Panel shown in Figure L8.31.

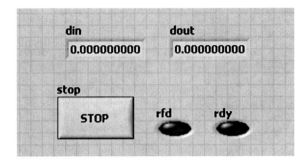

Figure L8.31: FP of the FPGA VI

L8.2.4 Finalizing Host VI

Now, let us create the remaining portion of the host VI, which allows writing the host data and reading the filtered data. Figure L8.32 helps one to better understand the host FPGA communication using DMA FIFO's.

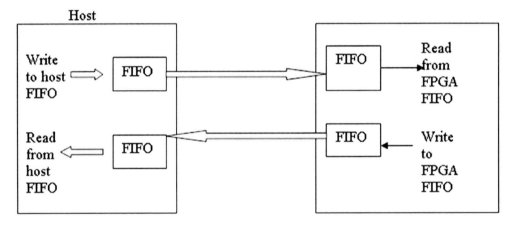

Figure L8.32: Host-FPGA VI communication using DMA FIFO

Figure L8.33 shows how to create the host part of the BD that writes to the host memory and reads from the host memory.

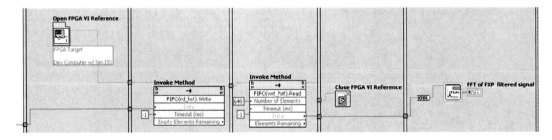

Figure L8.33: BD of FIFO part of host VI

The steps needed to create this communication are:

- Add more frames to the Block Diagram as shown in the figure.
- Place **Open FPGA VI reference** on the Block Diagram, and configure it to refer to the FPGA VI.
- Place **invoke method (FPGA Interface>>Invoke method)** in the next frame. Wire the **FPGA Reference Out** terminal of **Open FPGA VI reference** to the **FPGA Reference In** terminal of the **invoke method** function. Right click on the function, and choose **(method>>FIFO (rd_hst)>>write)**. The node resizes to the shape as shown in Figure L8.36. Thus, effectively a reference is created to the **host to target FIFO** on the FPGA. Wire the fixed-point input data array (Figure L8.36) to the **data** input of this function, and set the timeout to -1. **Timeout (ms)** specifies the number of milliseconds the Invoke Method function waits before timing out. The Invoke Method function times out if the host part of the FIFO does not contain enough space to write **Data** by the time the number of specified milliseconds elapses. The default is 5000 milliseconds. A value of –1 causes the Invoke Method function to wait indefinitely. By choosing **method,** a number of options are made available that can be used to invoke various methods such as read, write, etc., on FPGA targets to control the FPGA VI.
- Similarly, create another invoke method node **FIFO (wrt_hst).Read,** and wire the **FPGA Reference Out** terminal of **FIFO (rd_hst).Write** to the **FPGA Reference In** terminal of **FIFO (wrt_hst).Read.** This will read from the host part of the DMA FIFO, the data that has the filtering result and is transferred from the FPGA **FIFO (wrt_hst)** to the host side. Since the input signal is 640 elements wide, create a numeric control with the value 640, and wire it to the **Number of elements** input of this function. Set the timeout to -1.
- In the next frame, place a **Close FPGA VI reference** to close the reference to the FPGA VI.
- Finally, in the last frame, place a **to double precision float** function and wire the **data** output of **FIFO (wrt_hst)** to **Read to this function** so that the fixed-point filtered data returned from the FIFO is converted to the equivalent floating-point representation. Finally, take the FFT as done earlier for the floating-point filter and display the result using an indicator.

Figure L8.34 shows the FP of the host and FPGA VIs simultaneously.

L8.2.5 FFT of FXP Filtered Signal

Now, run the host VI. As the coefficients file has already been saved, there is no need to save the coefficients again. So, set the **generate Coe File** control to **false,** and run the host VI. As the program runs, the results as indicated in Figure L8.34 are obtained. The input data in fixed-point are written to the host FIFO, then transferred through the DMA to the FPGA FIFO where the IP Node gets the filtered fixed-point data element by element. As **rdy** goes high, the filtered data is written to the FPGA FIFO, the FPGA VI stops, and the filtered data is transferred through the DMA to the host FIFO where it is converted to real representation and the FFT is displayed. As seen in the FP of the complete host VI (see Figure L8.34), the FFT of the fixed-point filtering implemented in the LabVIEW FPGA matches closely with that of the floating-point filtering result.

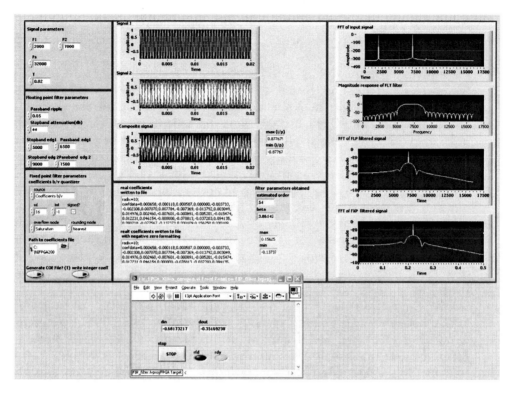

Figure L8.34: Complete FP of the host and FPGA VIs after execution is complete

L8.3 Lab Experiments

- Repeat the lab in the chapter by creating various subVIs using the filter specifications of a passband between 4000 to 7000 Hz, stopband edge frequencies of 2000 Hz and 9000 Hz, a sampling frequency of 32000 Hz, a passband ripple of 0.05, and a stopband attenuation of 20dB. The number of coefficients would be less than 55. Instead of using the Xilinx Core Generator, design the filter in the SOS direct form-2 representation in the LabVIEW FPGA. Use memory to load the filter coefficients from a text file as done in the lab example using .coe files.

- In this experiment, it is desired to see the coefficient quantization effects. Redo the lab in the chapter but this time create a case structure so that the coefficients are written in a file to be read by the Xilinx Core Generator after selecting the coefficient quantization. For this purpose, add a subVI to the design to set the coefficients using VIs such as DFD Set Quantizer. Then, simulate FIR filtering using VIs such as DFD FXP Simulation.

APPENDIX L8:

An Overview of Xilinx Core Generator

Xilinx Core Generator is an easy to use design tool which provides parameterizable COREs optimized for Xilinx FPGAs. As this Core Generator is started, the main window or GUI opens up as illustrated in Figure L8.A.1.

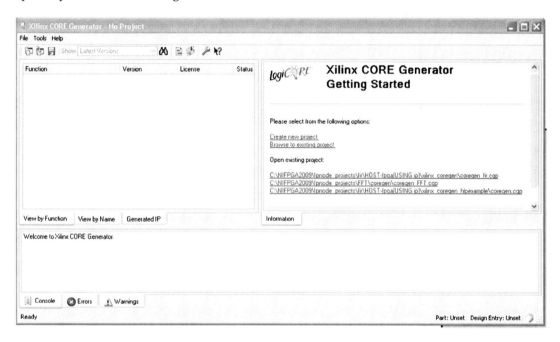

Figure L8.A.1: Xilinx Core Generator opening window

The first step in implementing any algorithm is to create a project as follows:

- Either click on **Create a new project** (upper right hand corner) or **select File>>new project.**
- Select the folder where it is desired to save the Xilinx Project file (has extension .cgp). It contains information on project-specific property settings, versions of the Core available in the project and user-specified output files. Click **save.**

As the project is saved, a new window opens up which gives the following options:

- **Part:** Here, the user can choose the FPGA family (Virtex, Spartan, etc.), the device type, package type and speed grade (see Figure L8.A.2).
- **Generation:** Here, one can choose the design entry method which can be **VHDL, Verilog** or **Schematic** (see Figure L8.A.3). It also gives setting for implementation files and simulation files and flow settings corresponding to the leading FPGA vendors.
- **Advanced:** This is for setting advanced features.
- Click **ok.**

In the GUI, a set of functions appears in the left side of the window as indicated in Figure L8.A.4.

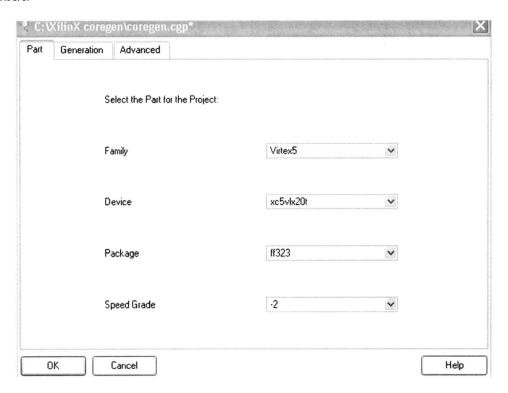

Figure L8.A.2: Implementation options in Coregen

These functions allow generating IP cores for various applications including Digital Signal Processing, Video Processing, etc. For using the core generator to design filters, or an FIR filter, choose **Digital signal processing>>Filters>>FIR compiler** as shown in Figure L8.A.5.

This opens up the FIR Compiler v4.0. Refer to the data sheet of the compiler for a detailed explanation of its features. An overview is provided next.

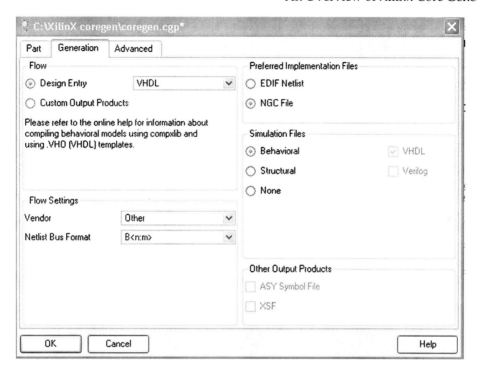

Figure L8.A.3: File and flow options in Coregen

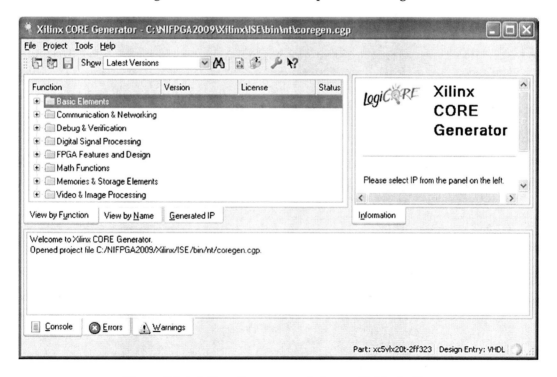

Figure L8.A.4: Functions or modules available in Coregen

Figure L8.A.5: FIR Compiler in Coregen

L8.A.1 FIR Compiler v4.0

The Xilinx® LogiCORE™ IP FIR Compiler core is optimized to provide highly parameteriz-able, area-efficient, high-performance FIR filters either via a Multiply-Accumulate (MAC) or Distributed Arithmetic (DA) architecture. Some of its salient features are:

- The core gives the flexibility to generate a wide range of filters including single rate, half band, Hilbert transform, polyphase filter bank and interpolated filters.
- Supports up to 256 sets of coefficients, with 2 to 1024 coefficients per set.
- Input data and filter coefficients can have up to 49-bit precision.
- Provides online coefficient reload capability.

The FIR Compiler can implement three different types of filter architectures: Distributed Arithmetic, Systolic Multiply-Accumulate and Transpose Multiply-Accumulate. For the full set of features and those that are supported for other FPGAs, the reader is referred to the data sheet. Figure L8.A.6 gives the pinout diagram for the Xilinx Core and Table L8.A.1 lists an explanation of some of the signals used in it.

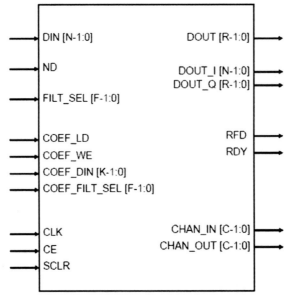

Figure L8.A.6: XILINX FIR CORE Pinout diagram [2]

Pin name	Used for	Description	Optional
SCLR	Input	**Synchronous clear**: Assert high synchronously with CLK to reset the filter internal state machines. It has priority over CE.	yes
CLK	Input	**Clock:** System clock for the CORE.	no
CE	Input	**Clock Enable:** Core clock enable (active high). Available for MAC-based FIR implementations	yes
DIN [N-1:0]	Input	**Data IN:** CORE accepts data input on this pin. Data is N bits wide.	no
ND	Input	**New data:** When this signal is asserted (high), the data sample presented on the DIN port is accepted into the filter core. ND should not be asserted while *RFD* is low, as those samples presented when RFD is low are ignored by the core.	yes
DOUT [N-1:0]	Output	**Data OUT:** CORE provides the filtering result on this pin, which is an R-bit-wide output sample bus. R depends on the filter parameters (data precision, coefficient precision, number of taps, and coefficient optimization selection) and acts as a full-precision output port to avoid any overflow. R = N (wordlength of input data) + K (wordlength of coefficients) + Bit growth (Log_2(number of coefficients))	no
RFD	Output	**Ready for data**: When active high, it indicates that core is ready to accept new data sample.	no
RDY	Output	**Ready**: High indicates new filter output sample available on Dout.	no

Table L8.A.1: Some of the important signals in Coregen

For the detailed timing and explanation of the handshake signals, refer to [2]. The first step in generating a filter core using the FIR Compiler is to load the coefficients file that is previously generated. The coefficient file has a COE extension. This is an ASCII text file with a single-line header that defines the radix of the number representation used for the coefficient data, followed by the coefficient values, as shown in Figure L8.A.7, either in a multiple-line or a single-line format.

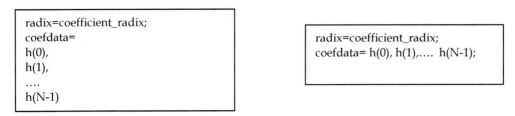

Multiple-line format   Single-line format

Figure L8.A.7: File formats for Xilinx Filter Coefficients file

The coefficients can be of the type integer or real. For integers, three different representations are possible: base-10, base-16 or base-2. These correspond to coefficient_radix=10, coefficient_radix=16 and coefficient_radix=2, respectively. For real coefficients, base-10 is the only radix allowed. If the user has saved the coefficients in real or floating-point format, in order to set the wordlength and fraction length for the fixed-point conversion of these coefficients, choose the option **Quantize only** in the **quantization** input of the filter configuration page of the FIR Compiler GUI. Although the core determines the best wordlength and fraction length based upon the real coefficients range and number, one can alter these parameters if desired. Similarly, if the coefficients are saved as integer coefficients, choose the option **Integer coefficients** in the **quantization** input of the filter configuration page. Since the coefficients are of integer type, the user can only change the wordlength as the fraction length is 0, which is indicated by the indicator on the filter configuration page.

L8.3 Bibliography

[1] Xilinx, *CORE Generator Guide* 3.1i, 2003.
[2] Xilinx, *FIR Compiler v4.0*, DS534, 2008.

CHAPTER 9
Adaptive Filtering

Adaptive filtering is used in many signal processing applications including noise cancellation and system identification. In most cases, the coefficients of an FIR filter are modified according to an error signal in order to adapt to a desired signal. In this chapter, a brief overview of system identification and noise cancellation is presented wherein an adaptive FIR filter is used.

9.1 System Identification

In system identification, the behavior of an unknown system is modeled by accessing its input and output. An adaptive FIR filter can be used to adapt to the output of the unknown system based on the same input. As indicated in Figure 9.1, the difference in the output of the system, $d[n]$, and the output of the adaptive FIR filter, $y[n]$, constitutes the error term, $e[n]$, which is used to update the coefficients of the filter.

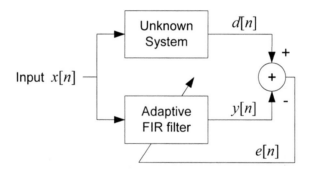

Figure 9.1: System identification

A popular algorithm which is widely used to update each coefficient of the FIR filter based on the error term, or the difference between the outputs of the two systems, is the Least Mean Square (LMS) algorithm [1]. The update is done as follows:

$$h_n\big[k\big] = h_{n-1}\big[k\big] + \delta e\big[n\big]x\big[n-k\big] \tag{9.1}$$

where h's denote the unit sample response or FIR filter coefficients, and δ a step size. This adaptation causes the output $y[n]$ to approach $d[n]$. A small step size will ensure convergence but results in a slow adaptation rate. A large step size, though faster, may lead to skipping over the solution.

9.2 Noise Cancellation

A system for adaptive noise cancellation has two inputs consisting of a noise-corrupted signal and a noise source. Figure 9.2 illustrates an adaptive noise cancellation system. A desired signal $s[n]$ is corrupted by a noise signal $v_1[n]$ which originates from a noise source signal $v_0[n]$. Bear in mind that the original noise source signal gets altered as it passes through the environment or channel whose characteristics are unknown. For example, this alteration can be in the form of a lowpass filtering process. Consequently, the original noise signal $v_0[n]$ cannot be simply subtracted from the noise corrupted signal as there exists an unknown dependency between the two noise signals $v_1[n]$ and $v_0[n]$. The adaptive filter is thus used to provide an estimate for the noise signal $v_1[n]$.

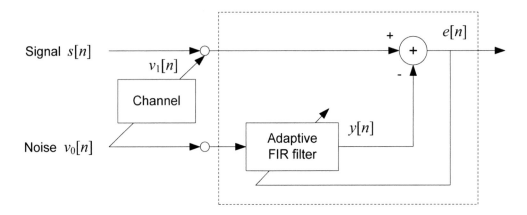

Figure 9.2: Noise cancellation

The weights of the filter are adjusted in the same manner stated previously. The error term of this system is given by

$$e[n] = s[n] + v_1[n] - y[n] \tag{9.2}$$

The error $e[n]$ approaches the signal $s[n]$ as the filter output adapts to the noise component of the input $v_1[n]$. To obtain an effective noise cancellation system, the sensor for capturing the noise source should be placed adequately far from the signal source.

9.3 Bibliography

1)   S. Haykin, *Adaptive Filter Theory*, Prentice-Hall, 1996.

Lab 9:

FPGA Implementation of Adaptive Filtering

L9.1 System Identification in LabVIEW FPGA

As shown in Figure 9.1, the transfer function of an unknown system is identified in this lab using the LMS algorithm. The following steps are carried out for this purpose:

- Create a sinusoidal signal or an input signal **xn** and filter it through an IIR Butterworth filter acting as the unknown system to get the desired output **dn**.
- On the host side, filter the input through an FIR adaptive filter and compare its output **yn** with **dn**.
- Simulate the same system using fixed-point settings to examine the fixed-point outcome. Create the same adaptive filter as an FPGA VI using the above fixed-point settings, and compare the FPGA outcome with both the floating-point and fixed-point outcomes.

Now, let us go through these steps in details.

L9.1.1 Creating a Project

Create a project named **LMS_on_FPGA** to run in simulation mode. The project explorer screen is shown in Figure L9.1 where the folder **subVI** is a virtual folder that contains three VIs: **sig_filt_dsgn.vi, LMS_lv_flp.vi, LMS_fxp_sim.vi, LMS_Read_FPGA.vi**. These subVIs are called from the top level host VI **LMS_host_top.vi**. To create such folder, right click on **My computer** and choose it as a subVI (**New>>Virtual folder>>rename**). A virtual folder provides a convenient place to keep similar VIs in a one place, reducing the clutter without using real physical disk space. On the virtual folder tab, right click and select **new>>VI** to create the subVIs. Similarly, **LMS_FPGA_VI** is the top level FPGA VI that uses the subVI **FXP LMS** created in the virtual folder **FXP LMS**. The design of the subVIs is explained in the sections that follow.

L9.1.2 Design of sig_filt_dsgn.vi

Figures L9.2 and L9.3 show the BD and FP of **sig_filt_dsgn.vi**, respectively, which include the creation of the input signal **xn** and the Butterworth filtered desired signal **dn**. On the BD, place the function **Sine wave.vi** and create its corresponding controls: **No. of input samples, amplitude** and **Input Signal Frequency (Hz)**. Since this function accepts at its input terminal a discrete frequency, create a control **fs[Hz]** for the sampling frequency and divide the

Input Signal Frequency by it. Create an array indicator **input (xn)** to store the input signal values.

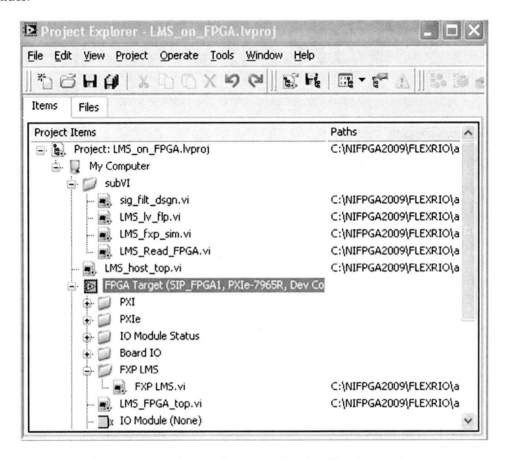

Figure L9.1: Project explorer for adaptive filtering project

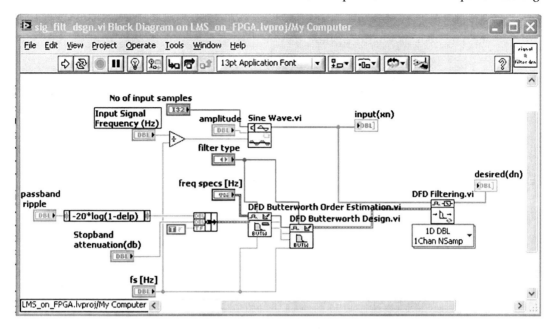

Figure L9.2: BD of sig_filt_dsgn.vi

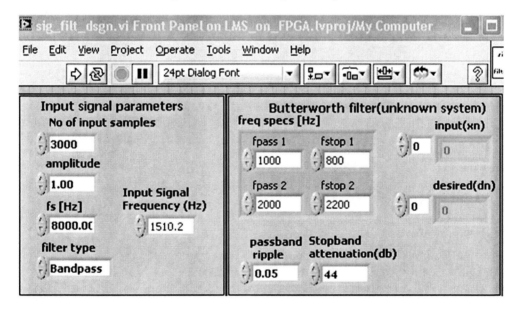

Figure L9.3: FP of sig_filt_dsgn.vi

Now, place the **DFD Butterworth Order Estimation.vi** (Signal processing>>Digital filter design>>filter design>>advance IIR filter design>>DFD Butterworth Order Estimation.vi) on the BD. This VI determines the minimum order of the Butterworth filter for meeting the given specifications. Wire the **freq specs[Hz]** cluster to its corresponding input. This cluster carries the frequency specification for the filter including the passband edge frequencies **fpass1** and **fpass2,** and the stopband edge frequencies **fpass1** and **fpass2** in Hz. The input terminal **ripple specs** accepts the passband ripple and the stopband attenuation in linear or db scale. Create a **passband ripple** numeric control in linear scale, convert it to the db scale using an expression node. Create a **Stopband attenuation [db]** numeric control, and a Boolean constant with value true. Wire these into a cluster and the output of this cluster to the **ripple specs** input of the function.

Now, place the function **DFD Butterworth design (Signal processing>>Digital filter design>>filter design>>advance IIR filter design>>Butterworth design.vi)** on the BD and wire its inputs of **order, low cutoff frequency[Hz]**, and **high cutoff frequency[Hz]** to the corresponding outputs of the DFD Butterworth order estimation function. This function designs the Butterworth filter. Wire the **fs[Hz]** input of this function to the **sampling frequency** control already created. Finally, filter the input **xn** with this filter by creating the function **DFD filtering.vi (Signal processing>>Digital filter design>>filter design>>processing>>DFD filtering.vi)**, and wire the output of the sine wave to the **signal in** terminal, and the **filter** terminal to the **filter out** terminal of the **Butterworth design** function. To ensure that the function operates on array data, choose the instance **single channel>>multiple samples>>1D double**. This way, the output of this function is the desired signal, which is shown by the indicator **desired (dn)**. To make this VI a subVI, wire the appropriate connectors and change the icon of the subVI to **signal & filter dsn**. This is illustrated in Figure L9.4.

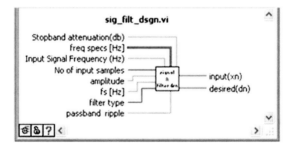

Figure L9.4: Terminals of subVI sig_filt_dsgn.vi

L9.1.3 Design of LMS_Lv_flp.vi

Figures L9.5 and L9.6 show the BD and FP of **LMS_Lv_flp.vi**, respectively, which include the creation of an FIR LMS filter on the host VI. Place the function **AFT create FIR LMS.vi (Signal processing>>adaptive filters>>AFT create FIR LMS.vi)** on the BD. This function creates an FIR LMS filter based on the step size δ (see Equation 9.1) and the filter length. Create the numeric controls **step size** and filter **length**, and wire them to the respective terminals of this function. This function also has an input leakage, which is used in a variant of the LMS called leaky LMS. Keep it unwired.

Now, place the function **AFT Filter signal and update coefficients.vi (Signal processing>>adaptive filters>> AFT Filter signal and update coefficients.vi)** on the BD. This VI filters the input signal based on the type of the adaptive filter connected to it and updates its weights in an iterative fashion. Wire the **adaptive filter in** terminal of this function to the **adaptive filter out** terminal of the **AFT Filter signal** function so that the function updates the weights as per the LMS equation. To ensure that the function operates on array data, right click on the function and choose **select type>>AFT Filter signal and update coefficients (Real array)**. Wire the array controls **x(n)** and **d(n)** to their respective inputs of this function, representing the input and the desired signal, respectively. The output filtered signal and the error signal are captured using the indicators **e(n)** and **y(n)**. Finally, to free the resources used by these VIs, wire the **AFT Filter signal** and **update coefficients** functions to **AFT Destroy Adaptive filter.vi (Signal processing>>adaptive filters>>AFT Destroy Adaptive filter.vi)**. Then, wire the respective **error in** and **error out** terminals of these VIs. Make this VI as a subVI with its connector terminals as shown in Figure L9.7.

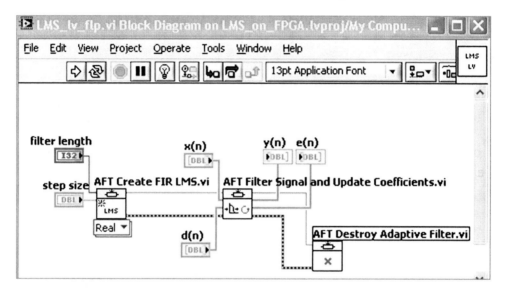

Figure L9.5: BD of LMS_Lv_flp.vi

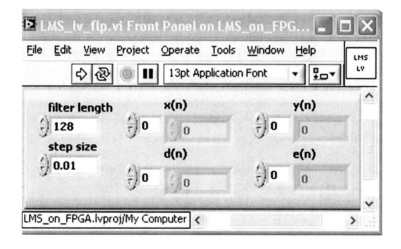

Figure L9.6: FP of LMS_Lv_flp.vi

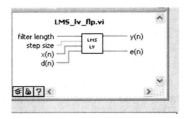

Figure L9.7: Terminals of subVI LMS_Lv_flp.vi

L9.1.4 Design of LMS_fxp_sim.vi

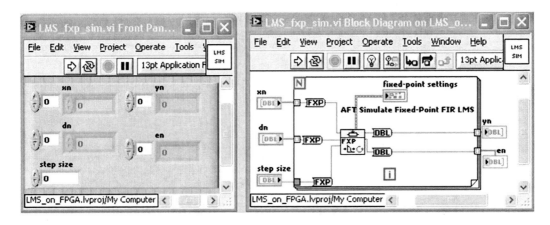

Figure L9.8: FP and BD of LMS_fxp_sim.vi

Figure L9.8 shows the BD and FP of **LMS_fxp_sim.vi** which simulates the fixed-point LMS filter version. Place the function **AFT Simulate fixed-Point FIR LMS.vi (Signal processing>>adaptive filters>>fixed point>>AFT Simulate fixed-Point FIR LMS.vi)** on the BD. This function allows configuring the fixed-point settings of the input signal x(n), the desired signal d(n), the step size and the weights w(n). Double click on it to open its configuration page (see Figure L9.9). Set the filter length to 30 (the default is 128). The step size gets automatically set to 0.00999832 based on the <+/16,-2> settings. The wordlength of the step size cannot be changed, but its integer wordlength can be. Keep the rest of the default settings at <+/-24,1>. Click **save FXP settings** to save these settings in a folder as an XML file **LMS_fxp_set.xml.**

Place this function in a For Loop. Create these corresponding numeric controls and indicators **xn, dn, step size, yn,** and **en** as shown in Figure L9.8. Convert the float values of xn, dn, and step size to match the fixed-point settings of Figure L9.9 via the function **to fixed point**. Then, create an indicator **fixed-point settings** to pass these values to the host VI. To reduce the FP space occupied by this indicator, hide it by choosing hide indicator. As done for the previous VIs, make this VI a subVI as shown in Figure L9.10.

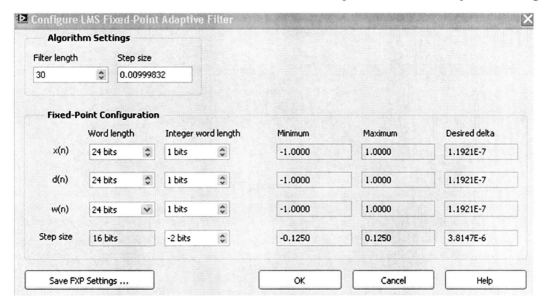

Figure L9.9: Configuring the fixed-point LMS filter for simulation

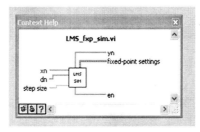

Figure L9.10: Terminals of subVI LMS_fxp_sim.vi

L9.1.5 Design of LMS_host_top.vi

Let us now design the portion of the host VI for bulding the LMS filter on the host for simulation purposes. Refer to Figures L9.11 and L9.12. Create a sequence diagram, and add two frames to it. Consider the second frame 2. Right click and choose **select VI>>browse to the folder where the subVIs were saved>>select sig_filt_dsgn.vi.** Create the indicated controls and indicators as shown in Figure L9.11. However, for the input signal and desired signal indicators, instead of creating array indicators, create the chart indicators **Input signal (xn)** and **Desired signal (dn)** (Figure L9.13) by going to the FP and choosing **Modern>>Graph>>Waveform Chart**.

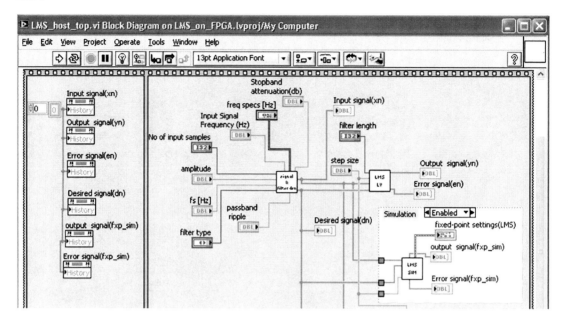

Figure L9.11: BD of portion of LMS_host_top.vi

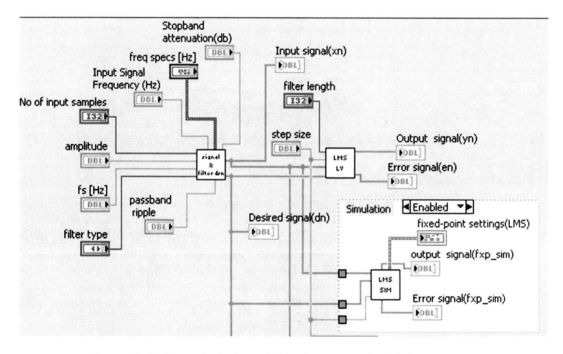

Figure L9.12: Expanded view of BD of portion of LMS_host_top.vi

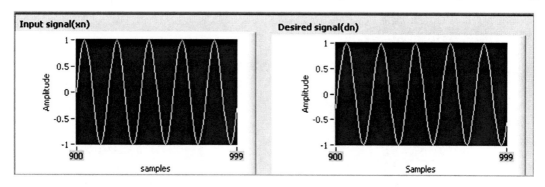

Figure L9.13: Chart indicators for input signal and desired signals

By default, the chart history length appears as 1024. Right click on these charts, select chart history, and enter the value 100.

Similarly, add the other subVI **LMS_Lv_flp.vi** to the BD, and create the corresponding controls: **filter length** and **step size**. Wire the input terminals **x(n)** and **d(n)** to **Input signal(xn)** and **Desired signal (dn)** from **sig_filt_dsgn.vi**. Wire the output terminals **y(n)** and **e(n)** to the chart indicators **Error signal(en)** and **output signal(yn)**, which are configured with the chart history length of 1000 and 100, respectively (Figure L9.14).

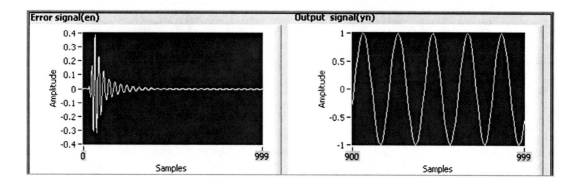

Figure L9.14: Chart indicators for LMS o/p signal and error signal

Now, create a **Diagram Disable structure** and in the enabled frame, add the subVI **LMS_fxp_sim.vi**. Wire the input controls as shown in Figure L9.12, and create the chart indicators **Error signal(fxp_sim)** and **output signal(fxp_sim)** wired to the terminals **e(n)** and **y(n)**, respectively (Figure L9.15).

Create the indicator **fixed point settings** to display the fixed-point LMS filter settings. Figure L9.16 shows the FP created so far. Set the parameters for the input signal to create a sine wave of 1000 samples with amplitude 1 and frequency 500 Hz at the sampling frequency of 10 kHz. Set the IIR filter (unknown system) as a bandpass filter with the passband from 300 to 800 Hz, and the stopband edge frequencies of 100 Hz and 1000 Hz. Set the stopband attenuation to 44 db, and the passband ripple to 0.05. For the LMS filter, set the filter length to 30 and the step size to 0.1. Save and run the VI.

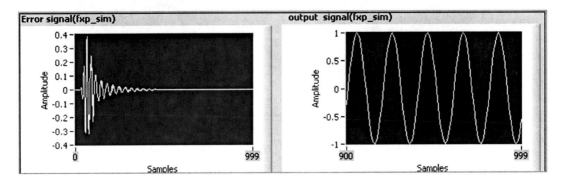

Figure L9.15: Chart indicators for LMS o/p signal and error signal for simulation VI

As seen from Figure L9.16, the LMS filter output for both the host VI and the simulation VI will match the desired signal, and so do the error signals. This figure also displays the fixed-point settings.

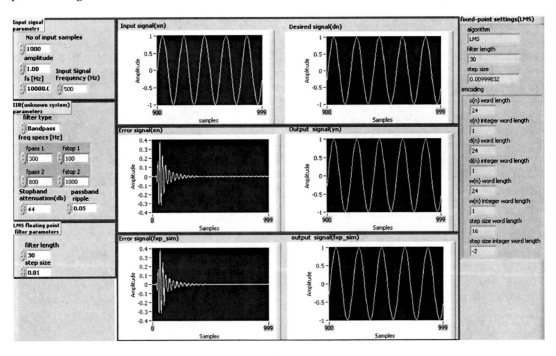

Figure L9.16: FP of portion of LMS_host_top.vi in run state

L9.1.6 Design of FXP_LMS.vi

Now, the stage is set to create a fixed-point FIR LMS filter on the FPGA side. In the project explorer, right click on the FPGA target and select **Start IP generator>>Adaptive>> LMS**

adaptive filter. Figure L9.17 shows the implementation screen that opens up. It shows the default parameters similar to the simulation filter. Select **Load FXP Settings** and browse to the file **LMS_fxp_set.xml** created in section 4 above. The window now changes to Figure L9.18. Select **Ok**. This creates the LMS filter as an FPGA VI, FXP_LMS.vi, under the virtual filter FXP LMS (Figure L9.1).

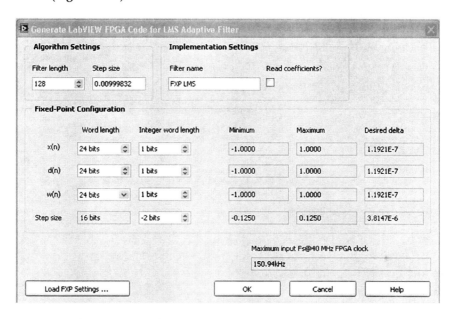

Figure L9.17: Fixed-point LMS filter settings (default)

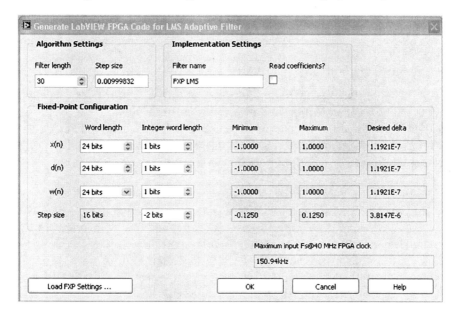

Figure L9.18: Fixed-point LMS filter settings (chosen from XML file)

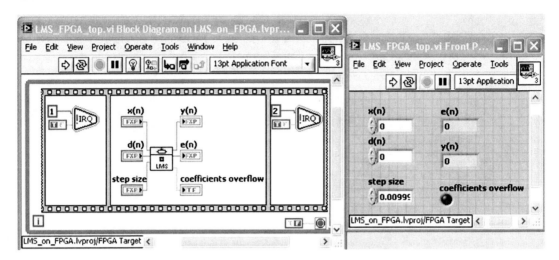

Figure L9.19: FP and BD of LMS_FPGA_top.vi

L9.1.7 Design of LMS_FPGA_top.vi

Refer to Figure L9.19. It shows the top level FPGA VI FP and BD, **LMS_FPGA_top.vi.** Create a sequence diagram with three frames. In the middle frame, add the subVI FXP_LMS. Create the corresponding controls and indicators.

To synchronize the host and the FPGA VI, interrupts need to be used. Add the function **Interrupt.vi** to the first frame. Wire the constant **1** to **IRQ number**, and **true** to **wait until cleared**. Similarly, create an Interrupt 2 in the third frame. Place the entire structure in a while loop with the constant **true** wired to the conditional terminal. This completes the FPGA VI design.

L9.1.8 Design of LMS_Read_FPGA.vi

Now, let us create the host VI that communicates with the top level FPGA VI just created. Refer to Figure L9.20. Create three numeric controls **input**, **desired** and **step size**. Place a sequence diagram with three frames in it. Create the function **Open FPGA VI reference** with the resource name control wired to it. In the middle frame, create a For Loop. Wire the count terminal to the size of the input data by using the **Array size** function.

To write the input signal, desired signal, and step size to the FPGA VI, create three **Read write** controls, and wire them to the FPGA VI reference. Choose the method as **write**, and wire them to the corresponding controls after the conversion to fixed-point (fixed-point settings need to match the FPGA VI). Place the **Invoke method** function on the BD, connect its input reference terminal to the output reference terminal of the control x(n), and choose **Acknowledge IRQ** with IRQ number 1. Refer to Figures L9.19 and L9.20. As soon as the FPGA VI runs, it sends the interrupt number 1 to the host, which writes the data and acknowledges it before the dataflow moves to the next element in the chain. Similarly, create the **Wait on IRQ** and **Acknowledge IRQ** Invoke Method functions, and wire the IRQ number **2** to them with the timeout of -1. Wire them as shown in Figures L9.19 and L9.20. As the FPGA computes the LMS output sample and the error sample, it sends the interrupt number

2 to the host, which is still waiting for the interrupt. It acknowledges the interrupt and captures the values read from the FPGA VI using **Read write Controls, y(n) and e(n)** which are configured in **read** Mode. Thus, the host does not send a new sample to the FPGA VI till it receives the computed sample from the FPGA VI synchronizing the two. Figure L9.20 also shows that the merge error is used to merge the errors, and the VI can be stopped on an error by capturing the error status and wiring it to the loop condition terminal (the For Loop is used with the conditional terminal option).

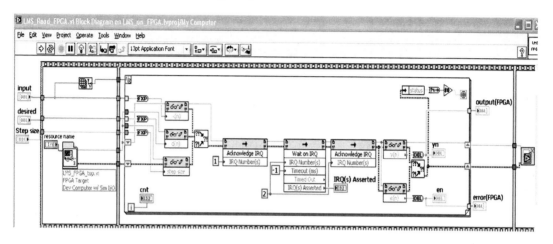

Figure L9.20: FP of LMS_Read_FPGA.vi

As indicated, the LMS filtered signal and the error signal are captured on the host sample by sample using the chart indicators **y(n), e(n)** and the numeric array indicators **output (FPGA)** and **error (FPGA)**. Finally, the reference to the FPGA VI is closed using **Close FPGA VI reference**. Make this VI a subVI as shown in Figure L9.21. The indicators y(n) and e(n) are not passed to the calling VI.

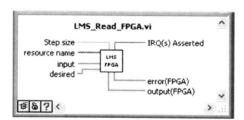

Figure L9.21: Terminals of subVI LMS_Read_FPGA.vi

L9.1.9 Completing Design of Top-Level Host VI

Refer to Figure L9.22. This figure shows the FP for the completed host VI **LMS_FPGA_top.vi.** Figure L9.23 shows the portion of this VI that uses LMS_Read_FPGA.vi. As seen from this figure, LMS_Read_FPGA.vi is placed in the enable case of the diagram disable structure. Wire the controls as indicated, and create two chart indicators for the terminals **output (FPGA)** and **error (FPGA).** Capture the state of IRQ number using **IRQ asserted.** To ensure that all the charts are initialized to zero at the start,

269

create a property node for the chart history **history data**, and wire array constants with values zero to it.

This completes the design. If the host VI is now run, one can see the LMS filter output and the error signal read into the host from the FPGA side. In simulation mode, this is a slow process. One can see that the values are updated point by point in the subVI LMS_Read_FPGA.vi. Alternatively, one can also add y(n) and e(n) as additional terminals in this subVI and capture them in the host VI. Also, one can run the LMS subVIs LMS_Read_FPGA.vi and FXP_LMS.vi together by enabling both of the Diagram disable structures. This option is shown in Figure L9.23. As seen in this figure, the FPGA implemented filter generates the same results as for the host and simulation versions.

This VI can be made to run in real-time by passing a small amount of input data as training data so that the filter weights get updated quickly and then these weights are read from the FPGA VI into the host to be used for the remaining real-time test data, which is a way of modeling the unknown system running in real-time.

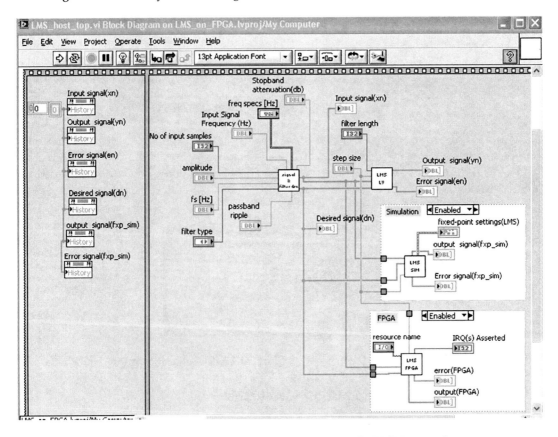

Figure L9.22: BD of completed host VI LMS_FPGA_top.vi

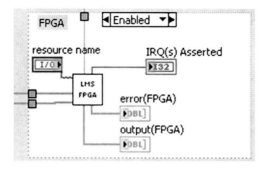

Figure L9.23: BD of portion of host VI that communicates with FPGA VI

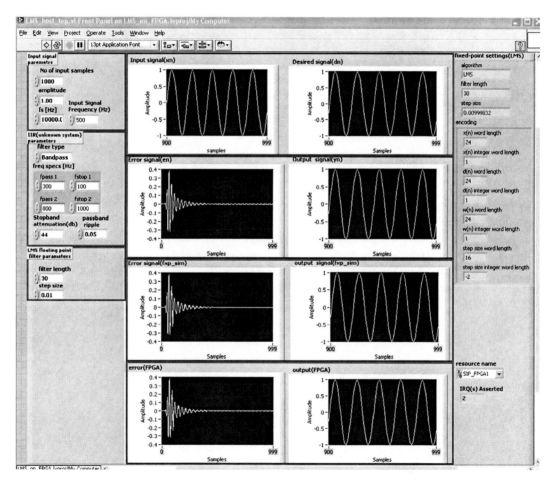

Figure L9.24: FP of host VI LMS_FPGA_top.vi after execution

L9.2 Lab Experiments

- Build a VI graphically to implement the inverse system identification problem shown in Figure L9.25 by modifying the system identification VI appearing in Figure L9.22. Generate the desired signal by setting the delay equal to one half the order of the unknown system. Create the LMS adaptive filter in the FPGA VI as done in the lab covered in the text. Verify the inverse system identification VI for the system orders of 12 and 16.

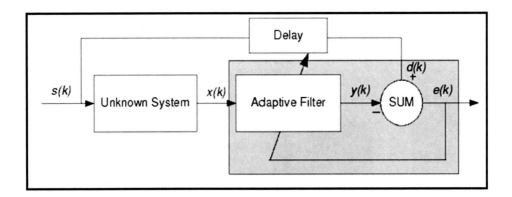

Figure L9.25: Inverse System Identification

- Implememt the system identification problem by creating a subVI that implements LMS filtering in an FPGA VI without using the LabVIEW IP generator. The subVI should have inputs corresponding to input signal, desired signal and step size, and outputs corresponding to output signal and error signal. Use the host-FPGA DMA communication for passing the inputs and getting back the results. The step size may be passed as a fixed-point array value or via read/write control (in that case, use interrupt handling).

CHAPTER 10
Frequency Domain Processing

Transformation of signals to the frequency domain is widely used in signal processing. In many cases, such transformations provide a more effective representation and a more computationally efficient processing of signals as compared to the time domain processing. For example, due to the equivalency of convolution operation in the time domain to multiplication in the frequency domain, one can find the output of a linear system by simply multiplying the Fourier transform of the input signal by the system transfer function.

This chapter presents an overview of three widely used frequency domain transformations, namely fast Fourier transform (FFT), discrete cosine transform (DCT), and short-time Fourier transform (STFT). More theoretical details regarding these transformations can be found in many signal processing textbooks, e.g. [1].

10.1 Discrete Fourier Transform (DFT) and Fast Fourier Transform (FFT)

Discrete Fourier Transform (DFT) $X[k]$ of an N-point signal $x[n]$ is given by

$$\left[\begin{array}{l} X[k] = \sum_{n=0}^{N-1} x[n] W_N^{nk}, \ k = 0, 1, ..., N-1 \\ x[n] = \frac{1}{N} \sum_{k=0}^{N-1} X[k] W_N^{-nk}, \ n = 0, 1, ..., N-1 \end{array} \right] \tag{10.1}$$

where $W_N = e^{-j2\pi/N}$. The above transform equations require N complex multiplications and $N-1$ complex additions for each term. For all N terms, N^2 complex multiplications and $N^2 - N$ complex additions are needed. As it is well known, the direct computation of (10.1) is not efficient.

To obtain a fast or real-time implementation of (10.1), a fast Fourier transform (FFT) algorithm is often used which makes use of the symmetry properties of DFT. There are many approaches to finding a fast implementation of DFT, that is there are many variations of FFT implementation. Here, the approach presented in [2] for computing a 2N-point FFT is stated to provide the idea behind FFT. This approach involves forming two new N-point signals $x_1[n]$ and $x_2[n]$ from a 2N-point signal $g[n]$ by splitting it into an even and an odd part as follows:

$$x_1[n] = g[2n] \qquad 0 \le n \le N-1$$
$$x_2[n] = g[2n+1] \tag{10.2}$$

From the two sequences $x_1[n]$ and $x_2[n]$, a new complex sequence $x[n]$ is defined to be

$$x[n] = x_1[n] + jx_2[n] \qquad 0 \le n \le N-1 \tag{10.3}$$

To get $G[k]$, the DFT of $g[n]$, this equation

$$G[k] = X[k]A[k] + X^*[N-k]B[k]$$
$$k = 0,1,...,N-1, \text{ with } X[N] = X[0] \tag{10.4}$$

is used, where

$$A[k] = \frac{1}{2}\left(1 - jW_{2N}^k\right) \tag{10.5}$$

and

$$B[k] = \frac{1}{2}\left(1 + jW_{2N}^k\right) \tag{10.6}$$

Only N points of $G[k]$ are computed from (10.4). The remaining points are found by using the complex conjugate property of $G[k]$, that is $G[2N-k] = G^*[k]$. As a result, a $2N$-point transform is calculated based on an N-point transform, leading to a reduction in the number of operations.

10.2 Discrete Cosine Transform (DCT)

Although Discrete Fourier Transform provides a useful means of analyzing the frequency content of signals, in applications such as image compression, Discrete Cosine Transform (DCT) is often used in place of DFT. Two-dimensional DCT is done on image blocks to gain improved energy compactness as compared to DFT. Although there are several DCT versions, here the so called DCT-2 formulation is given

$$\left[\begin{array}{l} X_{DCT}[k] = \sqrt{\frac{2}{N}}\alpha(k)\sum_{n=0}^{N-1} x[n]\cos(\frac{\pi k(2n+1)}{2N}), \quad 0 \le k \le N-1 \\ x[n] = \sqrt{\frac{2}{N}}\sum_{k=0}^{N-1}\alpha(k)X_{DCT}[k]\cos(\frac{\pi k(2n+1)}{2N}), \quad 0 \le n \le N-1 \end{array}\right] \tag{10.7}$$

where $\alpha(k) = \begin{bmatrix} \dfrac{1}{\sqrt{2}}, k = 0 \\ 1, k = 1, 2 .... N - 1 \end{bmatrix}$.

10.3 Short-Time Fourier Transform (STFT)

Short-time Fourier transform (STFT) is a sequence of Fourier transforms of a windowed signal. STFT provides the time-localized frequency information for situations when frequency components of a signal vary over time, whereas the standard Fourier transform provides the frequency information averaged over the entire signal time interval.

The STFT pair is given by

$$\begin{bmatrix} X_{STFT}[m,n] = \sum_{k=0}^{L-1} x[k] g[k-m] e^{-j2\pi nk/L} \\ x[k] = \sum_{m} \sum_{n} X_{STFT}[m,n] g[k-m] e^{j2\pi nk/L} \end{bmatrix} \qquad (10.8)$$

where $x[k]$ denotes a signal, $g[k]$ an L-point window function. From (10.8), the STFT of $x[k]$ can be interpreted as the Fourier transform of the product $x[k]g[k-m]$. Figure 10.1 illustrates computing STFT by taking Fourier transforms of a windowed signal.

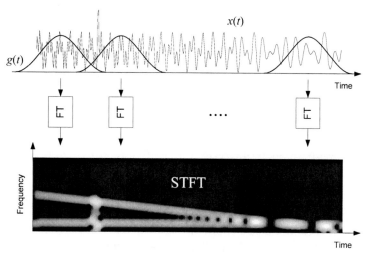

Figure 10.1: Short-time Fourier transform

There exists a trade-off between time and frequency resolution in STFT. In other words, although a narrow-width window results in a better resolution in the time domain, it generates a poor resolution in the frequency domain, and vice versa. Visualization of STFT is often realized via its spectrogram, which is an intensity plot of STFT magnitude over time. Three spectrograms illustrating different time-frequency resolutions are shown in Figure 10.2. The implementation details of STFT are described in Lab 10.

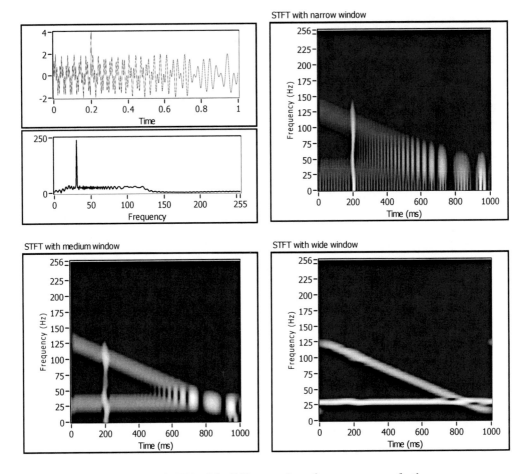

Figure 10.2: STFT with different time frequency resolutions

10.4 Signal Processing Toolset

Signal Processing Toolset (SPT) is an add-on toolkit of LabVIEW that provides useful tools for performing time-frequency analysis [3]. SPT has three components: Joint Time-Frequency Analysis (JTFA), Super-Resolution Spectral Analysis (SRSA), and Wavelet Analysis (discussed in Chapter 12).

The VIs associated with STFT are included as part of the JTFA component. The SRSA component is based on the model-based frequency analysis normally used for situations when a limited number of samples is available. The VIs associated with the SRSA component include high-resolution spectral analysis and parameter estimation, such as amplitude, phase, damping factor, and damped sinusoidal estimation. The VIs associated with the Wavelet Analysis component include 1D and 2D wavelet transform as well as their filter bank implementations.

10.5 Bibliography

[1] C. Burrus, R. Gopinath, and H. Gao, *Wavelets and Wavelet Transforms A Primer*, Prentice-Hall, 1998.

[2] Texas Instruments, *TI Application Report SPRA291*, 1997.

[3] National Instruments, *Signal Processing Toolset User Manual*, Part Number 322142C-01, 2002.

Lab 10:

Frequency Analysis in LabVIEW FPGA

L10.1 FFT in LabVIEW FPGA

This section covers the FPGA implementation of Fast Fourier Transform or FFT. A sinusoidal signal is first generated and corrupted by noise. FFT is then used to separate the signal and noise components. This is done in two ways: using the FFT function provided in LabVIEW and also using the FFT function as part of the FPGA toolkit. It is seen that both ways generate the same outcome. The FPGA implementation is first done without SCTL – single cycle time loop - (or without any handshake signals) and then with SCTL (or with handshake signals).

L10.1.1 FFT in LabVIEW FPGA Outside SCTL

Figure L10.1 shows a snapshot of the FFT project. This project **FFT_fpga.lvproj** is created in the LabVIEW FPGA by setting the FPGA target mode to **Execute VI on the PC with simulated I/O** and **random IO for FPGA IO read**. On **My Computer,** create a subVI named **FFT_sub.vi,** and open its Block Diagram as shown in Figure L10.2.

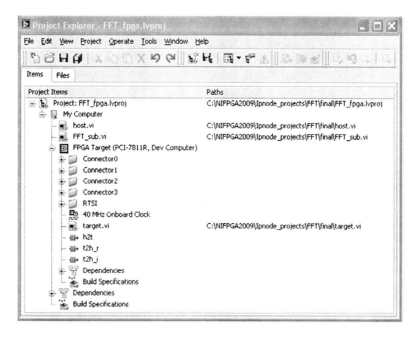

Figure L10.1: LabVIEW FPGA project FFT_fpga.lvproj

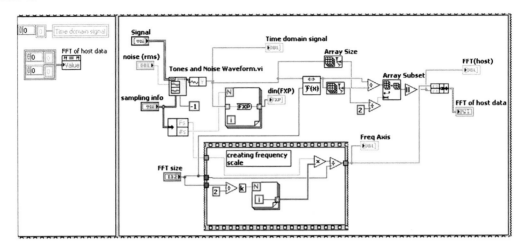

Figure L10.2: FFT_sub.vi BD

Let us go through the creation of the items shown in this figure one by one:

- Create a **sequence diagram (Structures>>Flat sequence diagram)** with two frames.
- In the second frame, add the function **tones and noise waveform (Signal processing>>Waveform generation>>tones and noise waveform).**

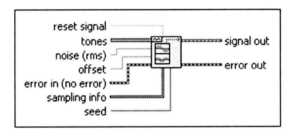

Figure L10.3: tones and noise waveform.vi

This function is shown in Figure L10.3 and its signals are:

▪ **Tones:** This is an array of clusters, where each cluster can be used to specify parameters of a sine or cosine waveform to be generated. The parameters include **Frequency, Amplitude and Phase** of the tone.

▪ **Sampling info:** This is a cluster of two elements: **Fs** the sampling frequency in Hz, and **#s** the number of samples.

▪ **Offset:** This is the DC offset of the signal. The default is 0.0.

▪ **Seed:** This is used to generate a new or different set of noise samples each time the VI is called. For values greater than 0, the noise sample generator is reseeded. The default is –1.

▪ **Noise:** It specifies the RMS level of the additive Gaussian noise. The default is 0.0.

- **Reset signal:** If this signal is **True**, it sets the time stamp to 0, the phase of each tone to the **phase** value used in the **tones** array, and the seed to the **seed** control value.
- **Signal Out:** This is the output signal generated by the function.

To obtain a single tone signal, create an array of single cluster which has three elements: frequency, amplitude and phase. To do so, on the FP of the VI, create three numeric controls and name them **Frequency, Amplitude,** and **Phase**. Create a **Cluster (right click on FP>>Array Matrix and cluster>>Cluster)**. Then, create an **array** control signal (**right click on FP>>Array Matrix and cluster>>array,** specify a **numeric control** and drag it into the **array**). Next, select the three numeric controls and place them inside the cluster control, and then select this single cluster and drag it into the array signal as shown in Figure L10.4.

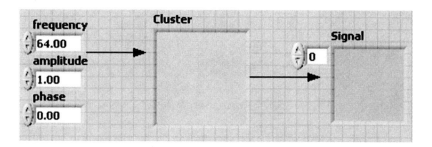

Figure L10.4: Creating sine wave signal

This creates the **signal** as depicted in Figure L10.2. Set the frequency to **64Hz,** amplitude to **0.75** and phase to **0** for the sine wave. Next, right click on **sampling info** input of the function, and select **create control** to specify **Fs** and **#s.** Next, create a constant with value **-1** and wire it to **seed.** Add the noise control **noise** as a numeric control and wire it to the **noise** input. Finally, right click on **signal out>>create indicator** and wire it. Create the array component out of it by wiring **get waveform components** to it. Wire its output to the graph indicator **time domain signal.**

- Use the **unbundle by name** function wired to the **Sampling info** control to get the individual elements sampling frequency **Fs** and number of signal samples **#s.** Create a For Loop, and wire **#s** to the **count terminal N**. This completes the signal generation part. Next, as done in the previous chapters, convert the signal into the Q16.15 fixed-point format using the **to fixed point** function and a For Loop, as shown in Figure L10.2.
- Wire the time domain signal output to the function **FFT (signal processing>>transforms>>FFT.vi)**. This function can be used to perform FFT on real or complex data, and on one- or two-dimensional data. When wired to an input, it determines the polymorphic instance to use, or the user can select the instance to use. By default, it accepts real data. Its inputs and outputs are:
  - **X:** It is the input vector whose FFT is to be computed. Wire the time domain signal output to this input

- **shift?**: It specifies whether the DC component is at the center of the FFT spectrum to show both the positive and negative frequencies if set to **True** (the default is False). Keep it unwired.
- **FFT size:** Specifies the size of FFT. If **FFT size** is greater than the size of **X**, the VI appends zeroes to the end of **X** to match the size of **FFT size**, but if **FFT size** is less than the size of X, it uses only the first **FFT size** elements in X to perform FFT. Create a numeric **FFT size** with the value 512, and wire it to this input.
- **FFT {X}**: It is an array representing the FFT result of the **input** X.

The FFT function is implemented in LabVIEW in such a way that the FFT result needs to get scaled or normalized by dividing it with the length of the input signal. This is done using the **array size** and **divide** functions as shown in Figure L10.2 in the expanded view exhibited in Figure L10.5.

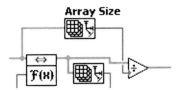

Figure L10.5: Normalizing the FFT result

- Discard the latter half of the FFT samples since it conveys the same information as the first half. For the present case, this corresponds to keeping 256 samples out of the 512 FFT samples. For this purpose, get the number of points by wiring the FFT{X} to the **size (Array>>array size)** function, wiring the function output to the **divide** function (along with constant 2), and finally connecting the divide output to the **length** input of the **array subset** function. The **index** input of this function is to be unwired (index 0), and the **array** input is to be connected to the array having the normalized FFT samples which were created earlier. Thus, its output terminal **subarray** is an array that starts at index 0, and has **length** elements out of the input array. Wire the subarray terminal to the **absolute (Numeric>>absolute)** function, and its output to a numeric indicator **FFT host**.

- Display the FFT. By definition, DFT is given by

$$X_k = \sum_{n=0}^{N-1} x_n e^{-\frac{j2\pi kn}{N}} \quad k = 0, ..., N-1 \tag{10.9}$$

Hence, the discrete time sampled frequency range is given by $f = k/N$ and the frequency range in Hz is given by

$$F = f * Fs = (k/N)*Fs \tag{10.10}$$

In other words, the frequencies are 0, Fs/N, 2*Fs/N... (N-1/N)*Fs Hz. For the present case, N=Fs=512. Therefore, the frequency scale is 0, 1, 2... 511 Hz.

However, as discussed earlier, one half of the FFT samples is kept, or the corresponding one half of the frequency samples, that is *0, Fs/N, 2*Fs/N, ((N/2)-1/N)*Fs* or 0, 1... 255 Hz. This is implemented as shown in Figure L10.2. To create the array named **k,** use a For Loop with **N** (loop count) wired to FFT/2 (the **divide** function with one input wired to the **FFT size** control, and the other input wired to constant **2**), and **iteration terminal** wired to the right border of the loop. Then, right click and choose **enable indexing** to create an array. Since the iteration terminal goes from 0 to loop count-1, with the loop count equal to *N/2*, the array generates **k** bins. Finally, implement Equation (L10.1) by wiring **Fs** and **array** to **multiply** and the output of the multiply and FFT size to the **divide** function. Place this portion of the BD within a smaller sequence diagram. Create an array indicator **freq axis,** and wire it to the output of the **divide** function, thus representing the frequency scale.

- Create a **bundle** function on the BD, and wire the first input to the array of frequency scale just created, and the other input to the FFT array absolute output created above. As a result, this creates a cluster of two elements, first element is an array of frequency axis, and the other element is an array of corresponding FFT values. On the FP, place a **XY graph (Graph>>XY graph)**, and on the BD, wire its cluster input terminal to the cluster just created. The XY graph plots data in Y cluster against X cluster. This completes the BD. The FP is shown in Figure L10.7.

Now, one needs to make this VI as a subVI. SubVIs are similar to functions in other programming languages, and thus provide the same advantages including modularity, easy debugging, and less cluttered design. On the upper right hand corner of the FP near **?,** right click on the VI icon and select **show connector**. It displays a number of square boxes. Click on each of the FP controls/indicators (to be created in any other VI when this VI is used as a subVI in it) and then click on the box, thus connecting the two. When all the controls and indicators are connected to the boxes (see Figure L10.6), save the VI. Now, this VI is ready to be used as a subVI in other programs.

Figure L10.6: Connecting items on Front Panel with boxes

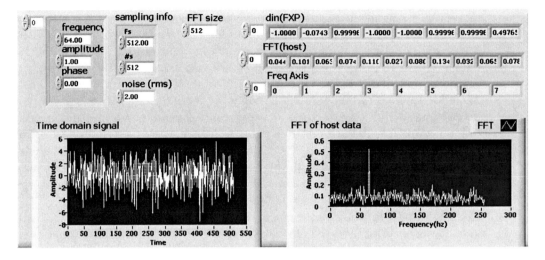

Figure L10.7: FP of FFT_sub.vi

Next, let us create the FPGA VI to communicate with the host VI. Create a new VI under the **FPGA** target, and save it as **target.vi.** Open its BD. The BD is shown in Figure L10.8. The steps of creating it are listed below:

- Create a **while loop** on the BD, and wire a **stop** control to the **loop condition** terminal.

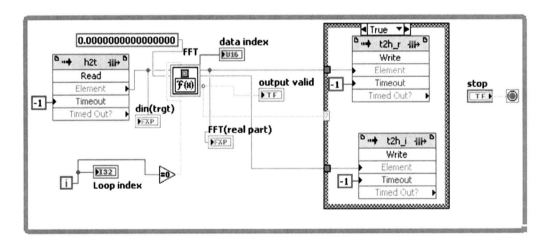

Figure L10.8: BD of target.vi

- The BD shows the target to the host DMA FIFO, **h2t**. Create this FIFO from the project explorer as done in the filtering chapter. Set its size to 2053 elements (see Figure L10.9) and the data type to fixed-point Q16.15, as this FIFO accepts the fixed-point composite signal from the host FIFO whose FFT is to be computed. The FIFO size is kept 2053 due to the reason that will become clear in the next section.

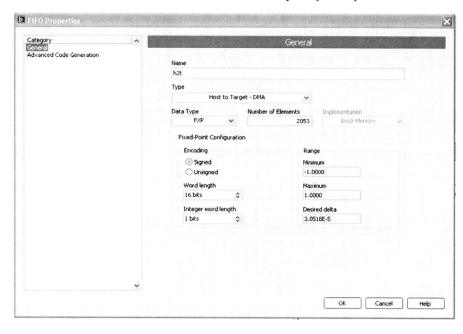

Figure L10.9: Properties page of h2t DMA FIFO

- Create the function **FFT (FPGA Math and Analysis>>FFT)** and place it on the BD. The FFT VI needs a detailed explanation. By double clicking on the VI, a window opens where one can **configure** the function (see Figure L10.11). If this function is placed **Outside the Single Cycle Timed Loop** (as done for this program), only the following terminals become visible:

  - **Reset:** It resets the VI when **True**. Thus, when the function is wired to a true value and called (within the while loop, each iteration of the while loop is equivalent to 1 call to the function), the function resets on the same call. Also, when true, the function **output valid** returns **False**. As shown in Figure L10.8, wire the **loop index** terminal of the while loop to the **equal to zero (Comparison>>equal to zero)** function, and wire its output to **reset terminal.** Thus, as soon as the while loop starts, the FFT VI is reset.
  - **Real data in:** The FFT VI can accept complex data. This terminal accepts the real part of the input data. As shown in Figure L10.8, wire the **element** terminal of the **h2t** FIFO to this terminal so that it accepts data from the FIFO.
  - **Imaginary data In:** This terminal accepts the imaginary part of the input data. Since using real data, wire the numeric constant with the value **0** to it, and set the data type to Q4.3 (kept minimum to save FPGA resources).
  - **Data index:** It refers to the output bin that is being provided by the VI. Right click on the control, create a corresponding indicator **Data index**, and wire it to the terminal data index.
  - **Real data out:** Returns the real part of the FFT result. Right click on the control, create a corresponding indicator and name it **FFT (real part)**, and wire it to the terminal data index.
  - **Imaginary data out:** Returns the imaginary part of the FFT result.

- **Output valid:** This terminal when true indicates that the FFT function has computed a valid output data sample. This point is made more clear by looking at the timing diagram displayed in Figure L10.10.

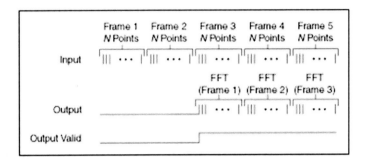

Figure L10.10: Timing diagram for FFT function when configured outside SCTL

Right click on the FFT function, choose **properties.** A page opens up as shown in Figure L10.11.

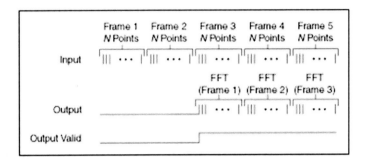

Figure L10.11: Configuring FFT function

The options appearing on this page include:

- **Length:** It specifies the length of the FFT. Choose 512, the same as in the host VI.
- **Direction: Forward** means FFT, **inverse** means inverse FFT or IFFT. Choose **forward.**

- **Output data type**: Here, one can specify the output data type, either explicitly using the options **wordlength (range [8-32])** and **integer wordlength**, or by letting the function decide the quantization using **adapt to source**. Choose the option **adapt to source**. Thus, the FFT result appears in the quantization format Q26.15 or <+- 26, 11>.
- **Execution mode:** It provides two options: **outside SCTL** and **inside SCTL**. Since the FFT VI was placed within a while loop, choose **outside SCTL**.
- **Clock rate**: It specifies the maximum clock rate (specified in the project explorer) for compilation of this function. If a faster clock is specified, the implementation of the function would require more pipelining stages which translates into a higher FPGA resource usage and latency for this function.
- **Throughput:** In general, throughput in the context of digital signal processing refers to the rate at which a new sample arrives at the processing unit. As the FFT function is configured outside SCTL, the throughput is disabled and has a value equal to 1 call/sample, which means that every time the function is called (as within the while loop), one sample of input data is accepted by the function to be processed.
- **Latency:** This refers to the time between an input sample is presented to the function and the time a valid output sample comes out. As seen in Figure L10.12, frames are presented to the function and there is a latency of 2*frame length before the first processed sample of frame 1 (first bin) comes out (and **output valid goes** high). After this initial latency, the function returns a valid output bin every time it is called. This assumes that a long data is divided into FFT frames, or streaming data. But for the input data of length 512, there is just one input frame, while the diagram shows that it takes 2*512 or 1024 function calls (while loop iterations) before the first bin comes out. Thus, as the loop starts, the first call to the function reads the first element of frame 1, and so on, till the first frame is finished. During this time, the function outputs garbage data (although the FFT real part indicator shows some values, they are not correct). Then, the data index reaches 512, meaning that the function is outputting the 513th bin if the second frame is fed to the input of the function. In the absence of this frame, the VI goes to the wait state (as the FIFO **h2t** is set to infinite timeout state, it keeps waiting for the data which does not exist). Thus, on the host side, one needs to append two more frames of zeroes at the end of frame 1 so that on the FPGA side, after latency of two frames, when frame 3 of zeroes is being read element by element, frame1 data is outputted as valid data.

Figure L10.12 provides an illustration of this process (1, 2, 3 refer to the order of data in and data out).

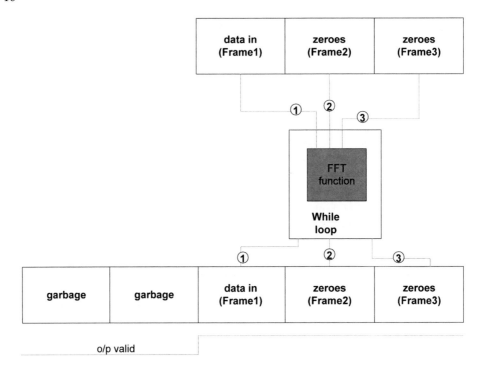

Figure L10.12: Frame processing sequence in FFT function outside SCTL

From this figure, it is seen that the input data of 3*frame size = 1536 elements is created. To hold this data, the **h2t** FIFO size is kept to 1536, which gets automatically coerced to 2053 elements.

- Now, create a **case** structure and wire its selector terminal to the **output valid** terminal of the FFT function. Although the FFT function does compute the output when the output valid is false, only the valid data starting from frame 1 is fed to the subsequent FIFO's. Thus, it is needed to create a target to the host DMA FIFO **t2h_r** (see Figure L10.13) on the FPGA VI to hold the FFT real part result for the entire three frames (1536 elements). This way, its size is coerced to 2047 elements. However, if the host program is run once, the target to host FIFO in the FPGA VI would hold the result of FFT for the first frame which gets transferred back by the DMA to the read FIFO on the host. As a result, t2h_r of length coerced to a value corresponding to one frame would be sufficient too. Also, set its data type to Q26.15 to match the FFT result. Similarly, create a FIFO **t2h_i** with the same settings for holding the imaginary part of the FFT. As shown in Figure L10.8, place these FIFO's within a case structure **true**, and wire them to the **Real data out** and **Imaginary data out** terminals of the FFT VI, respectively. Keep the **false** case empty.

This completes the FPGA VI. The completed FP is shown in Figure L10.14.

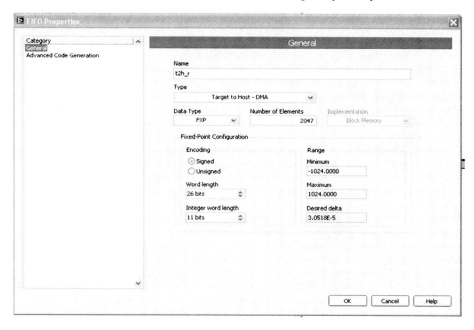

Figure L10.13: Properties page of t2h_r DMA FIFO

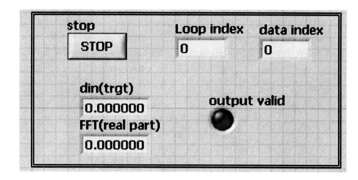

Figure L10.14: FP of target.vi

Now, let us create the host VI. Create the VI **host.vi** under **My Computer**, and open its BD. Upon completion, it looks as displayed in Figure L10.15.

Let us go through all the steps to create this diagram:

- Add a sequence diagram with five frames. In the first frame, add the subVI **FFT_sub.vi (select a vi>>browse>>FFT_sub.vi)**. Create the controls and indicators as done earlier while creating the subVI. A snapshot of the BD and FP is shown in Figures L10.15 and L10.19, respectively.

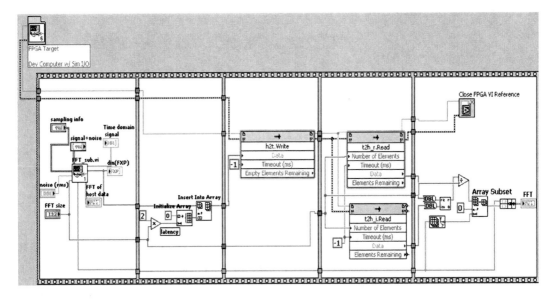

Figure L10.15: BD of host VI

- Open a reference to **target.vi** by using the **open Fpga VI reference** function. As was explained in the previous section, it is necessary to create two frames of zeroes, and append them to signal data (see Figure L10.16). To do this, in the second frame of the sequence diagram, multiply the **FFT size** output with constant **2**, and wire it to the **multiply (numeric>>multiply)** function so that its output is **latency**.

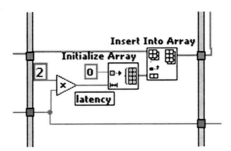

Figure L10.16: Latency setup

- Create two frames of zeroes of size equal to the latency using an **initialize array (Array, cluster and variant>> initialize array)** with the terminal **element** wired to constant **0**, and the terminal **index** wired to the **latency** array. Finally, append this array to the **din (Fxp)** output of the subVI by using the function **insert into array (Array, cluster and variant>>insert into array)** and wiring its terminal **element** to **din(Fxp)** as well as **new element** to the **initialized array** terminal of the **initialize array** function.
- In the next frame, create the function **invoke method**, and wire the **Fpga Vi reference in** terminal to the **Fpga Vi reference out** terminal of **open Fpga Vi reference**. Also, wire their **error in and error out** terminals. As was done in the filtering chap-

ter, configure it for write operation so that fixed-point input data with noise is written into the host FIFO, and then to the target FIFO **h2t** by the DMA. Wire its **timeout** terminal to **constant -1** (infinite) and the **data** terminal to the **insert array** terminal of **output array.** This is shown in Figure L10.17.

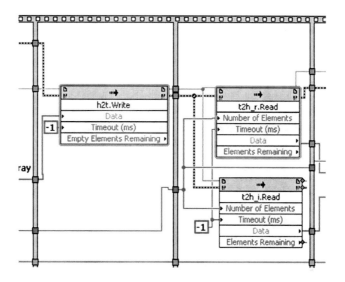

Figure L10.17: Configuring FIFOs on host VI

- Next, as the FPGA FIFO holds the FFT result, the first frame data (=FFT size, as depicted in Figure L10.12) needs to be captured. By using **invoke method,** as in the first step, create **t2h_r.read** and **t2h_i.read** (see Figure L10.17); one to read the real part of the FFT and the other to read the imaginary part. Wire the **FFT size** to the **number of elements** terminal of both nodes, and set their timeouts to -1.
- In the last step, convert the real and imaginary data to the equivalent floating-point representation using the function **to double.** Place the function **Re/Im to Polar (Numeric>>Re/Im to Polar)** on the BD. It has two inputs **x** and **y** which accept the real and imaginary part of the complex input and provides the magnitude and phase on the output terminals **r** and **theta.** Then, wire the **double** outputs to the input terminals of the function as shown in Figure L10.18.
- Last, as was done in the **FFT_sub.vi subVI,** perform the normalization of the result and display it using the same frequency scale as for the host VI floating-point FFT, refer to Figure L10.18. The XY graph indicator **FFT** shows the FFT result. Close the FPGA VI reference using the function **Close FPGA VI reference.**

The FP of the host VI is shown in Figure L10.19.It clearly shows that FFT separates the input sine wave frequency of 64 Hz from the noise by using the host only approach and the host–target DMA communication. Also, as shown in Figure L10.20, when the output valid goes high, it stays high and the data from the first frame is sent out (for loop index 1536, 512th element or the last element of the first frame).

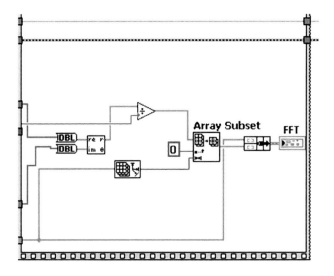

Figure L10.18: Displaying FFT result

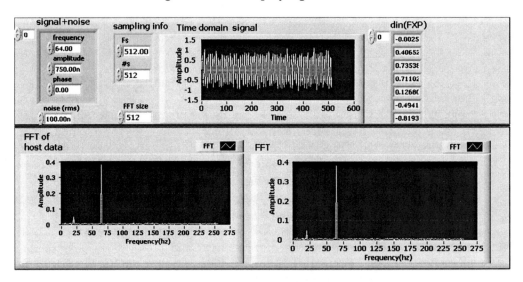

Figure L10.19: FP of host.vi

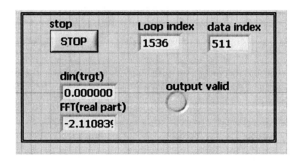

Figure L10.20: FP of target VI after execution

L10.1.2 FFT in LabVIEW FPGA Inside SCTL

Through this second method, the handshake mechanism is presented. First, create the target VI **target_hk.vi**, the FP of which is shown in Figure L10.21, and the BD in Figure L10.22.

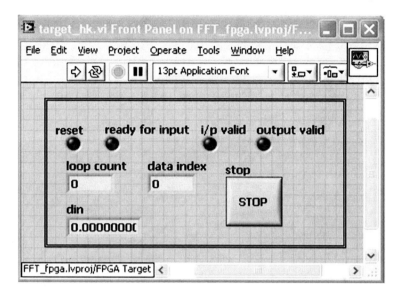

Figure L10.21: FP of target VI target_hk.vi

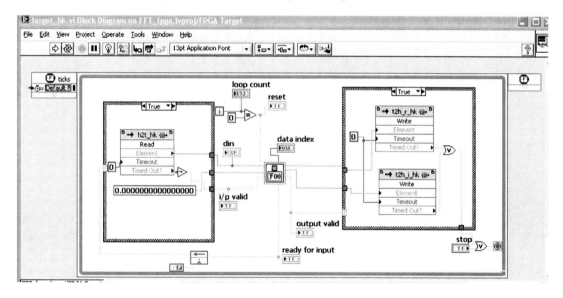

Figure L10.22: BD of target VI target_hk.vi with both case structures 'true'

To understand the way this BD is created, let us first understand how the FFT VI works when configured within a SCTL. As was stated earlier, this VI can be configured for operation inside a SCTL by choosing the option **inside single- cycle timed loop** on its properties page. As shown in Figure L10.23, the parameters **length, direction, Output data type,** and

Clock rate are set based on the same values used in the configuration **outside single-cycle timed loop** shown in Figure L10.11. However, the **latency** field now shows a value of **2564 cycles,** with **throughput** chosen to be **5.51 cycles/sample** (selecting **Inside single-cycle timed loop** automatically enables the **Throughput** option). As the Throughput field is clicked, it shows two options: **1 cycle/sample** (higher throughput rate) and **5.5 cycles/sample** (lower throughput rate). If 1 cycle/sample is chosen as the throughput, it translates to higher hardware resource utilization. This is because SCTL denotes that timing of the loop matches exactly the FPGA clock rate and with the throughput of 1 cycle/ sample, each iteration of the loop takes 1 clock cycle, and 1 sample of the data is inputted within this time. Hence, the function within the loop must process the data within one clock cycle, which translates into higher FPGA resources. Choose the throughput of **5.5 cycles/sample**.

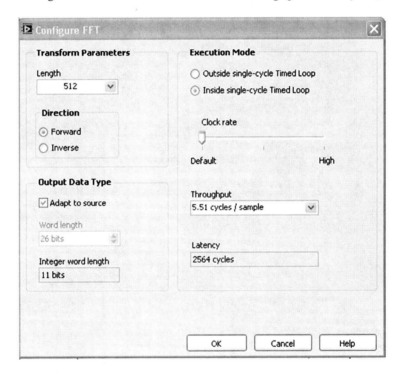

Figure L10.23: FFT VI properties within SCTL

Also, LabVIEW automatically computes the latency to be 2564 cycles. With this latency and SCTL mode, the timing diagram for the throughput of 1 cycle/sample and the throughput of 5.5cycles/sample are displayed in Figure L10.24.

It is to be noted that the timing diagram for the throughput of 1 cycle/sample is the same as that of the FFT VI (the same latency of 2*FFT size) configured outside the SCTL, while for the throughput greater than 1, the latency is not the same, and **output valid** is a pulsed output with the FFT frame result coming out of the function when this pulse is high.

In addition, within the SCTL, the FFT VI does not accept or return values while computing the FFT. So any data sent to this VI is discarded during the computation time. To prevent data loss, the host to target FIFO should be large enough to hold data until this function accepts values again. A rough estimate of the size of the FIFO is equal to the latency divided by the average system throughput. Thus, for the present case, the FIFO size should be 2564/5.51= 465. Hence, the **h2t_hk** FIFO size is coerced to **512** which is also the frame size.

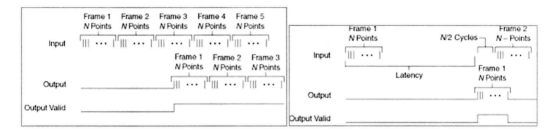

Throughput=1 cycle/sample   Throughput=5.5 cycles/sample

**Figure L10.24: Timing diagrams with throughput of
1 cycle/sample and 5.5 cycles/sample**

Now, place this function within a SCTL. The expanded view of the function when configured within the SCTL is shown in Figure L10.25. As can be seen, it has the same terminals as in the Outside SCTL configuration, but now it has new handshake terminals which are listed below.

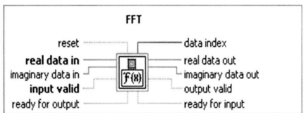

Figure L10.25: FFT VI within LabVIEW FPGA

- **Ready for input:** It is an output signal. When the FFT VI is ready to start processing data, the **ready for input** line goes **high**, which signals to the upstream node (a FIFO or whatever precedes this node) that the function can accept input samples at the next clock cycle. Create a **LED** indicator for the signal and name it **ready for input.** Now, as shown in Figure L10.22, create a **case** structure and wire this node terminal to the **case select** terminal of the structure by using a **feedback node** (a feedback node stores data from one VI execution or loop iteration to the next). Wire a **false** constant to the initializer terminal of the feedback node so that the first time the loop runs, or in the first clock cycle, the **false case** structure gets selected. Within **false case,** create a fixed-point constant (Q16.15) with value 0, and a **false** constant. Wire the fixed-point constant to the **real data in** and **imaginary data in** terminals of the function. This ensures that when the function is not ready to accept any input data, zeroes are fed to the input, which is like feeding no data.

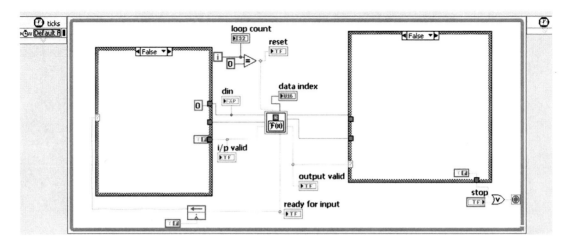

FIgure L10.26: BD of target vi target_hk.vi with both case structures 'false'

- **Input valid:** This is an input signal, which when high signals the FFT function that the next data sample has arrived to get processed. In general, any node x must send this signal to the downstream node (that follows x) before it can start reading data. Create a **LED** indicator for the signal, and name it **Input valid.** Wire the **false** constant created above (within **false** case) to this indicator, and onto the **input valid** terminal of the function. When **reset** is **high**, **input valid** and **ready for output** is ignored, while **output valid** and **ready for input** are **low**.

To better understand how the handshake signals work for a few iterations, let us assume that the host VI is made such that when it runs, the FPGA VI executes with Figures L10.27, L10.28, and L10.29 exhibiting the run state. When loop count = 0 (see Figure L10.27), **reset** goes **high**, **ready for input** goes **low** (and is stored for the next cycle by the feedback node), which selects **false case**, and **input valid** goes low too (though it is ignored). When, the loop count becomes 1 (see Figure L10.28), the **ready for input** signal goes **high**, but this value is stored for the next iteration, and the last stored value which was **low** is applied to the case structure select terminal, thus selecting the false case again, and generating a low **output valid** to the function. Next iteration, **ready for input** stays **high**, and the last stored **high value** is fed to the case structure, thus selecting **true case** (see Figure L10.29). This way, the handshake ensures that the function has to tell the upstream node that it is ready to accept data samples (ready for input high), then the upstream node has to acknowledge that input valid is high before sending data samples. Thus, even if **ready for input** goes **high**, the function would ignore input data samples unless **input valid** goes **high.**

Now, let us create the BD for the case when **ready for input** selects a true case. As seen in Figure L10.22, similar to the first example, create a host to target DMA FIFO **h2t_hk** on the BD with the properties as shown in Figure L10.30. Comparing Figure L10.30 and Figure L10.9, one can see that the **h2t_hk** FIFO has less elements since unlike the previous case, it is not required to continuously feed input frames till one gets one frame output (for throughput not equal to 1, see Figure L10.24). Hence, there is no need to append any zero frames.

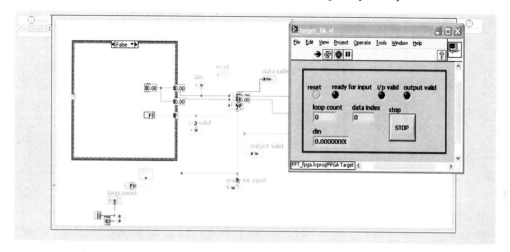

Figure L10.27: BD of target VI target_hk.vi in debug mode in the first loop iteration

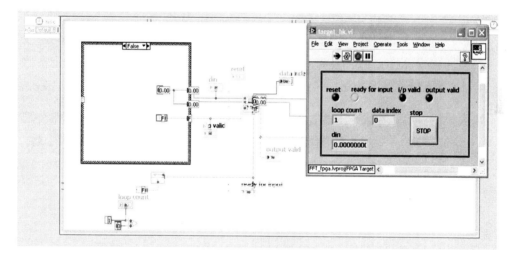

Figure L10.28: BD of target VI target_hk.vi in debug mode in the second loop iteration

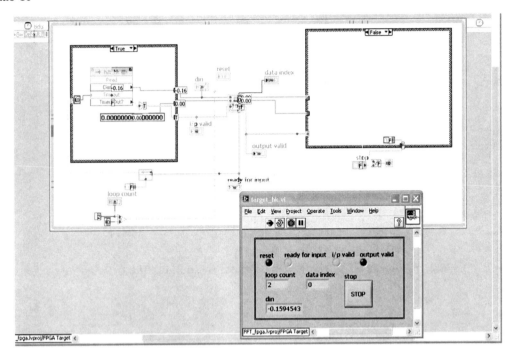

Figure L10.29: BD of target VI target_hk.vi in debug mode in the third loop iteration

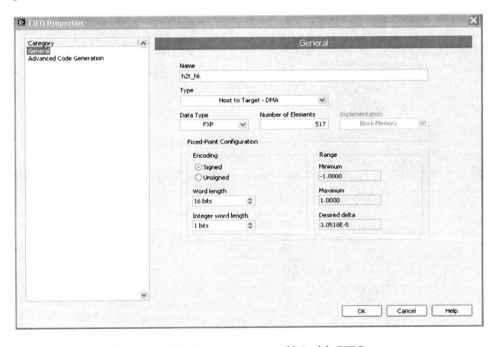

Figure L10.30: Property page of h2t_hk FIFO

Wire the **element** output of the FIFO to the **real data in** input of the function and the **zero** fixed-point constant (Q16.15) to the **imaginary data in** input (as data is real). If the FFT function is ready to accept data, but if the FIFO has not acquired data within the specified **timeout** (0 or no wait), it will be empty and the function should not attempt to read the input line. This is implemented by wiring the **timed out** terminal of the FIFO to the **input valid** indicator after inversion with a **not** function. When the DMA controller does not transfer the input samples to the FPGA FIFO fast enough, the FIFO times out giving a **high,** which turns **input valid** low. This way the VI ignores any data that is inputted to the system during this period.

Now, data samples are fed to the FFT function, point by point, and when one frame of data has been read, **ready for input** goes **low** (see Figure L10.32), though it is still computing the FFT result. However, as indicated by the timing diagram, the FFT result will come out of the function after latency. Next, let us go through the remaining handshake signals:

- **Output valid:** This terminal becomes **high** if this node/function has computed a result that downstream nodes can use.
- **Ready for output:** It specifies whether downstream nodes are ready to accept data samples that this node/function has computed. If this node was followed by a downstream node, the **ready for input** output of that node would need to be wired to this input of the current node using the **feedback** node. However, one needs to pass the FFT result from the target to the host FIFO directly by keeping this terminal unwired.

The FFT properties page shows that FFT result is Q26.15. Therefore, similar to **t2h_r** and **t2h_i,** or the FIFO's in the first example, create two DMA FIFO's **t2h_r_hk** and **t2h_i_hk** with the same configuration and place them in the **true** case of the second case structure as shown in Figure L10.22. The FIFO's will hold one frame of the FFT result, so the size is 512, which coerces to 1023 (see Figure L10.31).

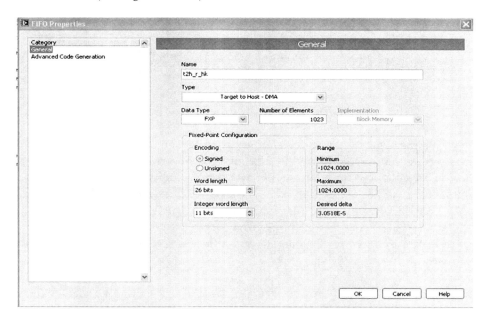

Figure L10.31: Properties of t2h_r_hk DMA FIFO

Lab 10

For the time **output valid** is low, one does not need to write anything in the FIFO's. Thus, in the **false** case of the case structure, create a **false** constant, and wire it along with the **Boolean control stop** to the respective inputs of an **or** function, the output of which is wired to the **SCTL** conditional terminal (see Figure L10.26).

Now, after the latency period, **output valid** goes high. Next, wire the **real data out** and **imaginary data out** terminals of the FFT function to the **element** input of the respective FIFO's, and set the Timeout to 0. Also, wire the corresponding **timeout** terminals of the FIFO's to one of the terminals of the **or** function as shown in Figure L10.22. If the DMA controller has not transferred the FFT result from the **t2h_r_hk** and **t2h_i_hk** FIFO's to the host FIFO memory within the specified timeout, the FIFO's fill up and data gets corrupted. Wiring as indicated stops the target VI and prevents this condition.

Hence, after initialization, till loop count/clock cycle reaches 513, **ready for input** stays high, **input valid** stays high, and **output valid** stays low. Ready for input goes low, after one frame of data gets passed to the FFT function, but since the frame data sample 1 was inputted to the FFT function from clock cycle/loop count 2 (as soon as input valid became high), **ready for input** stays high for two 513 (512 counts), and then goes low at the loop count of **514** (see Figure L10.32).

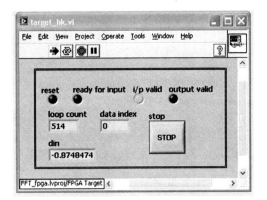 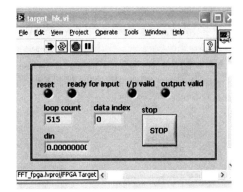

Figure L10.32: Ready for input and input valid going low

In the next cycle, after one clock delay, **input valid** also goes low (see Figure L10.32). During all these cycles, **data index** is 0, indicating that the FFT result has not been computed yet. **Output valid** stays low for a latency of 2560 samples (approximately 5 cycles/sample*512 samples) and then it becomes **high** when the loop index reaches 2566 (see Figure L10.33). At this point, the FFT result for the first frame sample 1 comes out of the respective terminals. The data index also starts incrementing from zero till it reaches 255 (N/2 or 256th sample), and **ready for input** goes high again, but no input sample is acquired as **input valid** is low (see Figures L10.24 and L10.34).

Finally, as soon as the data index reaches 511, the FFT sample is fed to the FIFO, and is read back to the host. After that, the process repeats as it did from the loop count 1.

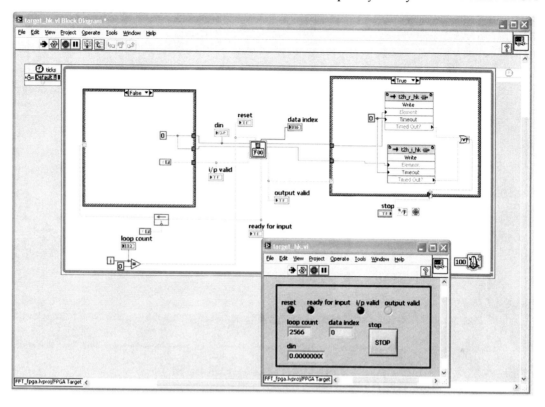

Figure L10.33: Output valid going high

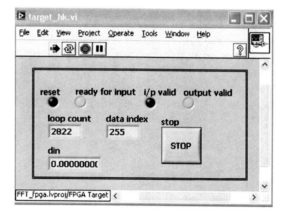

Figure L10.34: Ready for input becoming high again

Now, let us create the host VI. Figure L10.35 shows the BD of it.

To create this BD, follow the steps as in the first example for **host.vi**. Comparison with Figure L10.15 reveals that the diagram is similar except for the second frame, which does not have the diagram for appending zeroes (not required as explained above). Also, the sequence diagram is enclosed in a while loop.

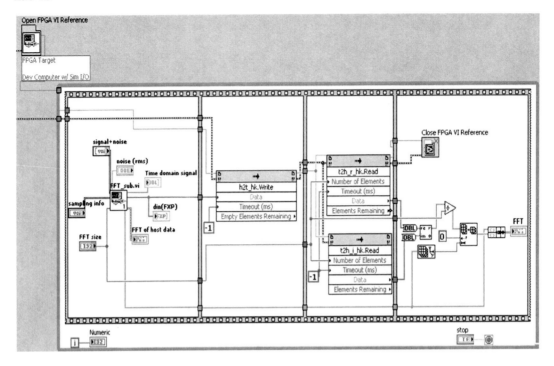

Figure L10.35: BD of host VI host_hk.vi

This completes the design. Now, run the host VI, and change the noise to see the result. It is verified that as more noise is added, the desired FFT response for both floating-point and fixed-point FFT shows a peak at the required 64 Hz frequency, but it starts getting buried in noise. Figures L10.36 and L10.37 for the FP of the VI show this effect. However, note that as more noise is added, since the combined signal and noise is fed to the FPGA VI in fixed-point format, Q16.15 may not be sufficient to effectively represent the input signal. This may show up in the result as quantization noise which cannot get distinguished from the RMS noise added in the system (this holds true for example 1 as well). In such cases, one can start the design with a higher dynamic range setting such as Q20.16 depending upon the expected noise level.

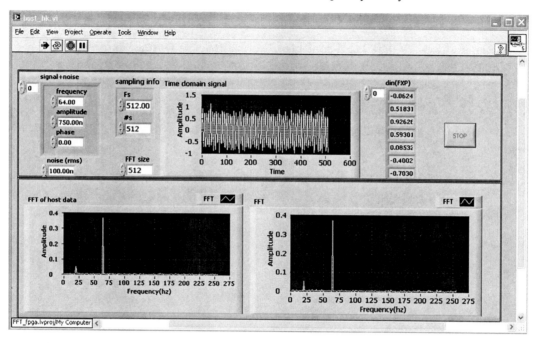

Figure L10.36: FP of host VI host_hk.vi in run mode with no noise

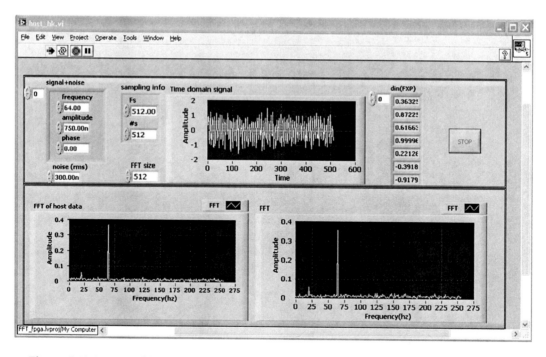

Figure L10.37: FP of host VI host_hk.vi in run mode with RMS noise of amplitude 1

L10.2 DCT in LabVIEW FPGA

This section discusses the FPGA implementation of Discrete Cosine Transform (DCT) as a frequency domain processing tool. A host VI is first created to generate a sinusoidal signal together with a ramp signal. The FFT and DCT magnitudes are then computed and displayed together with their number of coefficients. An FPGA VI is also created to perform the DCT computation using FFT on the same data that is used in the host (after fixed-point conversion), and then confirming the result on the host. Let us create the VIs in a step by step manner.

Create the project **DCT.lvproj** using the host VI **host.vi**, and then add the FPGA VI **target_hk.vi** to the project, as shown in Figure L10.38. Set the FPGA target mode to **Execute Vi on development computer with simulated I/O** with **random IO data for FPGA IO read.**

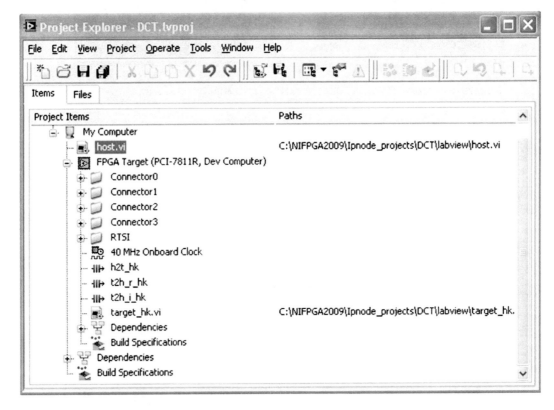

Figure L10.38: LabVIEW FPGA project DCT.lvproj

L10.2.1 Host VI

Follow the steps below to create the host VI:

1. On the BD (see Figure L10.39), create a **sequence diagram (Structures>>Flat sequence diagram)** with six frames in it.
2. In the first frame, go to **(view>>tools>>edit text)** and add text 'signal generation' to have a name for frame1. In this frame, create three numeric constants **frequency,**

amplitude and phase. Cluster them together using a **cluster control** (as outlined in the FFT example), and name it **Sine parameters**. Using the **unbundle by name** function, unbundle the individual numeric controls out of the cluster.

3. Place the function **sine Waveform.vi (Signal processing>>Waveform generation>>sine waveform.vi)** on the BD. This function is shown in Figure L10.40.

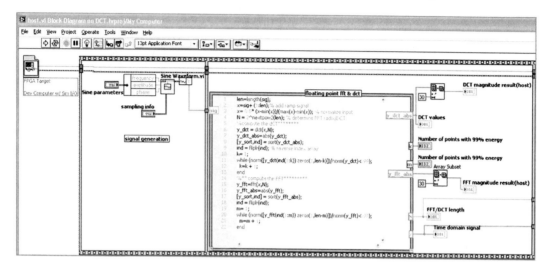

Figure L10.39: BD of host.vi

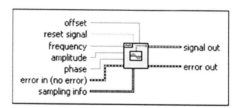

Figure L10.40: Sine waveform.vi

The signals in this function include:

- **Offset:** This is the DC offset of the signal. The default value is 0.0. It is optional. Keep this unwired.
- **Reset signal:** If this signal is True, it sets the time stamp to zero, the phase of each tone to the **phase** value used in the **tones** array, and the seed to the **seed** control value. Keep this unwired.
- **Frequency**: Specifies the frequency of the sinusoidal signal.
- **Amplitude**: Specifies the amplitude of the sinusoidal signal.
- **Phase**: Specifies the phase of the sinusoidal signal.
- Wire the controls out of the **unbundle by name** function to the corresponding terminals of frequency, amplitude and phase, and set the values to 100, 50, 90, respectively.

- **Sampling info:** This is a cluster of two elements: **Fs** for the sampling frequency in Hz, and **#s** for the number of samples. Create the Cluster Control with two elements **Fs** and **#s** with values 400 Hz and 256, respectively.
- **error in (no error):** This describes any error that propagates from any previous node. Keep this unwired.
- **Signal Out**: This is the output signal generated by the function.
- **Error out**: This contains error information. Keep it unwired.

This way, a cosine waveform **sig(n)=50* cos(2*pi*100/4000*[1:256])** is created. Convert this waveform signal into **double array** using **get waveform components**.

4. Name the second frame as **floating point fft &dct.** In this frame, compute the FFT and DCT of the signal using a **Mathscript node (Mathematics>>Script and formula>>Mathscript node).** One can use this node to write the text based .m files. Lab-VIEW also has the Matlab Script node with the same functionality, but the MathScript node has the advantage that it does not require Matlab (Matlab Command Window) to be running in the background while the Matlab Script node does. On the other hand, the Matlab Script node works as a complete Matlab system, allowing full functionality of Matlab, while the Mathsript node may not allow certain functions to be used in the node. To allow larger codes to fit in the Mathscript node, one can include scrollbar in the node, and also include line numbers for easy debugging. To do so, right click on the node and select **visible items>>scrollbar** and **visible items>>Line numbers.** Right click on the left border of the node, and select **add input.** A blank box appears as shown in Figure L10.41. Here, enter the name (can be a scalar, array, string types) and wire it to the signal to be used in the node. The signal name then becomes a variable. For the **input,** the variable automatically adapts to the signal wired to it. Enter the name **sig** and wire it to the input sinusoidal signal.

Figure L10.41: Creating variables on the border of MathScript node

Let us go through the code line by line.

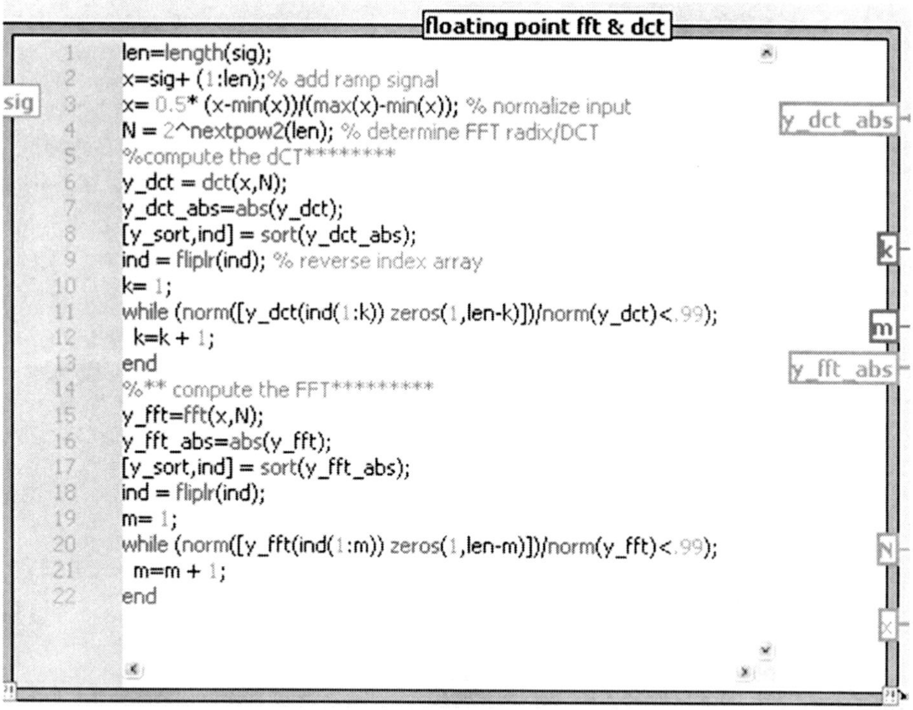

```
                        ┌─────────floating point fft & dct─┐
  1   len=length(sig);
  2   x=sig+ (1:len);% add ramp signal
  3   x= 0.5* (x-min(x))/(max(x)-min(x)); % normalize input
  4   N = 2^nextpow2(len); % determine FFT radix/DCT
  5   %compute the dCT********
  6   y_dct = dct(x,N);
  7   y_dct_abs=abs(y_dct);
  8   [y_sort,ind] = sort(y_dct_abs);
  9   ind = fliplr(ind); % reverse index array
 10   k= 1;
 11   while (norm([y_dct(ind(1:k)) zeros(1,len-k)])/norm(y_dct)<.99);
 12     k=k + 1;
 13   end
 14   %** compute the FFT*********
 15   y_fft=fft(x,N);
 16   y_fft_abs=abs(y_fft);
 17   [y_sort,ind] = sort(y_fft_abs);
 18   ind = fliplr(ind);
 19   m= 1;
 20   while (norm([y_fft(ind(1:m)) zeros(1,len-m)])/norm(y_fft)<.99);
 21     m=m + 1;
 22   end
```

sig → (inputs)
y_dct_abs, k, m, y_fft_abs, N, x → (outputs)

Figure L10.42: Floating-point FFT and DCT of input signal using MathScript node

- In line 1, the length of the input signal is computed using the function **length.**
- In line 2, a ramp signal (with sample values 1,2.3...256) of length equal to the input signal **sig** is created, and added to the input signal generating the composite signal **x.**
- As the sinusoidal signal is of amplitude 50, when converted to fixed-point before being fed into the FPGA VI, the wordlength requirement may be large, which translates to larger size FIFO's in the FPGA. So, in the third line, normalize the input for it to lie between -0.5 to 0.5.
- In the forth line, the FFT radix/DCT radix is computed using the function **nextpow2.** This function works as follows: N= Nextpow2(len) is the same as N>= abs(log$_2$(len)). That is, it computes the smallest power of two that is greater than or equal to the absolute value of the argument. Here, the length 256 is used, so N=256. But if the length was say 200, still the FFT size would be 256.
- Line 6 computes the DCT using the function **DCT.**
- Line 7 line computes the absolute value of the DCT using the function **abs.**
- Line 8-13 computes the number of DCT samples that have energy within 99% of the total energy in the DCT samples. Line 9 sorts the absolute DCT magnitudes from lower to higher magnitudes, stores them in an array **y_sort,** and their index numbers (corresponding to the y_dct_abs array) in the variable **ind** using the function **sort.** Next, the indices are flipped using the function **fliplr** (similar to the LabVIEW function **reverse 1-Darray**).

307

Next, consider the test example provided below in Matlab

```
y_dct= [7 -20 -5  -11];
y_dct_abs= [7 20 5 11];{index = 1 2 3 4}
y_sort= [5 7 11 20]; ind= [3 1 4 2];
ind= [2 4 1 3];{after flipping}
k=1;
Norm (y_dct) = sqrt( 7^2+-20^2+ -5^2+ -11^2);{energy of signal)
norm([y_dct(ind(1:k)) zeros(1,len-k)]) = norm([ y_dct(ind(1)) zeroes(1,3)])
= norm([y_dct(2)] 0 0 0])
= norm([-20 0 0 0]) = 20
```

This continues till the norm /energy of the samples becomes less than 99% of the total energy. The DCT magnitudes are thus arranged from lowest to highest, the indices are arranged from those which correspond to the highest DCT magnitude to the lowest magnitude, and then the norm of the DCT sample with the highest magnitude is found (the first element of the index array) with respect to the signal norm. This procedure continues till the energy of the DCT samples chosen becomes equal to 99% of the total energy. One starts from the highest magnitude so that a faster convergence is obtained. For example, in a signal of length 100, there might be 2 samples that have 99% of the signal energy. So it is better to start the algorithm from those samples.

- In lines 15 to 22, the FFT of the signal is computed, and in a manner similar to the DCT, the number of FFT samples with 99% of the total energy in the samples are computed.

Now, to create the output signals for the Mathscript node, right click on the right border of the node, and select **add output**. Write the name of the signal in the blank box that appears. The signal type should match the corresponding script variable. This can be done by right clicking on the signal, selecting **Choose data type**, and then selecting the data type from the options displayed. Thus, as shown in Figure L10.42, create the signals **y_dct_abs** and **y_fft_abs** (double array) that correspond to the magnitude of the DCT and FFT results, and the signal **N** that refers to the FFT/DCT size, the signal **x** that refers to the composite time domain signal, and finally, the signals **k** and **m** that represent the count of the samples with 99% energy for the DCT and FFT results, respectively.

5. Create an array indicator, and name it **DCT values**. Wire the Mathscript variable **y_dct_abs** to this indicator. This array then represents the values of the DCT transform applied onto the signal x. Similarly, create a graph indicator to display the first 30 samples (for better visibility, use fewer number of samples for display), and name it **DCT magnitude result (host)**. Wire the Mathscript variable **y_dct_abs** to this graph via the function **array subset**, whose parameter **length** is set to 30. Right click on the graph, and select **visible items>>plot legend**. Click plot (**legend>> Common plots>>stem plot**) (see Figure L10.43). To hide the plot legend, click on **visible items>>plot legend**.

6. Create two numeric indicators, name them **number of points with 99% energy**, and wire each of them to the Mathsript variables **k** and **m**.

7. Create a graph indicator, with the same properties as the earlier graph, and name it **FFT magnitude result (host)**. Wire the Mathscript variable **y_fft_abs** to this graph via the function **array subset**, whose parameter **length** is set to 30.

8. Create a numeric indicator, name it **FFT/DCT length**, and wire it to the variable **N** in the Mathscript node. This represents the FFT/DCT size.

9. Finally, create a graph indicator, name it **time domain signal**, and wire it to the Mathscript variable **x** for giving a graphical view of the signal as shown in Figure L10.43.

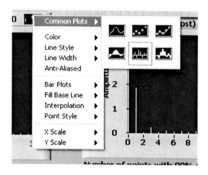

Figure L10.43: Creating stem plot in LabVIEW graph

This completes the second frame.

10. Name the third frame as **reformatting data for dct (using fft)**. Here, a signal twice the length of the signal **x** is created by first flipping the signal and then appending it with the unclipped version. Thus, in general, a signal y(n) is created as

$$y(n) = \{ \begin{array}{ll} x(n) & \text{for} \quad 0<=n<= N\text{-}1, \\ x(2{*}n\text{-}1\text{-}N) & \text{for} \quad n<=N<=2N\text{-}1, \\ 0 & \text{otherwise} \} \end{array}$$

Wire the signal **x** to the **array** input terminal of the **reverse input array (array>> reverse input array)** function for the **reversed array** terminal of the function to return the flipped array (see Figure L10.44). Wire this terminal with the array input terminal of the **insert into array** function for the **output array** terminal of this function to return **y(n).** Create a graph indicator for y(n) as **Time domain signal (double frame) for DCT computation.** Figure L10.45 shows the FP of the host VI (when fully completed). The graph of y(n) will show the flipped version of x(n) appended with x(n). Finally, use a For Loop and the function **to fixed point**, as done in the FFT example, to convert this signal into fixed-point data type for utilization in the FPGA VI. As the signal amplitude is limited to the absolute value of 0.5, the quantization chosen is < 18,-1> (unsigned as signal is always positive) with the wordlength of 18, the fraction length of 19, and the integer wordlength of -1. This provides a range of 0 to 0.49999 with the precision of 1.9073E-6. Note that in an actual target, it would be better to use wordlengths that match the target accumulator wordlength to speed up the datapath. This completes the first part of the host VI.

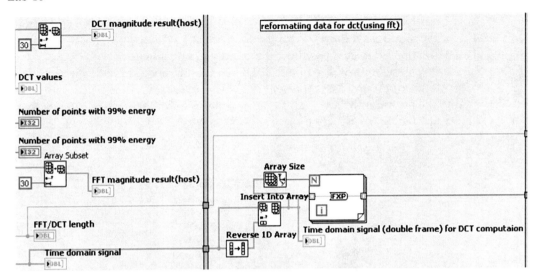

Figure L10.44: Formatting input data for DCT using FFT

L10.2.2 Target VI

Open the BD of the target VI **target_hk.vi**. Figures L10.46 and L10.47 show that the BD is the same as that of the VI **target_hk.vi** of the FFT project **FFT_fpga.lvproj**. To create this VI, either copy the BD of the target_hk.vi from the FFT_fpga.lvproj into the empty BD of target_hk.vi in the current project, or add the file from the Project menu so that it is referenced in both the projects. One can also copy that file into the current folder and add it into the project from the current folder.

As seen in Figures L10.46 and L10.47, the host to target DMA FIFO **h2t_hk** and the target to host FIFOs **t2h_r_hk** and **t2h_i_hk** are used for communication between the host and the target. The properties of these FIFOs are different from those in the **FFT** project. When this target file is added into the **DCT.lvproj**, the **run** arrow sign on the FP of the VI would be broken, as the target is added to the project without first creating the FIFOs. So, on the **h2t_hk** FIFO, select **new FIFO** and a new FIFO named **FIFO** appears on the project explorer. Rename it **h2t_hk** and set its properties to the ones shown in Figure L10.48.

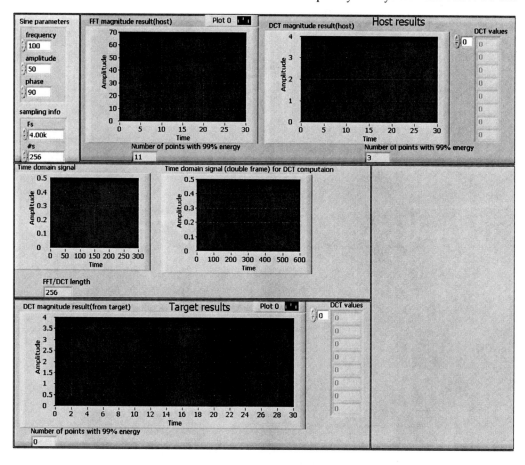

Figure L10.45: FP of host.vi

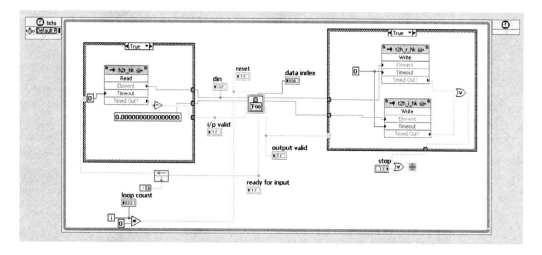

Figure L10.46: BD of target_hk.vi when case structures are true

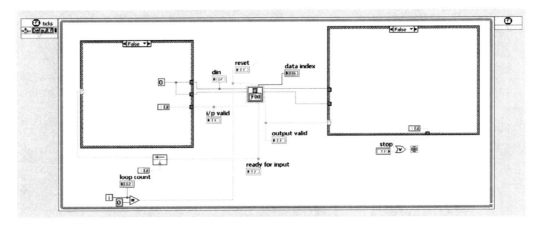

Figure L10.47: BD of target_hk.vi when case structures are false

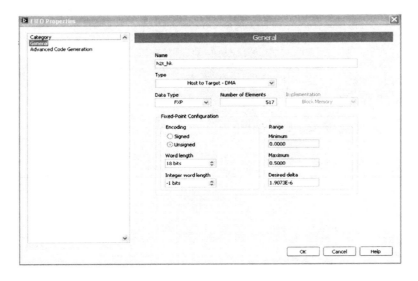

Figure L10.48: Setting properties of h2k_hk FIFO

As OK is pressed, the FIFO appearing in the FPGA VI automatically gets its broken wires fixed and gets connected as shown in Figures L10.46 and L10.47. Once the **element** output of this FIFO is connected to the **real data in** terminal of the FFT.VI, set its properties as shown in Figure L10.49. It can be seen that the latency, throughput and

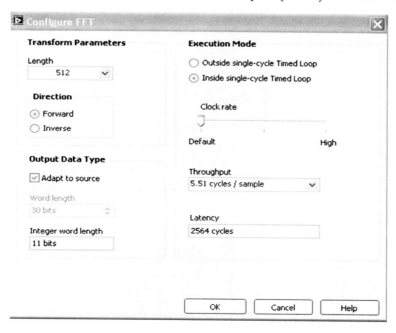

Figure L10.49: Property page of FFT.vi

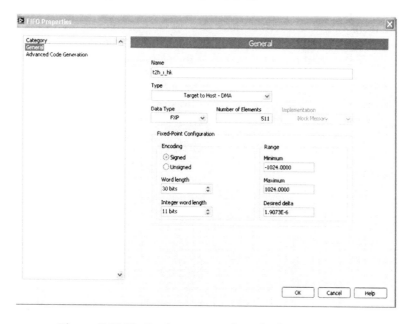

Figure L10.50: Setting properties of t2h_i_hk FIFO

integer wordlength of the result is the same as in the FFT project, the only value that changes is the wordlength, which is 30 bits (it was 26 bits in Figure L10.23). Note that the FFT of size **2*N** is computed, where N is the FFT size of the original data x(n). Now, in a similar manner, create the **t2h_r_hk** and **t2h_i_hk** FIFOs with the quantization that matches the FFT

313

function quantization, as shown for the **t2h_i_hk** FIFO in Figure L10.50. Figure L10.51 illustrates the FP of the target VI.

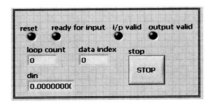

Figure L10.51: FP of target_hk.vi

L10.2.3 Completion of Host VI

Having created the target VI, let us now complete the host VI by adding the portion that communicates with the target VI by carrying out the following steps:

1. Open a reference to **target_hk.vi** by using the function **open Fpga VI reference**.
2. Name frame 4 as **FP data written to target**. In the frame, create the function **invoke method**, and wire the terminal **Fpga VI reference in** to the terminal **Fpga VI reference out** of **open Fpga VI reference**. Also, wire their **error in and error out** terminals (see Figure L10.52). Configure it for write operation so that fixed-point input data from frame 3 is written into this FIFO, and then to the target FIFO **h2t_hk** by the DMA. Wire its **timeout** terminal to **constant -1** (infinite) and the **data** terminal to the **To fixed point** function output at the right side of the For Loop.

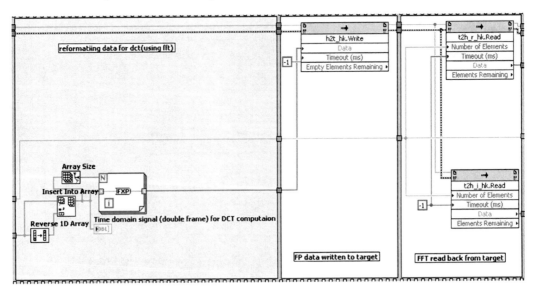

Figure L10.52: BD of host.vi communicating with target_hk.vi

3. Name frame 5 as **FFT read back from target**. Capture half of the FFT frame, computed by the target VI using the **invoke method** node. Create a **read** method to the

t2h_r_hk and **t2h_i_hk** FIFOs with the timeout of -1, and the number of elements equal to one frame (wired to N in the Mathscript node).

4. Finally, add a Mathscript node in the last frame (see Figure L10.53), and **close the FPGA VI reference** outside the frame. The expanded view of the node is shown in Figure L10.54.

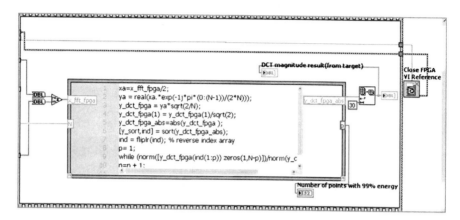

Figure L10.53: Computing DCT from FFT

5. The real and imaginary FFT result, which are fixed-point data, are converted to double format using the function **to double precision float**, and the output is passed to the **x** and **y** inputs of the function **re/Im to complex**. The output of this function **x+iy** is the complex FFT result, which is wired to the input variable **x_fft_fpga**. The other input variable is **N,** which is the same as N in the previous Mathscript node for the FFT radix.

```
1   xa=x_fft_fpga/2;
2   ya = real(xa.*exp(-1j*pi*(0:(N-1))/(2*N)));
3   y_dct_fpga = ya*sqrt(2/N);
4   y_dct_fpga(1) = y_dct_fpga(1)/sqrt(2);
5   y_dct_fpga_abs=abs(y_dct_fpga );
6   [y_sort,ind] = sort(y_dct_fpga_abs);
7   ind = fliplr(ind); % reverse index array
8   p= 1;
9   while (norm([y_dct_fpga(ind(1:p)) zeros(1,N-p)])/norm(y_dct_fpga)<.99);
10  p=p + 1;
11  end
```

Figure L10.54: DCT MathScript node

For the DCT computation from FFT, the signal x(n) whose N-point FFT is taken in the host VI is expanded to form the signal y(n) by taking its 2*N-point FFT in the target VI, and reading back N points of the FFT result back into the host. Here is the equation that computes DCT from FFT

$$y\_dct\_fpga(k)=b(k)*cos((x\_fft\_fpga/2)*exp((-j*\pi*(0:N-1))/(2*N));$$

where $b(k)=1/sqrt(N),k=0$

$$b(k)=2/sqrt(N),k=1,2....N-1$$

Lines 1 to 4 of the Mathscript node (see Figure L10.54) implements this equation. The rest of the code is similar to the one in the previous Mathscript node (see Figure L10.5), which computes the number of points in the DCT with 99% energy (p). The first 30 values of the DCT result based on the FFT computation done in the target VI are displayed by the graph indicator **DCT magnitude result** (connected to the variable y_dct_fpga_abs) and the numeric indicator **DCT values.**

L10.2.4 DCT Outcome

Figure L10.55 shows the DCT outcome on the FP of the **host.vi** and **target_hk.vi** when the **host.vi** is run. It can be seen that the DCT of the input data, done in the host VI, using the command **DCT** matches with the DCT of the same data, computed through FFT in the target VI. Moreover, both display the energy compaction property of the DCT as compared to the FFT.

It should be noted that the signal length is kept the same as the FFT length. If the two are not the same (suppose the signal is of length 200), the FFT and DCT need to be computed in the host VI using N=256, and the FFT in the target VI using N=512. However, when reading back the first 256 values in the host VI and using N=256 in the above equation, the DCT values may not match with those of the host VI , but the shape will remain similar with the energy compactness preserved.

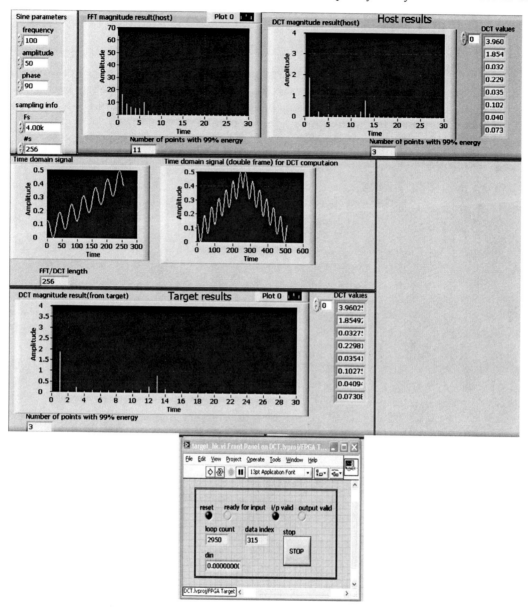

Figure L10.55: Host.vi and target_hk.vi results after one run

L10.3 Lab Experiments

- Redo Lab 10.1 without using the LabVIEW FPGA subVI FFT.vi. Instead, use Equation (10.1), where $W_N = e^{-j2\pi/N}$ can be passed to the FPGA VI from the host, or loaded from a text file into the memory. Note that N or number of frequency bins is to be kept as variable so that this value is passed as a read/write control or an array element. In essence, design the VI with this flexibility.
- Implement the DCT algorithm in an FPGA VI without using the FFT but by using Equation (10.7).

CHAPTER 11

FPGA Hardware Implementation

11.1 NI FlexRIO

In this chapter, the specific FPGA target board of NI FlexRIO, which contains a Xilinx Virtex-5 FPGA processor, is used to run the previously discussed FPGA codes. This target board has 512 MB of memory and fits into a PXI Express bus chassis with a PCI-e card interface placed inside the host PC. Here are the steps for installing this target as stated in [3]:

- Power off the computer.
- Place the PCI-e host card in one of the empty PCI express expansion slot.
- Connect the NI PXI-e 1073 chassis to the host card via a MXI express cable.
- Install the PXI-7965R adaptor module in the PXI-e 1073 chassis so that it fits into the backpane.
- Turn on the chassis and then the PC in this order. While turning off, use the reversed order, that is the PC first and then the chassis.
- If installation is successful, the power and link lights on the chassis are turned on, indicating correct power and correct connection between the chassis and the host PC.
- **Start>>National instruments>>Measurement and Automation Explorer (Max).** A window opens up as shown in Figure 11.1.

Figure 11.1: Measurement and Automation Explorer

- Expand **Devices and Interfaces** to display the chassis connected to the system. As seen in Figure 11.2, the option PXI-1073 chassis needs to be selected for the above hardware and if it is the only one, chassis-1 also get displayed. Expand **RIO Devices** to display the FlexRIO adaptor together with its connection properties. Under **RIO Allias on my system**, enter a name (**SIP_FPGA1** used here) for the adaptor.

- Close MAX.

This target is used next to perform actual hardware FPGA implementation.

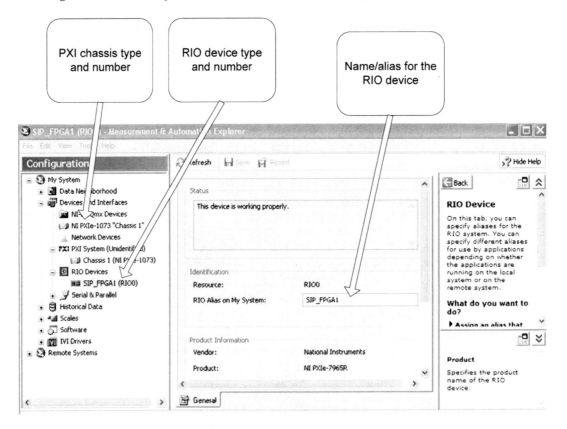

Figure 11.2: Different items in the MAX window

Lab 11 – Part 1:

Real-Time FPGA Hardware Implementation of FIR Filters

L11.2 Real-Time FIR Filters

In Chapter 8, an FIR filter was used to filter one sine tone from a composite signal consisting of two sine tones. The FPGA VI was run in the simulation mode. Here, the implementation is done on an actual FPGA target NI-FlexRIO. The FPGA and host VIs remain the same. However, for better visibility of the host VI BD, some portions of it are converted into sub-VIs.

First, create the project **FIR_on_target.lvproj** with the host VI **FIR_host.vi** and the subVIs **FXP_set_filter.vi** and **Sig_and_FLP_flt_dsgn.vi**. The FPGA VI **FIR_FPGA_on_trgt.vi** is then added to the project. Now, unlike the FPGA VIs discussed in the previous chapters that were designed to run in the simulation mode, this FPGA VI is to be executed on an actual FPGA target, that is the NI-FlexRIO board containing a Xilinx Virtex-5 FPGA. In the project explorer, choose **execute VI on>> FPGA target**, as depicted in Figure L11.1.

Figure L11.2 shows that the FPGA VI gets executed on the target board **SIP_FPGA1** (named at the time of connecting the board to the PC) corresponding to the NI FLEXRIO 7965R hardware board.

L11.2.1 Design of Sig_and_FLP_flt_dsgn.vi

Figures L11.3 and L11.4 show the BD and FP of this subVI, respectively. The BD is similar to the ones shown in Figures L8.2, L8.4, L8.6, L8.11, L8.12 and L8.13 of Chapter 8. Basically, this subVI creates a composite sine wave signal and an FIR filter based on the Kaiser window. The only difference from Chapter 8 is that frequency 1 is kept at a random number between 0 to 14000Hz with the sampling frequency of 32000 Hz, where within the while loop, it changes based on a **Time delay**. This ensures that while F2 is at 7000Hz, within the pass-band of the filter (6500 to 7500 HZ), F1 sweeps across the passband and stopbands, thus providing a comparison of the floating-point and fixed-point filtering versions. Figure L11.5 displays the connections of this subVI.

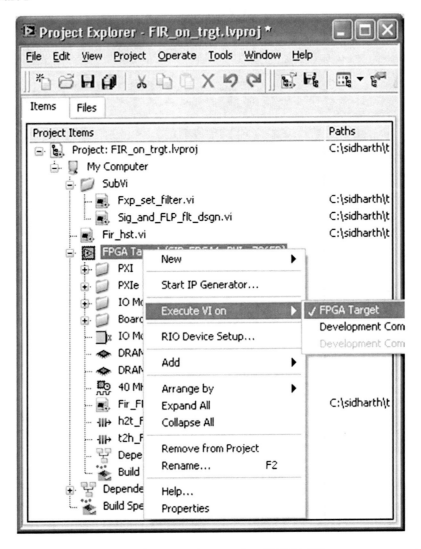

Figure L11.1: Choosing the execution mode of FPGA VI to run on target

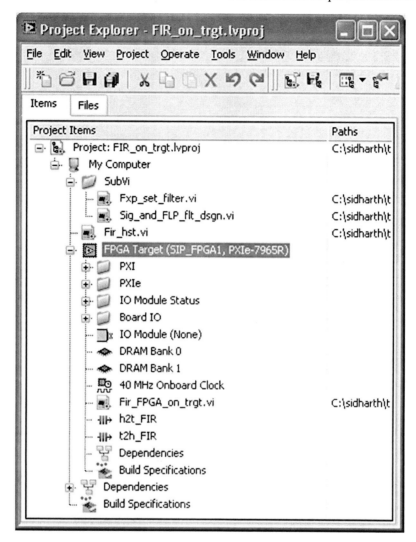

Figure L11.2: FPGA project FIR_on_trgt.lvproj

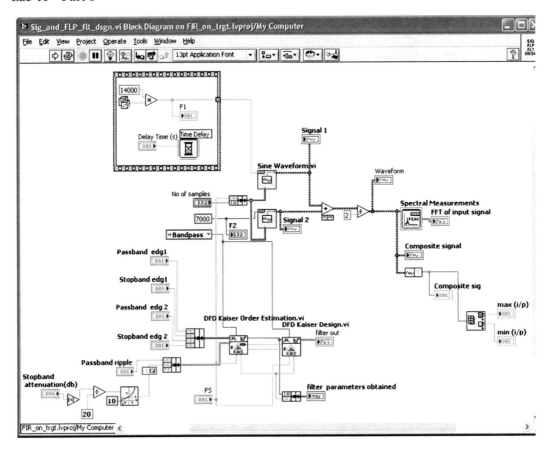

Figure L11.3: BD of Sig_and_FLP_flt_dsgn.vi

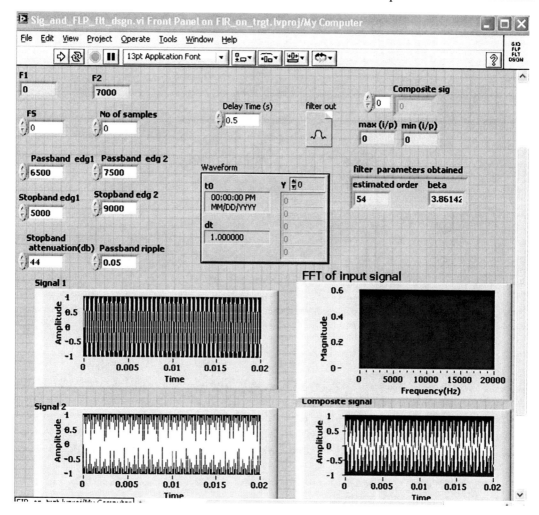

Figure L11.4: FP of Sig_and_FLP_flt_dsgn.vi

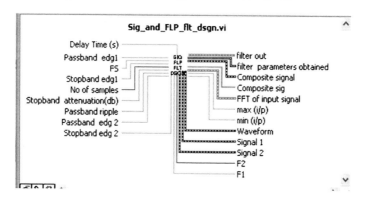

Figure L11.5: Connections of Sig_and_FLP_flt_dsgn.vi

L11.2.2 Design of FXP_set_filter.vi

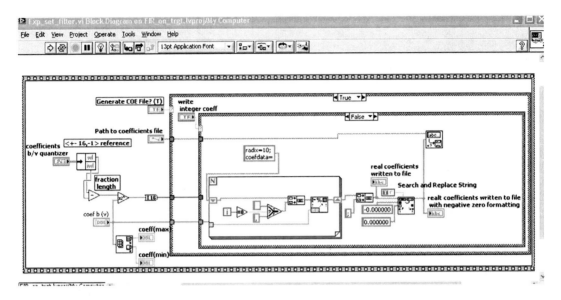

Figure L11.6: BD of FXP_set_filter.vi

Figure L11.6 dsiplays the BD of the subVI **FXP_set_filter.vi.** This subVI is similar to the ones shown in Figures L8.20 and L8.21. The purpose of this subVI is to generate a COE file or co-efficient file for the fixed-point filter to be used by the Xilinx FIR Compiler in order to gener-ate fixed-point filtering outcome. As done in Chapter 8, one can select to write the coeffi-cients file and choose the coefficient data format between integer and non-integer. Figure L11.7 displays the FP of this subVI and Figure L11.8 its connections.

L11.2.3 Design of FIR_FPGA_on_trgt

This is the FPGA VI that realizes the FIR filter. After the coefficients file is saved, the FIR core is generated using the FIR Compiler, which is then used in the FPGA VI. The BD of this VI is shown in Figure L11.9, which is similar to that shown in Figure L8.33.

L11.2.4 Design of FIR_hst.vi

Now, let us design the host VI **FIR_hst.vi.** Figure L11.10 displays a portion of the host VI that uses the subVIs in sections L11.2.1 and L11.2.2, while Figure L11.11 displays a portion of the host VI that communicates with the FPGA VI. Again, it is noted that these diagrams are similar to the ones shown in Chapter 8. Open a reference to **FIR_FPGA_on_trgt.vi** by using the **Open Fpga VI reference** function (refer to Figure L11.10). Until now, Open FPGA VI reference was being configured by selecting FPGA VI. This needs to be done differently now. Right click on the input **resource name** and create a control **resource name**. Then, choose the FPGA target that is used for the project. For our case here, **SIP_FPGA1** needs to be chosen which refers to the name of the FPGA target when it was first installed in the FlexRIO chassis and recognized in the Measurement and Automation Explorer software. Figure L11.12 shows the FP of the modified host VI.

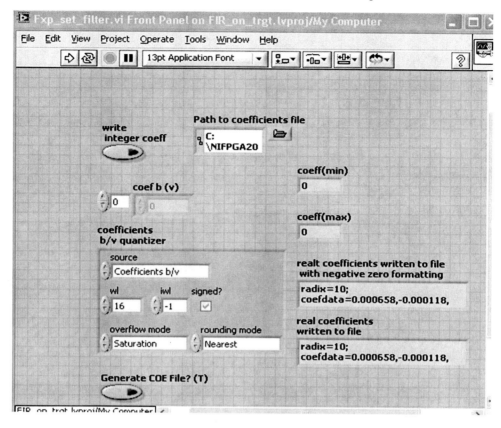

Figure L11.7: FP of FXP_set_filter.vi

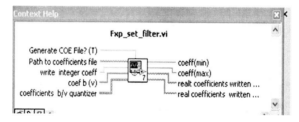

Figure L11.8: Connections of FXP_set_filter.vi

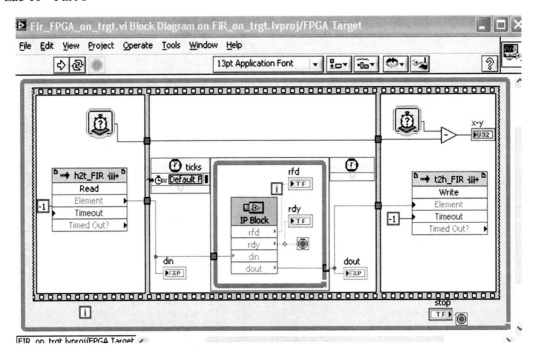

Figure L11.9: FP of FPGA VI

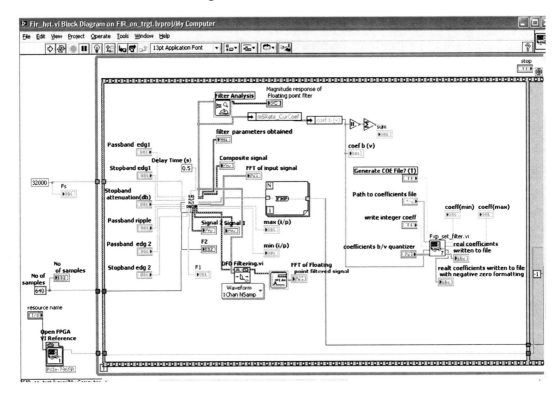

Figure L11.10: FP of a portion of FIR_hst.vi

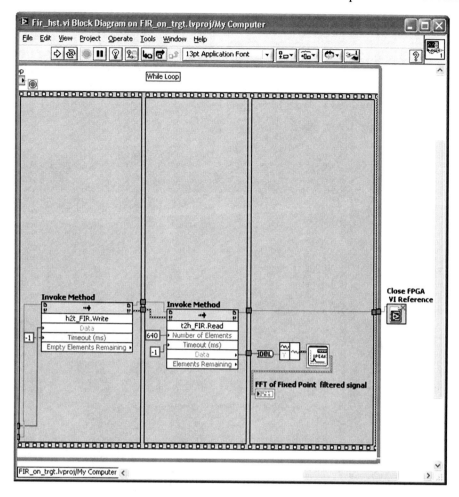

Figure L11.11: Portion of host VI that communicates with FPGA VI

L11.2.5 FPGA VI Compilation

Before the host VI can be run, it is necessary to compile the FPGA VI. Either run the FPGA VI, or right click on the FPGA VI name in the project explorer and choose **Compile**. The compilation goes through the following steps:

- **Generation of intermediate files:** At this stage, LabVIEW converts the FPGA VI into intermediate files, which are sent to the LabVIEW FPGA Compile Server.

- **Compilation status window:** The compile server then uses the Xilinx code generator to convert the intermediate files into a binary bitstream, which is stored as a bit file in a subdirectory, named **FPGA Bitfiles,** of the project directory. The Compilation Status window (see Figure L11.12) appears as soon as the code generation completes with no errors. In this window, the field **progress bar and status output** shows the percentage of the compilation process completed. The field **reports** allows users to choose reports such as **summary**, denoting a summary of the device utilization and timing details, and **estimated timing (synthesis)** providing a display of the

329

clock frequency used in the design and the maximum clock frequency for which the design would still be synthesizable. **Output window** displays the result of the choice made in the **reports** field. As shown in Figure L11.12, as per the device utilization, the design consumes 14.9% of the total slices on the Virtex-5 FPGA, 10.1% of the flip flops, 11.5% of the LUTs, 2.5% of the DSP blocks DSP48E, and 4.1% of the Block RAM. If desired, this report may be saved.

- Close the compilation status window.

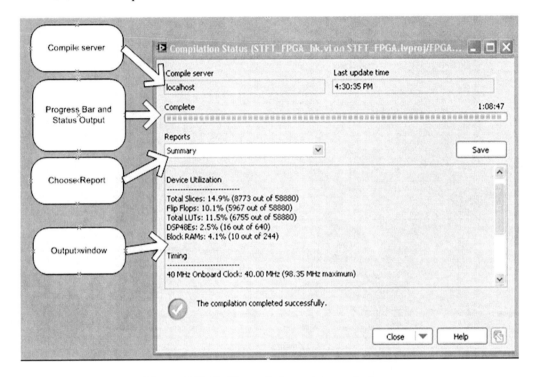

Figure L11.12: Compilation status window

If it is desired not to change the FIFOs in the FPGA VI, the FPGA VI needs to be compiled again. To address this issue, in the BD of the host VI, right click on **Open FPGA VI reference>>Configure Open FPGA VI reference>> Bitfile** and select the **Bitfile** generated during the compilation rather than the FPGA VI. This ensures that next time the FPGA VI runs directly from the bit file. That is, with the configuration option as shown in Figure L11.13, when one runs the FPGA VI, the bitfile corresponding to the VI is automatically downloaded to the FPGA target.

L11.2.6 Running Host VI

After having compiled the FPGA VI, run the host VI. It can be verified that as the frequency of sinewave 1 sweeps across the passband of the filter, the FFT of the FPGA filtered data follows this change consistent with the FFT of the floating-point filtered data. This is exhibited in Figures L11.14 and L11.15.

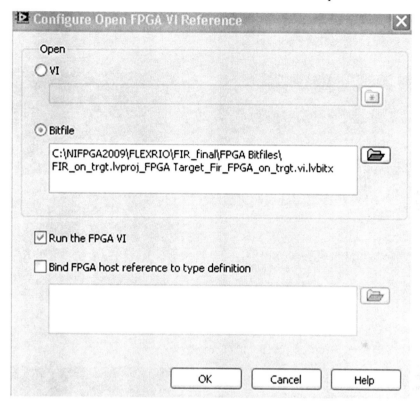

Figure L11.13: Configuring FPGA VI reference as bitfile

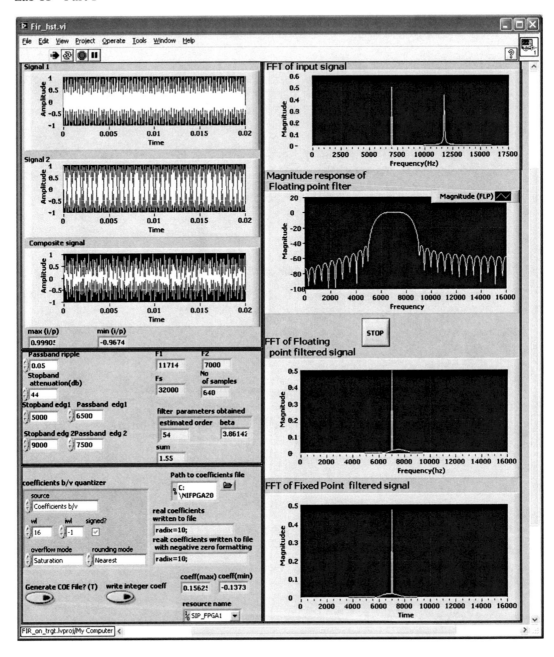

Figure L11.14: Host VI in run mode with one sinusoid within the filter passband

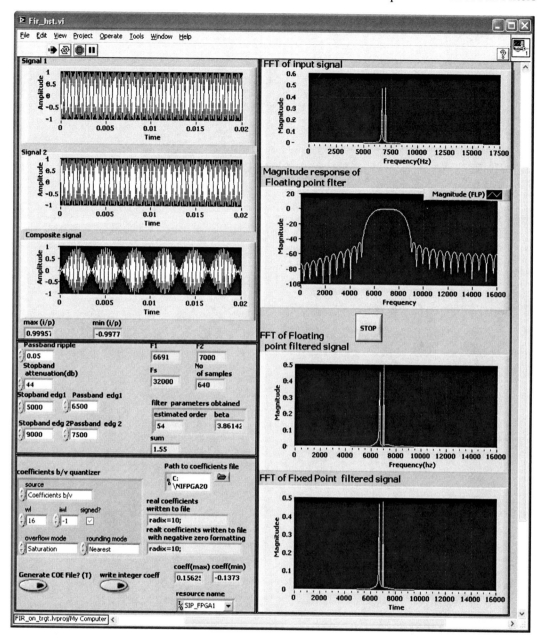

Figure L11.15: Host VI in run mode with both sinusoids within the filter passband

Lab 11 – Part 2:

Real-Time FPGA Hardware Implementation of IIR Filters

L11.3 Real-Time IIR Filters

As another example of FPGA hardware implementation, in this section an IIR Butterworth bandpass filter is implemented on the FPGA target hardware. Figure L11.16 gives an overview of the VIs and subVIs for this implementation as they are added in the project explorer **IIR_on_target.lvproj**.

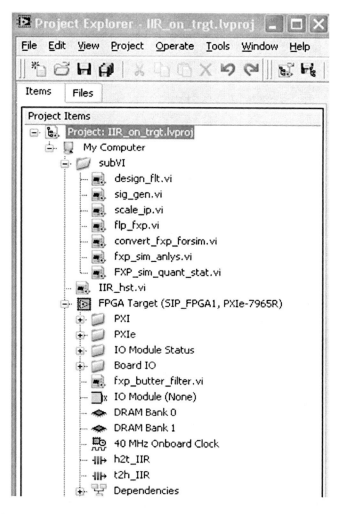

Figure L11.16: Project explorer for IIR filter implementation

Let us begin by creating the needed subVIs on the host side, which are placed in the virtual folder **subVI**.

L11.3.1 Design of design_flt.vi

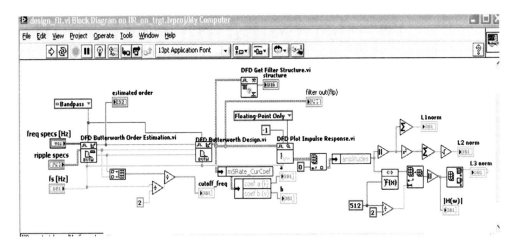

Figure L11.17: BD of subVI design_flt.vi

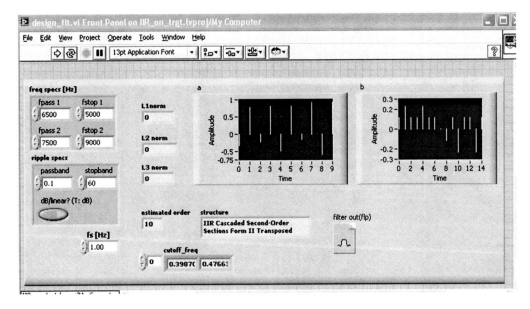

Figure L11.18: FP of subVI design_flt.vi

Figure L11.17 and Figure L11.8 display the BD and FP of the subVI, respectively. Use the function **DFD Butterworth Order estimation** to estimate the order of the filter, and then use the function **DFD Butterworth design** to design the filter in the same manner as done in Lab 9. In order to determine the structure of the filter to implement, use the function **DFD Get Filter structure.vi (Signal processing>>Digital filter design>>Utilities>>DFD Get Filter** structure.vi) and display the indicator **structure**. The default structure in LabVIEW is that of **Cascaded Second Order Sections (SOS) Form 2 Transposed.** To get the coefficients from the terminal **Filter out terminal of**, use an unbundle by name function similar to Lab 8. In order to scale the input as per the discussion in section 8.3, samples of the impulse response and magnitude response are needed. To obtain these, place the function **DFD Plot Impulse response.vi (Signal processing>>Digital filter design>>Filter Analysis>>DFD Plot Impulse response.vi)** on the BD. Its output terminal impulse response is a 1D array of a three-element cluster where the third element includes an array of impulse response samples. Use an **index array** to get this cluster element and then use the **unbundle by Name** function to get the impulse response h(n), as shown in the expanded view appearing in Figure L11.19.

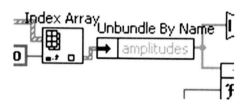

Figure L11.19: Obtaining impulse response samples

The L1 norm (refer to Equation (8.3)) is obtained as indicated in Figure L11.17 by using the **Absolute Value** function. Similarly, the L2 norm (refer to Equation (8.4)) is obtained by using the **Absolute Value, Square, Add array elements** and **Square Root** functions. Finally, the L3 norm (actually, L∞ norm) is obtained by using the **FFT, absolute value, array subset,** and **array max and min** functions. Finally, make this VI a subVI with its terminals as shown in Figure L11.20. This completes the design.

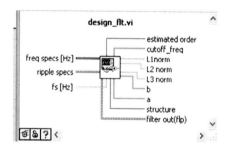

Figure L11.20: Connections of design.flt subVI

L11.3.2 Design of sig_gen.vi

Figure L11.21 shows the FP and BD of the VI **sig_gen.vi** that is used to create a composite signal consisting of two added sine waveforms. As seen in this figure, the control **Freq2** is a numeric control that is kept fixed (though it can be changed), and the control **Freq1** is a **vertical pointer slide** control for sweeping across the passband and stopband of the filter to see the filtering output. The default scale of the vertical pointer slide control is from 0 to 10, which is changed from 500 to 10000 by setting the scale property.

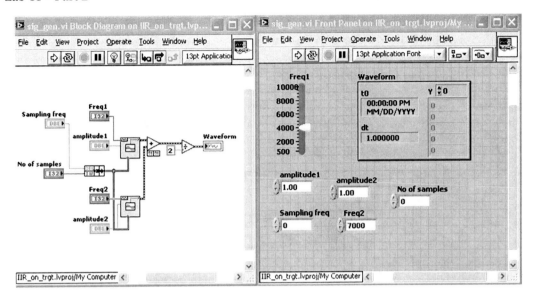

Figure L11.21: BD and FP of sig_gen.vi

The signal amplitudes are also divided by two after addition to keep the input between -1 to 1, thus allowing Q16 settings for the input. Figure L11.22 shows the connections for this subVI.

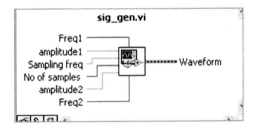

Figure L11.22: Connections of sig_gen subvi

L11.3.3 Design of scale_ip.vi

Figure L11.23 shows the FP and BD of the subVI scale_ip.vi, which scales the input **signal_in** to **signal_out**. This subVI uses a case structure to select scaling (**scale i/p**=true) or no scaling (**scale i/p**=false). **Scale type (i/p)** is an enum or enumeration control (on the FP, select **Modern>>Ring and enum>>enum**). To be able to choose between various norms for scaling the input, it is required to add the norm equivalent numbers and wiring them to a case structure for selecting a particular norm control. To do so, right click on the control and choose **Properties>>Edit items**, as shown in Figure L11.24. Click **insert**, add the name L1 norm under the heading **items** with the digital display 0. Similarly, add the other items. On the FP again, wire this control to the case structure and place the respective numeric controls for various norms. The diagrams also display the indicators for the maximum and minimum input values before and after scaling. Figure L11.25 shows the subVI connections.

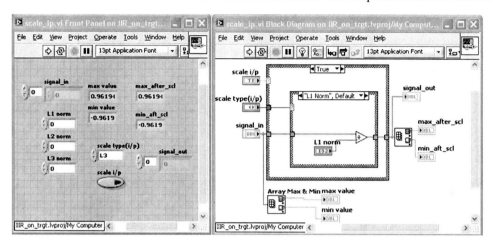

Figure L11.23: FP and BD for scale_ip.vi

Figure L11.24: Editing or adding items to enum control

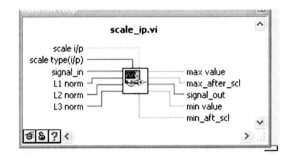

Figure L.11.25: Connections of scale_ip subvi

L11.3.4 Design of convert_fxp_for _sim.vi

Figure L11.26 dsiplays the BD of **fxp_for_sim.vi** and and Figure L11.27 displays its FP. This VI scales the filter coefficients, quantizes them with default settings, reads the settings, and then sets them as needed.

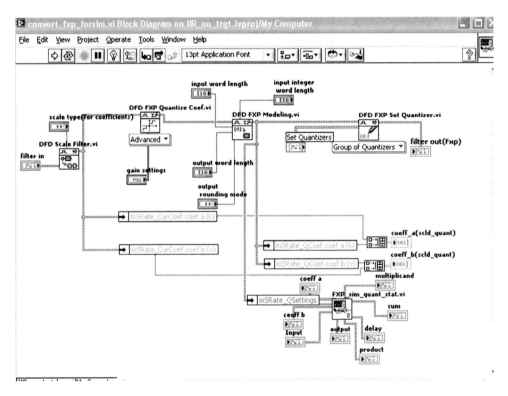

Figure L11.26: BD of convert_fxp_for _sim.vi

As shown in Figure L11.26, place the function **DFD Scale filter (Digital filter design>>Conversion>>DFD Scale filter)** on the BD. This VI scales the coefficients of the filter without changing its characteristics or properties. Right click on the input terminal **scale type** and create the corresponding enum control to set the scaling as No Norm (0), Time domain1 Norm(1), Time domain2 Norm(2), or Time domain inf-Norm(3). The Time domain norm1 and 2 are similar to the L1 and L2 norms that are used for scaling the input, while the Time domain norm 3 is defined as max $|h_i|$.

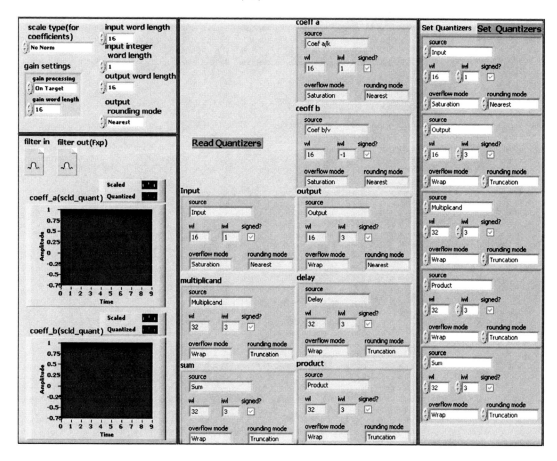

Figure L11.27: FP of convert_fxp_for _sim.vi

Place the function **DFD FXP quantize coef.vi** on the BD. Wire it to the controls and indicators as shown in the figure. This function can be used to set the input data word length and integer word length as well as the output data word length and rounding mode. It also sets the coefficient quantization parameters. However, it does not set the other quantization parameters including sum, multiplicand, which are set using the function **DFD FXP set quantizer.** This function is configured to work on a group of quantizers. Although Figure L11.27 does not show all the quantizers due to a lack of space, the quantizers for coefficients **a** and **b** are also set using this quantizer. To display the default quantization settings from the FXP modeling function so that they can be displayed in one window (and if the settings are not sufficient, they can be changed by the DFD FXP set quantizer), the subVI **FXP_sim_quant_stat.vi** is created. The BD of this subVI is shown in Figure L11.28.

Finally, the scaled floating-point coefficients and quantized fixed-point coefficients (after scaling) for both **a** and **b** coefficients are displayed in their respective graphs (co-

341

eff_a(scld_quant) and coeff_b(scld_quant)) after using the build array function, see Figure L11.26. Figure L11.29 shows the connections of this VI as a subVI.

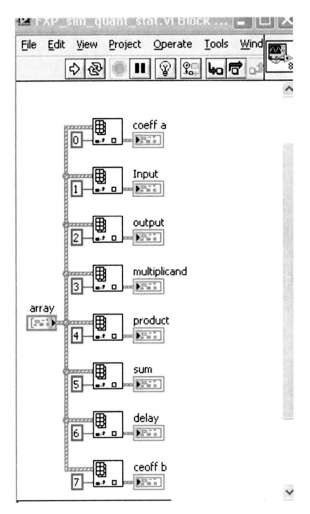

Figure L11.28: BD of FXP_sim_quant_stat.vi

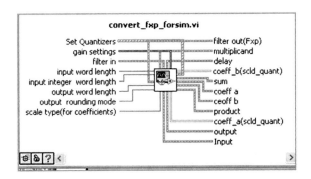

FIGURE L11.29: Connections of convert_fxp_for _sim.vi

L11.3.5 Design of fxp_sim_anyls.vi

This VI filters the input data using both floating-point coefficients and fixed-point coefficients and displays the output along with some other options. To filter the input data using the floating-point filter, **DFD Filtering.vi** is used. Both the filtered time domain data and the frequency domain data are plotted in the graphs **Filtered data (FLT)** and **FFT (FLT filtered)**, respectively.

To simulate the fixed-point filtering outcome, **DFD FXP simulation** is used. The **input range** for this function specifies the maximum absolute value of the input signal that can be represented in the chosen quantizer specifications. **Filtering statistics in** is used to provide the statistical information of the quantizers in the **filter in** input before the simulation. Wire it to the array constant as shown in Figure L11.30. More information on this can be obtained from [2]. The floating-point filtered data (in integer format simulation) comes out at the terminal **Integer signal out** and is displayed in both the time and frequency domains as shown by the indicators **Filtered Data (FXP) and FFT (FXP filtered)**. Note that the scale of the graphs (see Figure L11.31) are not normalized so that the scale is different from the floating-point graphs. The **filtering statistics report** terminal of the function carries the statistical information of the quantizers corresponding to the **filter in** input (except the coefficients quantizers) after the simulation in terms of **maximum value**, **minimum value**, **number of overflows** and **underflows**, and **number of operations**. This is displayed in the indicator **Filtering text report**. For the coefficients stastiscal information, **Coefficients report** is used. Now, to compare the floating-point and fixed-point filters, a number of graphs are used which include **Magnitude plot**, **Pole zero plot** and **impulse response**.The transfer function of the filter is also displayed using the function **DFD Render Transfer Function Equation.vi (DFD Toolkit>>Utilities>>DFD Render Transfer Function Equation.vi).** To display the equation for the fixed-point coefficients, choose the option FXP in the ring constant and connect it to the terminal **type (FLP)**. To choose the transfer function as a cascade of SOS sections, choose the option **Second order sections** in the input terminal **display**. The coefficients get captured in the indicator **transfer function (FXP)**.

This completes the design. Figure L11.32 shows the connections of this subVI.

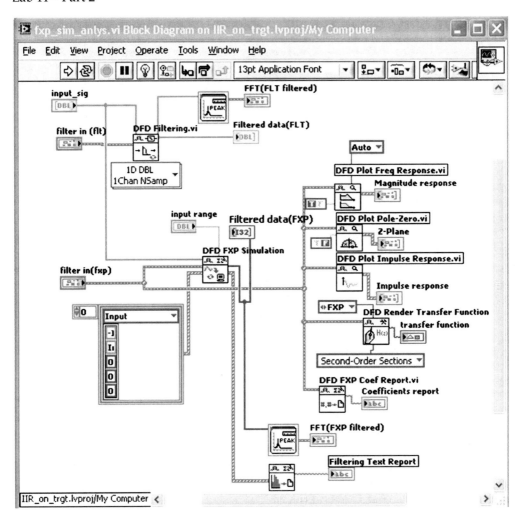

Figure L11.30: BD of fxp_sim_anyls.vi subVI

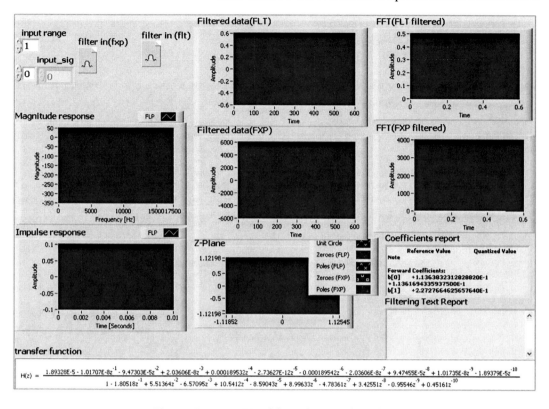

Figure L11.31: FP of fxp_sim_anyls subVI

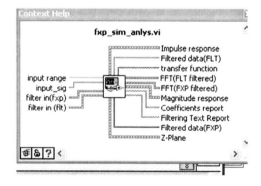

Figure L.11.32: Connections of fxp_sim_anyls.vi

L11.3.6 Design of IIR_hst.vi

Let us now design the top level host VI where the above subVIs are used, and also let us create the fixed-point filter on the FPGA side. Create a tab contol on the FP of this VI. Name the default page as **IIR filtering on target**. Right click and choose **add page after**, name the page as **Fixed point filter design and settings**. Similarly, add the third page **Fixed Point simulation on host**. This is shown in Figure L11.33.

IIR filtering on target	Fixed Point Filter Design and settings	Fixed Point simulation on host

Figure L.11.33: Naming of pages in tab control

On the first page, the controls and indicators corresponding to the signal, floating-point filtering, and fixed-point filtering on the target are provided. On the second page, the controls and indicators corresponding to the fixed-point filter is displayed. On the third page, the simulation results for the fixed-point filtering and its comparison with the floating-point filtering is shown.

Let us create this VI. As shown in Figure L11.34, a frame of the sequence diagram is obtained where the subVI design_filt.vi is added. Right click on the terminals of this subVI and add its various controls and parameters for the floating-point filter design. Here, the number of samples for the signal generation is set to 512.

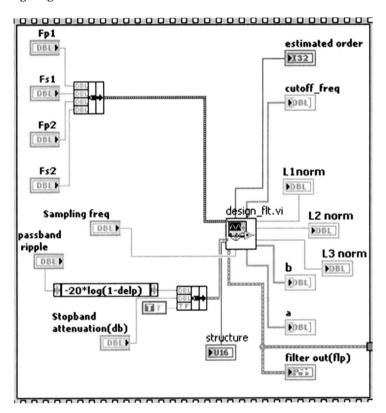

Figure L11.34: Portion of host VI for floating-point filter design

Now, on the FP, drag the controls and indicators created in this frame onto the first page of the tab control. The FP then looks like as indicated in Figure L11.35. Figure L11.36 shows another sequence diagram where in the first frame, a reference to the FPGA VI is opened with the resource name set to the current target. In the second frame, a While Loop is added and within it, the subVIs **sig_gen.vi** and **scale_input.vi** are added. The controls and indicators are smiliar to Figures L11.21 and L11.23. However, as the sampling frequency control is already created in the previous frame (see Figure L11.36), a local variable to this control is created by right clicking on the control and choosing **Create>>Local variable>>Right click on the local variable created>>Change to read**. Similarly, the local variables corresponding to the **L1, L2 and L3** norms are also created. Note that the FFT of the input data is displayed using the two graphs **FFT(i/p)** and **FFT (i/p)2**, one on the first page of the tab and the other on the third page. Also, note that the number of samples of the input signal is fixed to 512.

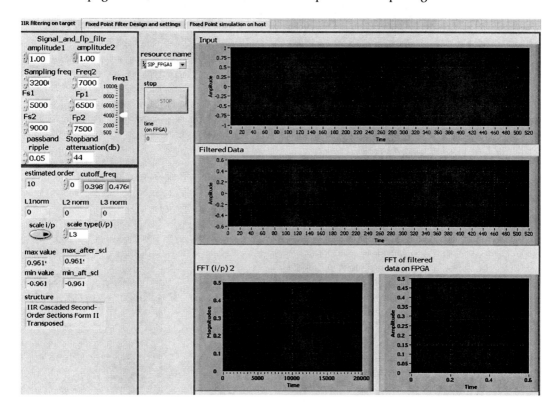

Figure L11.35: FP of host VI showing the first page of tab control

Refer to Figure L11.37. This figure shows the subVIs **convert_fxp_for _sim.vi** and **fxp_sim_anlys.vi** that are added to the BD with their respective controls and indicators. These are the same as in Figures L11.26 and L11.30. Note that for the fxp_sim_anlys subVI, the **input range** terminal is connected to the output of the scale_ip subVI, **max_after_scl**, so that the input range is automatically determined from the maximum value of the input. Figures L11.38, L11.39 and L11.40 show the FP diagrams created so far.

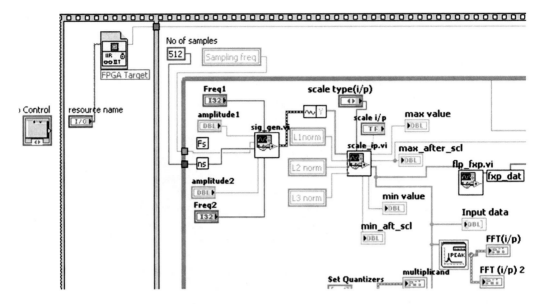

Figure L11.36: BD of a portion of host VI

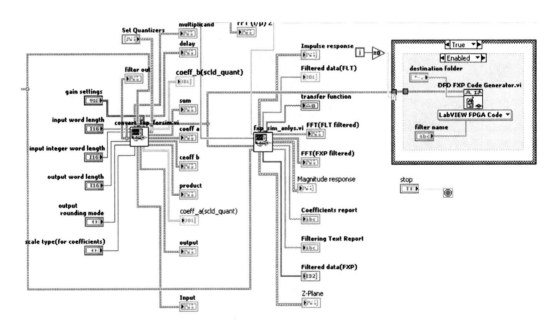

Figure L11.37: BD portion of host VI for fixed-point filtering

Let us now set the parameters and see the fixed-point simulation results for the purpose of deciding the fixed-point settings of the filter to be realized on the actual FPGA target hardware. As seen in Figure L11.38, a composite sine wave signal is generated with amplitude 1, sinewave frequencies of 3796Hz and 7000Hz, and the sampling frequency of 32000Hz. A passband from 6500Hz to 7500Hz with the stopband edges of 5000Hz and 9000Hz are specified. A passband ripple of 0.05 and a stopband attenuation of 60 dB are also considered. The control scale i/p is not pressed so that no scaling is applied to the input. The filter order gets estimated to be 10 with the L1, L2 and L3 norms appearing as 1.78, 0.28, and 1, respectively.

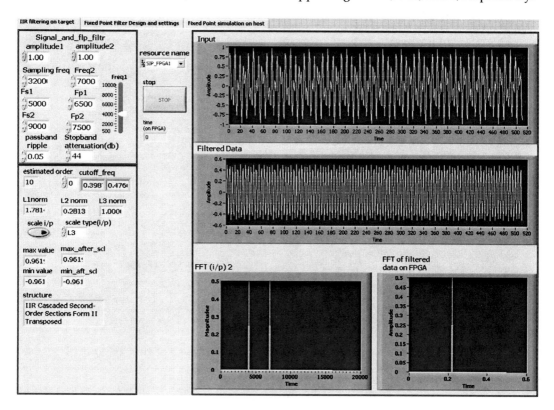

Figure L11.38: Host VI in run mode with first page of tab control displayed

Next, refer to the page shown in Figure L11.39. This page shows the fixed-point settings of the filter. The input is set to be Q15, and the output to be Q[3,13] or <+/- ,16,3>. The coefficients are also set as Q16. The unscaled and floating-point coefficients are plotted along with the scaled and quantized coefficients. From these graphs, the quantization settings are found adequate. However, the actual verification is done on page 3 of the tab control, see Figure L11.40.

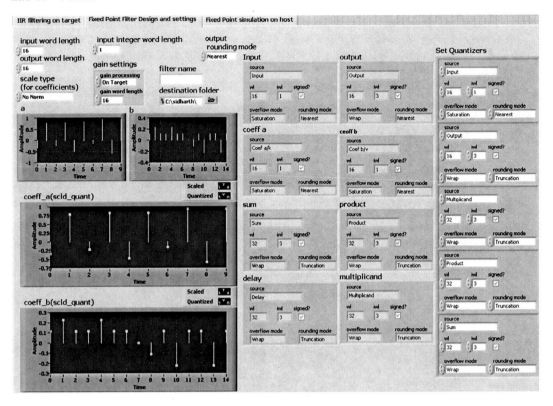

Figure L11.39: Host VI in run mode with second page of the tab control

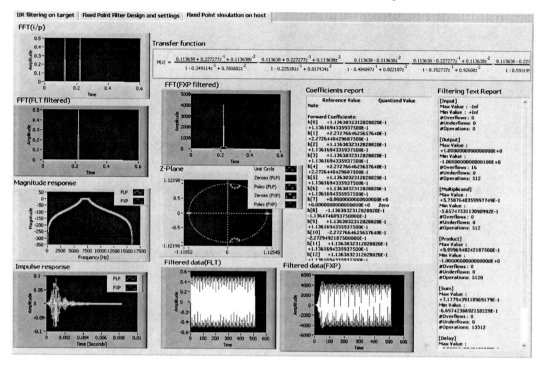

Figure L11.40: Host VI in run mode with third page of the tab control

It is worth pointing out the following points here. The filtering and coefficients report show no overflows and underflows. Both the floating-point and fixed-point results match in terms of the filtered data FFT outcome, the filter magnitude and impulse response plots, and the pole zero plot. The pole zero plot shows that even after quantization, the poles remain within the unit circle indicating the stability of the fixed-point filter. Thus, this allows us to proceed with the design of the fixed-point filter on the actual target.

Figure L11.37 shows the portion of the host VI for this design. The fixed-point filter is designed once. There is no need to design again during run time. Thus, a case structure and a diagram disable structure are used as shown in this figure. Place the function **DFD_FXP_code generator.vi (Signal processing>>Digital filter design>>Fixed point tools>> DFD_FXP_code generator.vi)** within the diagram disable structure, and enable the diagram. This VI generates the fixed-point code from the fixed-point filter based on the fixed point settings obtained by wiring the **filter out** terminal of the convert_fxp_for_sim subVI (corresponding to the DFD set quantizer) to the **filter in** terminal of the VI. The filter name and folder where the VI is automatically generated is created by wiring the corresponding controls **filter name** and **destination folder**. This creates a new project having the same name with the filter VI added to it, see Figure L11.41. Figure L11.42 shows the BD of this VI. However, in order to use it with the host VI, this VI needs to be modified by adding FIFOs. Also, the name of this VI needs to be changed to fxp_butter_filter. Close this project and add this VI to the current project **IIR_on_target**.

Figure L11.41: FPGA VI automatically created for fixed-point filter

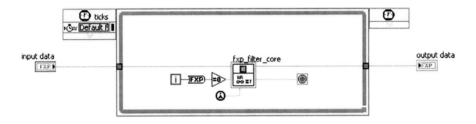

Figure L11.42: BD of the default FPGA VI

The BD of the modified VI is shown in Figure L11.43.

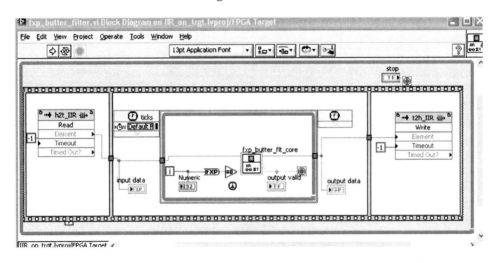

Figure L11.43: BD of the FPGA VI modified to communicate with host VI

To capture the input data from the host, the FIFO **h2t_IIR FIFO** is used, and to transfer the filtered result back to the host, the FIFO **t2hIIR FIFO** is used. The fixed-point settings for these are <+- 16,1> and <+- 16,3>, respectively, in accordance with the the simulation settings observed from page 2 of the tab control. Figure L11.45 shows the host side of the VI that communicates with the FPGA VI in the diagram disable structure. It shows that the input data is first converted to fixed-point via **subVI flp_fxp.vi** (see Figure L11.44), then transferred to the host via the target FIFO, and then the filtered result is obtained on the target via the host FIFO, whose time domain and FFT plots are obtained through the indicators **Filtered Data** and **FFT of filtered data on FPGA**.

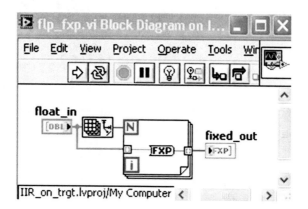

Figure L11.44: BD of SubVI flp_fxp.vi

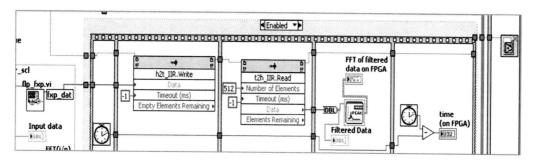

Figure L11.45: Portion of BD of host VI that communicates with FPGA VI

Figure L11.46 shows the host VI running with the real-time filtered data processed on the FPGA target.

353

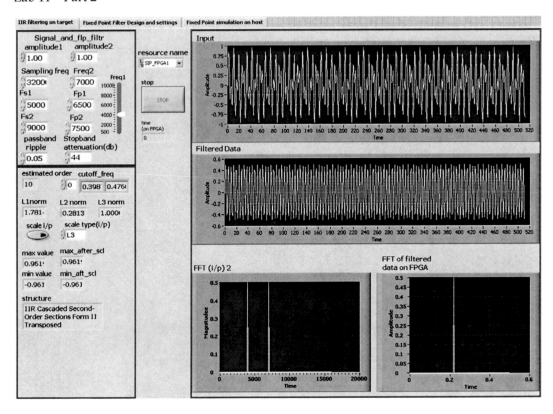

Figure L11.46: Host VI in run mode with filtering done on FPGA target

Lab 11 – Part 3:

Real-Time FPGA Hardware Implementation of Short Time Fourier Transform

This section discusses the real-time FPGA implementation of Short Time Fourier Transform (STFT) on the FPGA board NI FlexRIO containing the Xilinx Virtex-5 FPGA.

L11.4 STFT Overview

When the spectral content of a signal changes over time, the variation of its frequency content can be captured by using STFT. STFT involves dividing the signal into frames, placing an appropriate window over the frames, and then computing the DFT or FFT for each windowed frame. A spectrogram is often used to display the frequency content over time, where the x or horizontal axis shows the time or frame duration, the y or vertical axis shows the frequency or the FFT bins, and the z axis shows the FFT magnitude. The number of frames depends upon the specified amount of overlap between them. For example, consider a signal consisting of 8 samples. To compute STFT, the signal needs to be divided into R windowed frames. If the length of each windowed frame is 3 samples and 50% overlap is considered, STFT is computed over R=3 frames: frame 1 with indices from 0 to 3, frame 2 with indices from 2 to 5, and frame 3 with indices from 4 to 7. In other words, STFT is computed by taking 4-point FFT of each of the 3 frames, getting a total of 12 frequency bins.

In the next section, a host VI is first created to read a 'wav' file. An FPGA VI is also created to perform the spectral analysis of the file using STFT (after windowing on the host side). An intensity plot of the spectral content similar to a spectrogram is then displayed via the host VI.

L11.4.1 STFT Design

Let us now create the VIs in a step by step manner. First, create the project **STFT_FPGA.lvproj** with the host VI **STFT_host.vi**. The FPGA VI **STFT_FPGA_hk.vi** is then added to the project, see Figure L11.47.

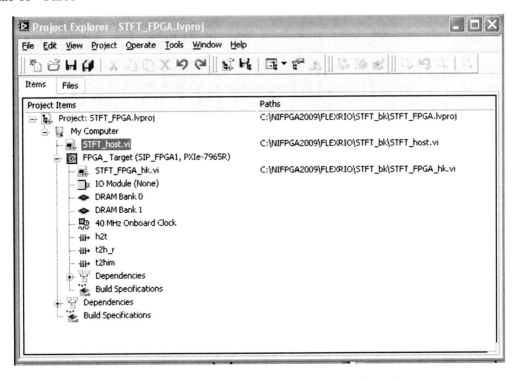

Figure L11.47: FPGA project STFT_FPGA.lvproj

L11.4.2 Host VI

The following steps need to be taken in order to create the host VI:

1. On the BD (see Figure L11.48), create a While Loop and label it as loop1.
2. Place a sequence diagram within the While Loop and label it as frame 1.
3. Place the function **Sound File Info.vi (Graphics & Sound>>Sound>>Files>> Sound File Info.vi)** in frame 1. This VI gets samples from the '.wav' file chosen. Right click on the input **path** and create a control **Path to Wave File.** This is the path to the '.wav' file to be read. The path can be seen by right clicking the control on the FP and choosing **browse options**. Wire this control to the function terminal. Browse to the '.wav' file and select it for the current program. An output of this function is **sound format.** It is a cluster that specifies the sampling rate, number of channels (mono/stereo) and bits per sample in the wave file to be read. Use **unbundle by name** to get the respective components of this cluster and create the indicators **sampling rate, number of channels** and **bits per sample.** The sampling rate in this wave file is **22.05 kHz**, bits per sample are **16**, and number of channels (stereo) is 2.
4. Next, configure the soundcard for reading and playing the '.wav' file. Place the function **Sound File Open.vi (Graphics & Sound>>Sound>>Files>>Sound File Open.vi)** in the BD. The function is a polymorphic VI that can be used to read a '.wav' file or write a new '.wav' file. Choose **read** for reading the file. Connect it to the left terminal of the shift register (see Figure L11.48). Wire the **path out** and **error out** terminals of the function **Sound File Info** to the terminals **pth** and **error in** of

the function **Sound File Open**. The output **sound file refnum** returns a reference to the sound file, which can be passed to other sound file VIs.

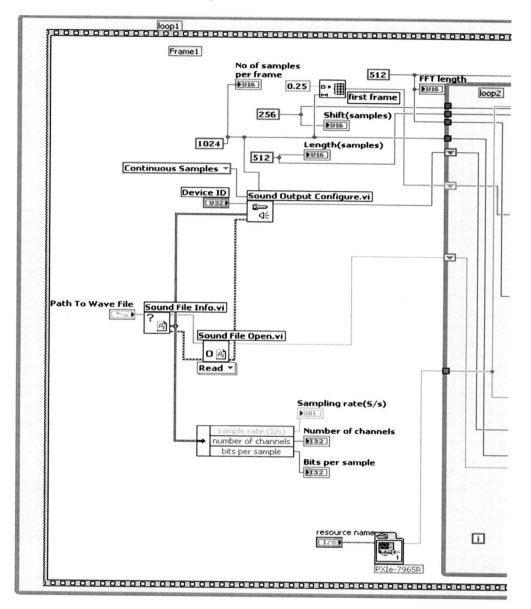

Figure L11.48: Snapshot of host VI for configuring soundcard

5. Place the function **Sound Output Configure.vi (Graphics & Sound>>Sound>>Output>>Sound Output Configure.vi)** in the BD. This function configures the soundcard output generating sound samples which can be written to a file using sound output write. Right click on the input terminal **Device ID** and create a control by the same name. Wire the control to this terminal. Set the ID to **0**. Right click on the terminal **sample mode** and create a ring constant with the option **Continuous**

Samples. This ensures that the sound write operation is performed repeatedly. Create a numeric constant with the value **1024** and wire it to the input terminal **number of samples/ch** that specifies the number of samples per channel in the buffer. Create an indicator **no of samples per frame** and wire it to this constant. Wire the output terminal **sound format** of the **Sound File Info** function to the input terminal **sound format** of the function **Sound Output Configure**. Connect the **error out** terminal of the **Sound File Open** function to the **error in** terminal of the **Sound Output Configure** function.

6. Create a while loop inside frame 1, and label it loop2. Then, create a sequence diagram with one frame inside loop2 and name it frame 2. Within this frame, create two additional sequence diagrams, each with three frames labeled as frames 3, 4, 5, and 6, 7, 8, respectively. This is shown in Figure L11.49.

7. As shown in Figure L11.50, wire the **refnum out** terminal of the function **sound file open** to the left edge of the while loop loop2. Right click and choose **replace with shift register**. The shift register terminals appear as indicated in this figure. Place the function **Sound File Close.vi (Graphics & Sound>>Sound>>Files>>Sound File Close.vi)** outside loop2 at the right side within frame1. This function closes an open '.wav' file. Wire its terminal **sound file refnum** to the right side terminal of the shift register. Similarly, create another shift register and wire the left terminal of the shift register to the **task id** output signal of the function **sound output configure**. Place the function **Sound Output Wait.vi (Graphics & Sound>>Sound>>Output>> Sound Output Wait.vi)** in frame1 and wire its **task id** input to the right terminal of the shift register. This function is placed outside the while loop as it waits till all data samples in the sound file are played by the output device. Place the function **Sound Output Clear.vi (Graphics & Sound>> Sound>> Output>> Sound Output Clear.vi)**. This VI stops sound playback from the output device, clears the buffer, returns the task to the default state, and clears the resources that are associated with the task. Wire the **task id** terminal of this function to the **sound output wait** function. Finally, place the function **Simple Error Handler.vi (Dialog and user interface>>Simple Error Handler.vi)** on the BD. This function displays the description of the error if an error occurs. Wire the **error out** and **error in** terminals of the functions **sound file close**, **sound output wait**, **sound output clear** and **simple error handler**.

8. Next, let us create the portion of the BD which actually reads and plays the wave file, see Figure L11.51.

Place the function **Sound File Read.vi (Graphics & Sound>> Sound>> Files>> Sound File Read.vi)** in frame 4. This function is a polymorphic VI which reads data from a '.wav' file into a waveform array. Right click on the function and choose **select type>>automatic** to generate waveform data of double precision type. Wire the **sound file refnum** input terminal of the function to the left terminal of the shift register (wired to **sound file open**, see Figure L11.48). To read 1024 samples from the file into the buffer, wire the terminal **number of samples/ch** to a constant with the value 1024. The input terminals **position mode** and **position offset** specify how the wave file read operation begins. If the position mode is set to **Absolute**, reading begins at the beginning of the file plus **position offset** relative to the beginning of the file. On the other hand, if the position mode is set to **Relative**, reading starts at the current location of the file plus **position offset**. The default is Relative noting that as all the samples in the file are read buffer by buffer, reading for each buffer should start at the position or index one plus the index where the last sample in the last buffer was read. Keep this terminal unwired so that the default option is selected. The output terminal **end of file** returns a true value when the last sample in the file is read. Wire this terminal and the terminal **sound refnum out** to the right border of frame 5, as shown in Figure L11.51. The **data** terminal reads sound data into a waveform array. Wire the **index array.vi (array>>index array.vi)** function to this terminal, and the output of the **index array** function to **get waveform components.vi**, the output of which is double precision sound data samples read from the file into the buffer each time loop2 runs. Wire this data signal to the right border of frame5, and also wire it to the respective indicator **time domain sound/speech signal (host VI)**.

Now, place the function **Sound Output Write.vi (Graphics & Sound>>Sound>> Output>>Sound Output Write.vi)** on the BD. This function writes data samples to the soundcard output. Wire the **task id** input terminal of the function to the left terminal of the shift register (which is connected to **task id** of the **sound output configure** function), and the **task id out** terminal to the right border of frame 5. The **data** terminal writes sound data samples to the internal buffers of the soundcard. In order to play the wave file, as it is read frame by frame, wire the **data** terminal of the **sound file read** function to the **sound output write** function.

Place the function **Tick count (ms).vi (Timing>>Tick count (ms))** in frames 3 and 5, as indicated in Figure L11.51. These timers return the time in milliseconds as the timer starts counting. In order to find the number of milliseconds that elapse for reading and playing 1024 samples from the wave file, wire the output of these timers to the **subtract.vi** function, and display the result by wiring to the indicator **sig_gen**. Figure L11.52 and Figure L11.53 exhibit snapshots of how the wires at the right border of frame 5 are connected. Now, loop2 should stop if either the **stop** control is pressed, or all the data samples in the wave file get read (**end of file** is true). Place the function **Compound Arithmetic (Numeric>>Compound Arithmetic)** in the BD, right click and choose **Change mode>>Or** to implement an **or function**. Create a Boolean control **stop,** and wire it to one of the inputs of the **compound arithmetic** function. Wire the other input to the end of the signal file. Wire the output to the **loop condition** terminal of the while loop.

As the wave file is finished reading, the functions **sound file close, sound output wait** and **sound output clear** execute in succession. However, one may wish to start the reading process again so that it is a continuous operation until finally stopped. For this purpose, the outer while loop or loop1 needs to have its **loop condition** set to **continue if true**. This way, the file reading by default starts again after file closing until **stop** is pressed, which is implemented by connecting the stop button to the loop condition using an **inverter (Boolean>>inverter),** as shown in Figure L11.53.

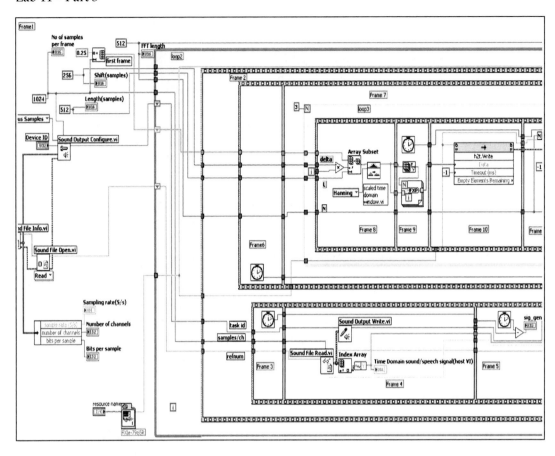

Figure L11.49: Creating while loop and sequence diagrams within frame 1

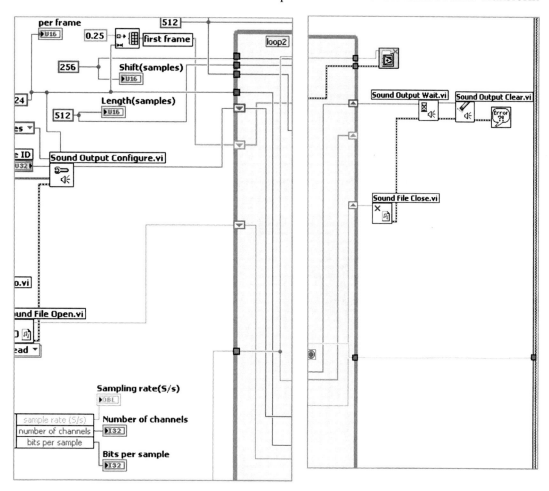

Figure L11.50: BD of sound close VI and shift register creations

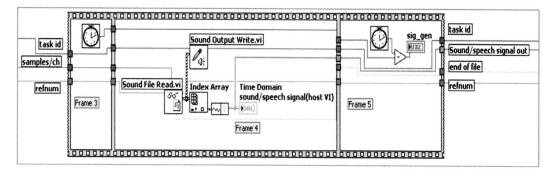

Figure L11.51: BD of host VI that reads and plays wave file

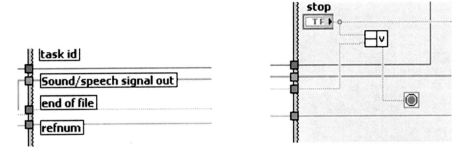

Figure L11.52: Snapshot of how wires at right border of frame 5 extend into loop2

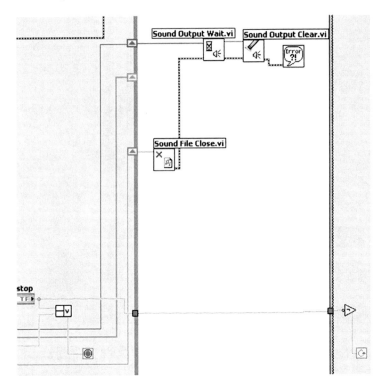

Figure L11.53: Snapshot of how wires at right border of frame 5 extend into loop2, and into frame 1

9. Since it is desired to read and play back 1024 samples of the wave file, within this time, the STFT of these samples needs to be computed. This is implemented by using a shift register so that as current 1024 samples are being read and played, the last 1024 samples get processed for the STFT computation. This requires wiring signal **out** to the right border of frame 2, as shown in Figure L11.53. Right click and choose **replace with shift register**. The other terminal of the shift register needs to be wired as shown in Figure L11.54. Since it is a good practice to initialize shift registers, this is achieved by wiring 1024 samples of value 0.25 each via the function **initialize array** with its **element** terminal wired to **constant 0.25** (chosen less than 1 to

fit in the quantization used) and the **dimension size 0** terminal wired to the constant **1024**. Figure L11.55 displays the FP portion of the host VI created so far.

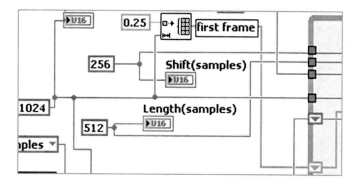

Figure L11.54: Wiring the left terminal of shift register carrying speech data

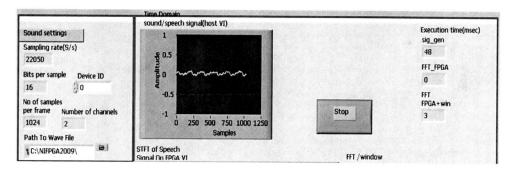

Figure L11.55: FP portion of host VI

10. As shown in Figure L11.56, it is desired to split each of the 1024 samples in the read buffer into three (R) frames, each with **L=512** samples, and then window them with a Hanning window. With 50% overlap between the windows, this corresponds to a shift of **delta=256** samples from the beginning of each frame. Thus, for the considered 1024 samples, three frames are created with the samples running from index 0 to 511, 256 to 767, and 512 to 1023, respectively. Within frame 7, create a For Loop and label it loop 3 with the constant 3 wired to its loop count terminal which controls the number of windows per buffer. Create five frames within the For Loop, and name them 8, 9, 10, 11, and 12. In frame 8, place the function **Array subset.vi** (array>>array_subset.vi) as shown in Figure L11.56. This function returns a subset or portion of the input array, beginning at the index specified by the input terminal **index** and the length specified by the **length** terminal.

Create a constant with the value **512**, and wire the **length** terminal of the **array subset** function to this constant (shown in frame 8 by L). Also, create an indicator **length (samples)** to display the length of each window in terms of the number of samples. Similarly, create a constant with the value **256**, and wire it to the indicator **Shift (samples)** to display the number of samples each window is to be shifted (shown in frame 8 by **delta**). Place a **multiplier.vi (numeric>>multiplier.vi)** in the BD, and wire its input terminals to the constant value 256 or delta, and to the for loop index terminal **i**. Wire its output to the index terminal of the **array subset** function. Also, as discussed above, wire the **data shift register** left terminal to the **array input** terminal of the function. This way, for every iteration of the while loop loop2, the previous loop speech/sound data of 1024 samples is fed into the **array subset** function, which splits it into three frames for three iterations of the for loop. Thus, when i=0, frame 1 has samples starting from index i*delta =0 and length 512, i.e. 0 to 511, frame 2 has samples starting from index i*delta=1*256=256 and length 512, i.e. 256 to 767, frame 3 has samples starting from index i*delta=2*256=512 and length 512, i.e. 512 to 1023.

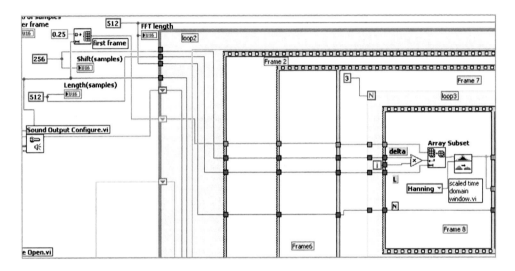

Figure L11.56: BD portion of host VI that windows speech signal

Place the function **scaled time domain window.vi (windows>>scaled time domain window.vi)** on the BD. This function applies a scaled window to data fed at the input terminal **X**. Wire the output subarray terminal of the **array subset** function to the **X** input of the window function. Right click on the input terminal window of the function and create a ring control. Choose **Hanning window** as window type. Thus, each of the successive frames of data is windowed by a Hanning window, and windowed data come out at the terminal **windowed X** of the function.

11. Now, one needs to compute the N=512-point FFT of these windowed frames in the FPGA VI one by one and read back the results to the host VI. To do so, in frame 9, convert the signal to the Q16 format (sufficient data precision for the discussed example) in a manner similar to the previous examples, see Figure L11.57.

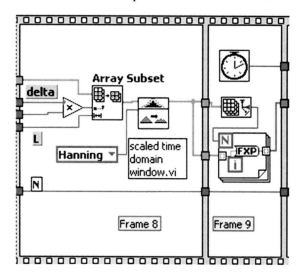

Figure L11.57: Converting signal to fixed-point format

L11.4.3 FPGA VI

Figures L11.58 and L11.59 show the BD and Figure L11.60 the FP of the FPGA VI **STFT_FPGA_hk.vi** when the case structure is false and true, respectively. Note that the BD is the same as that of the VI **target_hk.vi** of the FFT project **FFT_fpga.lvproj.** The previously discussed steps are used to create the BD for the current FPGA VI. For performing 512-point FFTs within a SCTL, the FFT function properties (see Figure L11.61) remain the same as those in FFT_fpga.lvproj.

In addition, the host to target FIFO h2t and the target to host FIFOs t2h_r and t2h_i have the same quantization and number of elements as the corresponding FIFOs in FFT_fpga.lvproj, see Figures L11.62 and L11.63.

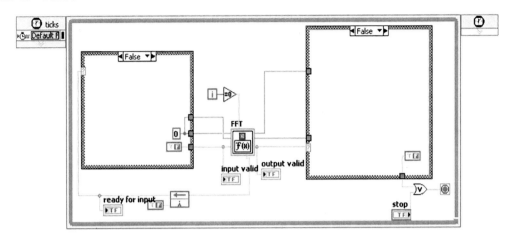

Figure L11.58: BD of STFT_FPGA_hk.vi when case structures are false

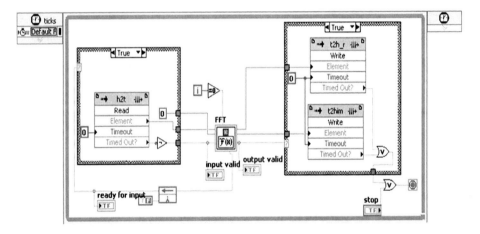

Figure L11.59: BD of STFT_FPGA_hk.vi when case structures are true

STFT_FPGA_hk.vi Front Pane...
File Edit View Project Operate Tools Win
13pt Application Font
input valid
ready for input
output valid
stop
STOP

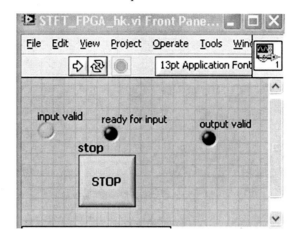

Figure L11.60: FP of STFT_FPGA_hk.vi

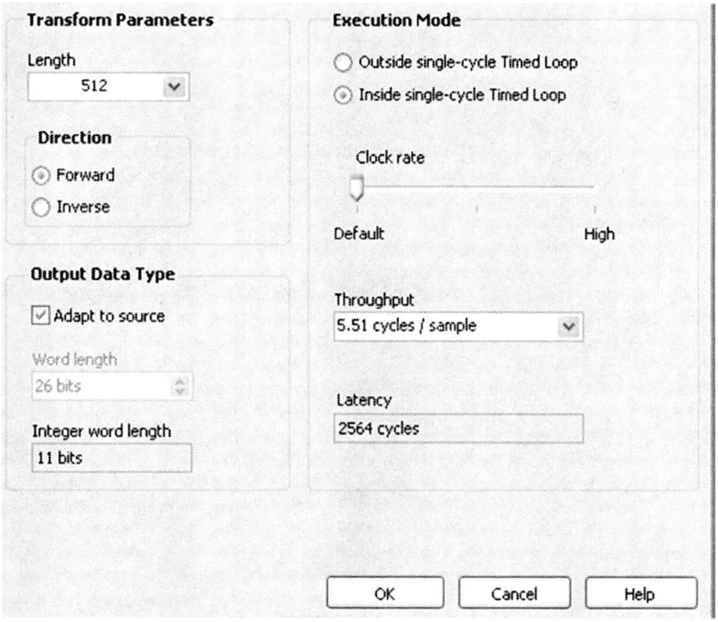

Transform Parameters
Length
512
Direction
Forward
Inverse
Output Data Type
Adapt to source
Word length
26 bits
Integer word length
11 bits
Execution Mode
Outside single-cycle Timed Loop
Inside single-cycle Timed Loop
Clock rate
Default
High
Throughput
5.51 cycles / sample
Latency
2564 cycles
OK
Cancel
Help

Figure L11.61: FFT function properties

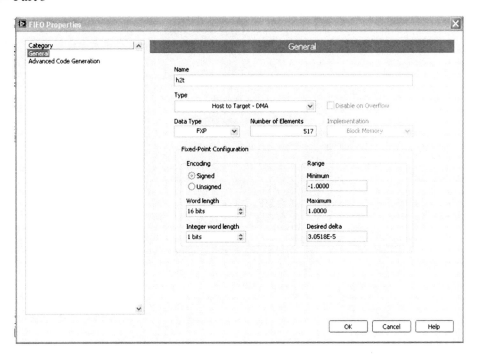

Figure L11.62: Properties of h2t FIFO

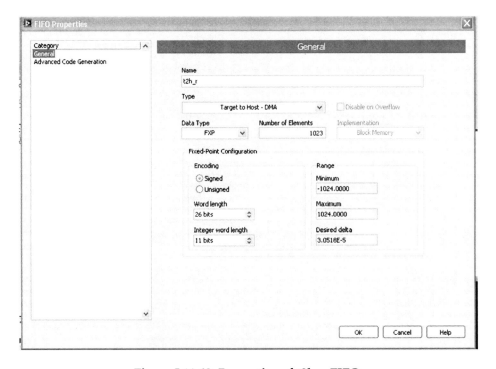

Figure L11.63: Properties of t2h_r FIFO

L11.4.4 Completion of Host VI

Having created the target VI, the host VI is completed next by adding the portion that communicates with the target VI by going through the following steps:

1. Open a reference to **target_hk.vi** by using the **Open Fpga VI reference** function (refer to Figure L11.48).
2. Figure L11.64 shows the BD of STFT_host.vi that communicates with target_hk.vi, and Figure L11.65 shows a zoomed view portion of this BD. In frame 10, create the function **invoke method**, and wire the **Fpga VI reference in** terminal to the **Fpga VI reference out** terminal of **Open Fpga VI reference**. Configure it for write operation so that fixed-point input data from frame 9 are written into this FIFO, and then to the target FIFO **h2t** by the DMA (Direct Memory Access). Wire its **timeout** terminal to **constant -1** (infinite) and the **data** terminal to the output of the **To fixed point** function at the right side of the For Loop that generates fixed-point data.

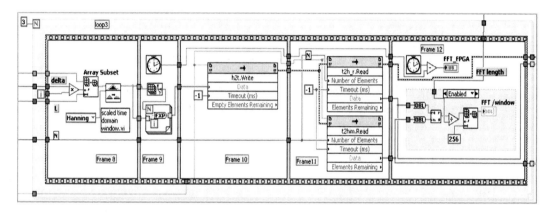

Figure L11.64: BD portion of STFT_host.vi that communicates with target_hk.vi

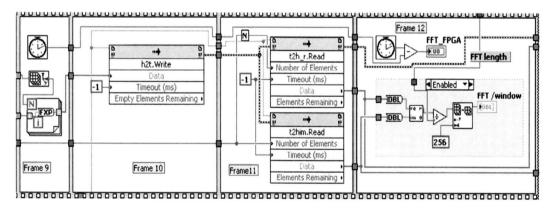

Figure L11.65: Expanded view of a portion of STFT_host.vi that communicates with target_hk.vi

3. In frame 11, the FFT frame computed by the target VI is obtained by using the invoke **method** node. Create a **read** method to the **t2h_r_hk** and **t2h_i_hk** FIFOs with the timeout of -1, and the number of elements equal to one frame (wired to the constant 512).

4. Compute the execution time of the FFT frame processing by using the Tick count (msec) function. Use an indicator **FFT_FPGA** for this execution time, see Figure L11.65 (the indicator type is unsigned 8-bit as this time it is within the unsigned 8 bit integer range; this is done to reduce memory usage).

5. In frame 12, the real and imaginary FFT values, which are of fixed-point data type, are converted to double format using the function **to double precision float**, and the output of this function is passed onto the inputs **x** and **y** of the function **Re/Im to Polar.vi (Numeric>>Re/Im to Polar.vi)** on the BD. Wire the output **r** that carries the FFT magnitude to the **divide** function along with the signal FFT length so as to normalize the FFT result. Wire the output of the **divide** function to the **array subset** function with the input **length** wired to the constant 256 in order to get only the positive frequency bins. Also, wire the output to the graph indicator **FFT/window**. This graph displays in real-time the FFT result of 256 frequency bins for each of the windowed frames of size 512 samples. To reduce the computation time, the FFT display may be disabled and only the spectrogram or the intensity plot may be displayed. As shown in Figure L11.65, enclose the graph along with the accompanying items in the structure **diagram disable (Structures>>diagram disable structure)**. Right click on the structure and choose **enable this diagram**. Keep the case with **disable** as empty. Later, if desired, the part that is enabled can be disabled. Now, consider the graph **FFT/window.** Right click on it and choose **properties.**As shown in Figure L11.66, change the name field to **frequency (Hz)** for x-axis to correspond to the frequency scale in Hz. Since by default the X scale indicates frequency bins, the scale can be converted to frequency scale in Hz by multiplying it with sampling frequency/FFT length = 22050/512= 43.06. Hence, in the field labeled **multiplier**, enter the value 43.06. Alternatively, one may use the graph property node to change the multiplier for the X scale to the desired value. As seen in Figure L11.66, the graph displays the FFT result up to the Nyquist frequency.

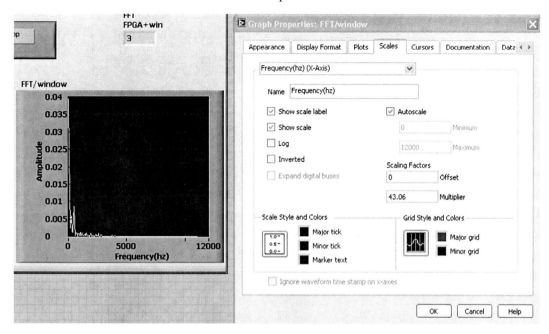

Figure L11.66: FP of host VI exhibiting FFT/window graph and its properties page

6. As displayed in Figure L11.67, in frame 7 outside the for loop, as done in the earlier FFT example, the fixed-point real and imaginary values are converted to double and then normalized with one half of the values taken. However, real and imaginary values coming out of the for loop is two-dimensional of size 3*512or R*N corresponding to 3 windowed frames of size 512. That is, the rows correspond to the number of time frames and the columns to the number of FFT frequency bins. The normalized and one half of the columns come out of the **array subset** function corresponding to 3 time frames and 256 frequency bins for each frame. These outcomes are wired to an intensity chart **STFT of Speech signal/Sound signal on FPGA VI**, see Figure L11.67. The intensity chart reverses the x and y scale for the input data. Thus, the input 2D array is specified as shown in Table L11.1. The intensity chart plots the points as indicated in Table L11.2. For more information about how to use intensity charts, refer to the LabVIEW 2009 help. Thus, the intensity chart will have the X scale with 3 elements and the Y scale with 256 elements. In order to represent the X scale in terms of time and the Y scale in terms of frequency in Hz, change the properties of the chart as shown in Figures L11.68 and L11.69. As seen in these figures, the multiplier for the X scale is set to 0.023 or 512/22050 or N/Fs to correspond to frame time/window time. Similarly, the Y scale has a multiplier of 43.06 as explained previously. Also, to be able to display the FFT values, choose **autoscale Z** and uncheck the **loose fit** property of the Z scale. Figure L11.70 shows the FP of the intensity chart, which is in a way equivalent to the spectrogram of speech/sound signals. The color corresponding to the Z values or the FFT outcome can be changed by using the property node and utilizing the property **color table** (refer to the LabVIEW help to see an example of how to do this). Also, change the chart history length to **64** to reduce the computation time.

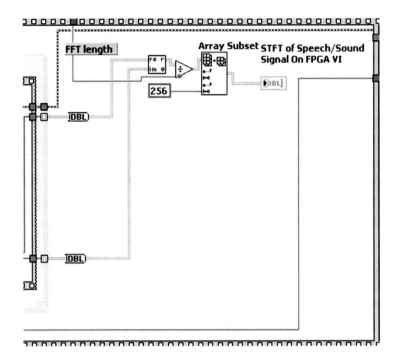

Figure L11.67: Data conversion

	0	1
0	a	b
1	d	e

Table L11.1: 2D array

	0	1
1	b	e
0	a	d

TableL11.2: Manner in which 2D array of Table L11.1 is displayed by intensity chart

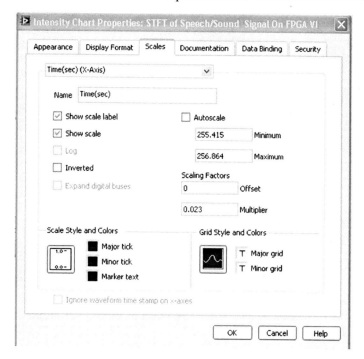

Figure L11.68: Changing properties of X scale

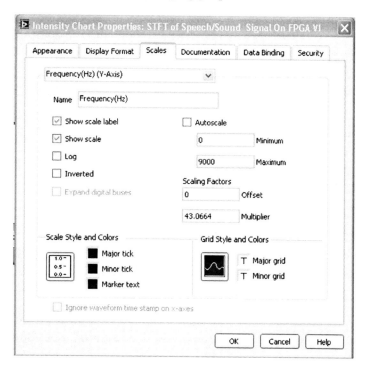

Figure L11.69: Changing properties of Y scale

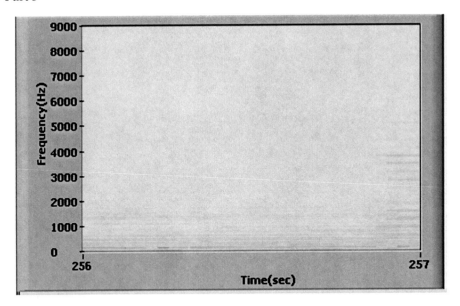

Figure L11.70: Intensity chart to display FFT outcome

7. Finally, create an indicator **FFT FPGA+win** to show the computation time for the windowing process on the host, the FFT computation on the FPGA, and the FFT and spectrogram displays on the host via the millisecond timers in frame 6 and frame 8, as shown in Figure L11.71. Close the FPGA VI reference by placing the function **close FPGA VI reference** in frame 1. This completes the design of both the host and FPGA VIs.

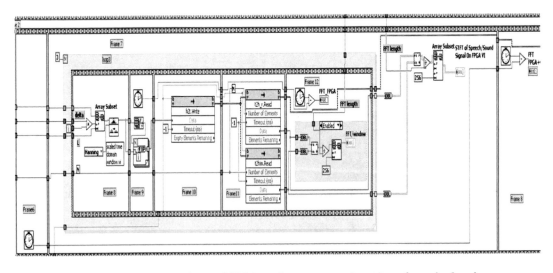

Figure L11.71: BD portion exhibiting the computation time for windowing, FFT and display

Figure L11.72 shows the FP of the host VI.

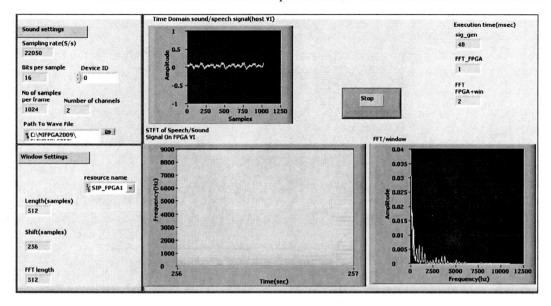

Figure L11.72: FP of STFT_host.vi

L11.4.5 FPGA VI Compilation

Before the host VI can be run, it is necessary to compile the FPGA VI. Either run the FPGA VI, or right click on the FPGA VI name in the project explorer and choose **Compile**. Run the host VI. As seen in Figure L11.73, the wave file is read and played back simultaneously together with the spectrogram of the windowed frames and the FFT of the frames are also displayed. It is seen that the file read and playback take about 48msec while the FFT computation on the FPGA VI takes about 1msec on average. Windowing of the sound buffered data into frames followed by the FFT and display on the host completes within 5msec, providing a real-time display without missing any speech/sound frames.

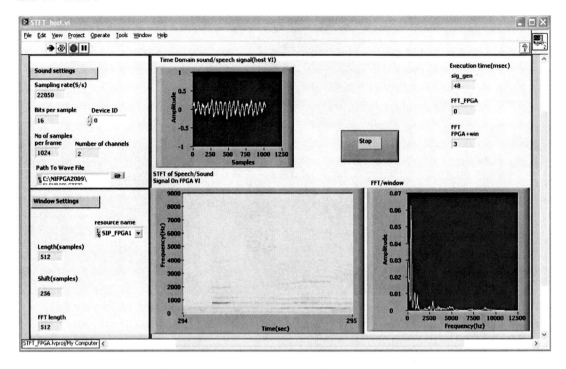

Figure L11.73: STFT_host.vi in run mode

L11.5 Lab Experiments

- Modify L7.3 to perform convolution graphically using fixed-point representation (for real numbers and not integers). Call this VI **convol_real.vi.** Run this VI on the FlexRIO target and verify the results. Next, redo the FIR lab by replacing the FIR_FPGA_on_trgt VI with the convolution VI, and extract the first p samples of the convolution result (p = signal length) as the filtered data outcome. Verify that the same results obtained by designing the FIR filter using the Xilinx Core Generator are obtained with the new design.

- Modify the STFT program so that the overlap between the windows is a variable between 25% to 75% in steps of 10. Verify the STFT of a speech signal by varying the window overlap on the FlexRIO target.

CHAPTER 12

Application Project 1: Discrete Wavelet Transform

In this chapter and the remaining two chapters, FPGA implementation of three application projects is discussed to provide a cross section of various issues one encounters when performing FPGA hardware implementation of DSP systems.

12.1 Discrete Wavelet Transform (DWT)

Wavelet transform offers a generalization of STFT. From a signal theory point of view, similar to DFT and STFT, wavelet transform can be viewed as the projection of a signal into a set of basis functions named wavelets. Such basis functions offer localization in the frequency domain. In contrast to STFT having equally-spaced time-frequency localization, wavelet transform provides high frequency resolution at low frequencies and high time resolution at high frequencies. Figure 12.1 provides a tiling depiction of the time-frequency resolution of wavelet transform as compared to STFT and DFT.

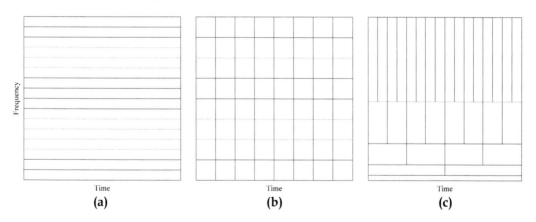

Figure 12.1: Time-frequency tiling for (a) DFT, (b) STFT, and (c) DWT

The discrete wavelet transform (DWT) of a signal $x[n]$ is defined based on so called approximation coefficients $W_\varphi[j_0, k]$ and detail coefficients $W_\psi[j, k]$ as follows:

$$W_{\varphi}[j_0,k] = \frac{1}{\sqrt{M}} \sum_n x[n] \varphi_{j_0,k}[n] \tag{12.1}$$

$$W_{\psi}[j,k] = \frac{1}{\sqrt{M}} \sum_n x[n] \psi_{j,k}[n] \qquad \text{for } j \geq j_0$$

and the Inverse DWT is given by

$$x[n] = \frac{1}{\sqrt{M}} \sum_k W_{\varphi}[j_0,k] \varphi_{j_0,k}[n] + \frac{1}{\sqrt{M}} \sum_{j=j_0}^{J} \sum_k W_{\psi}[j,k] \psi_{j,k}[n] \tag{12.2}$$

where $n = 0,1,2,\ldots,M-1$, $j = 0,1,2,\ldots,J-1$, $k = 0,1,2,\ldots,2^j-1$, and M denotes the number of samples to be transformed. This number is selected to be $M = 2^J$, where J indicates the number of transform levels. The basis functions $\{\varphi_{j,k}[n]\}$ and $\{\psi_{j,k}[n]\}$ are defined according to

$$\varphi_{j,k}[n] = 2^{j/2} \varphi\left[2^j n - k\right] \tag{12.3}$$

$$\psi_{j,k}[n] = 2^{j/2} \psi\left[2^j n - k\right]$$

where $\varphi[n]$ is called a scaling function and $\psi[n]$ a wavelet function.

For an efficient implementation of DWT, the filterbank structure is often used. Figure 12.2 shows the decomposition or analysis filterbank for obtaining the forward DWT coefficients. The approximation coefficients at a higher level are passed through a highpass and a lowpass filter followed by downsampling by a factor of two to compute both the detail and approximation coefficients at a lower level. This tree structure is repeated for a multi-level decomposition.

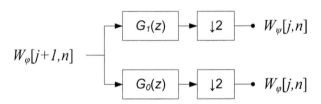

Figure12.2: Discrete wavelet transform decomposition filter bank,
G_0 lowpass and G_1 highpass decomposition filters

Inverse DWT (IDWT) is obtained by using the reconstruction or synthesis filterbank shown in Figure 12.3. The coefficients at a lower level are upsampled by a factor of two and passed through a highpass and a lowpass filter. The results are added together to obtain the approximation coefficients at a higher level.

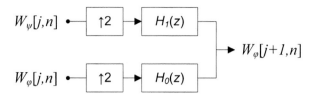

Figure12.3: Discrete wavelet transform reconstruction filter bank,
H_0 lowpass and H_1 highpass reconstruction filters

12.1.1 Higher Level Decomposition

Starting from the configuration of Figure 12.1, one may have higher levels of decomposition by passing the approximation coefficients through a number of filterbanks. Figure 12.4 shows an example for three level decomposition where three sets of detail coefficients at the three stages H, LH and LLH and a single set of approximation coefficients LLL at the third stage are resulted. For reconstruction, similar to Figure 12.3, the reverse of the decomposition process using muti-level synthesis filterbanks is performed.

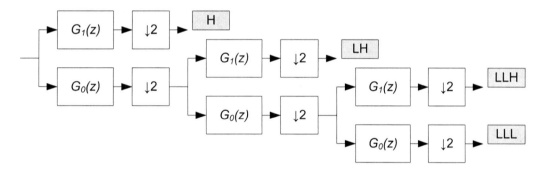

Figure12.4: Three-level wavelet decomposition

Lab 12:

FPGA Implementation of Discrete Wavelet Transform

This lab covers the real-time FPGA hardware implementation of discrete wavelet transform. Speech signal from a wave file is used as the input signal from which approximation and detail wavelet coefficients are computed in real-time. Signal reconstruction is then performed on the host PC and the reconstructed signal is compared to the original input signal.

L12.1 Project Features and Items

Figure L12.1 displays the project **project.lvproj**. The top level VI on the host or My Computer is **host_soundfile.vi**. A virtual folder named **subVIs** is created containing the subVI **gen_coeff.vi** and **configure_sound.vi**. Similarly, under the FPGA target, the top level VI is **dwtd4fxp.vi**. This VI uses the subVI **dwt_d4conv.vi** that is placed in the folder **subVIs**. The FPGA target also includes two other folders, namely **memories** and **FIFOs,** corresponding to the VI defined memory elements and DMA FIFOs, respectively.

Let us go through the design of the subVIs involved.

L12.2 Design of configure_sound.vi

Figure L12.2 displays the BD and FP of the subVI **configure_sound.vi**. This subVI reads sound samples from a specified wave file on the host VI using the function **Sound file read (DBL).vi**. A buffer of length 1024 is used for the samples. The buffer is passed out of this subVI as an indicator **Y**. This subVI also displays the Boolean indicator **end of file?** to indicate that all the data samples are read. Figure L12.3 shows the terminals of this subVI.

L12.3 Design of gen_coeff.vi

The subVI **gen_coeff.vi** is used to calculate the analysis filter coefficients on the PC in floating-point format, followed by their conversion into fixed-point format. Figure L12.4 shows the Front Panel and Block Diagram of this sub VI along with its terminals.The coefficients of the decomposition or analysis filters are generated by the subVI **WA Wavelet Filter.vi (Signal processing>>Wavelet analysis>>Discrete wavelet>>WA Wavelet Filter.vi)**. The decomposition filters are displayed by the indicator **analysis filters**. Using the **Unbundle by name** function, the elements of this cluster are obtained as lowpass and highpass coefficients which are converted to the fixed-point format <+/- 12,4> or an integer wordlength of 4 and a fractional length of 8. The fixed-point coefficients are displayed by their respective indicators **Low Pass Coefficients** and **High Pass Coefficients**.

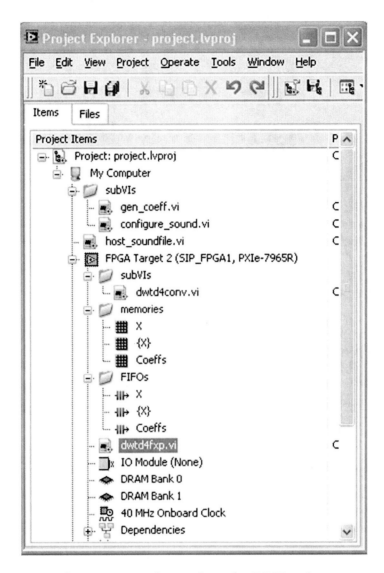

Figure L12.1: Project explorer for DWT project

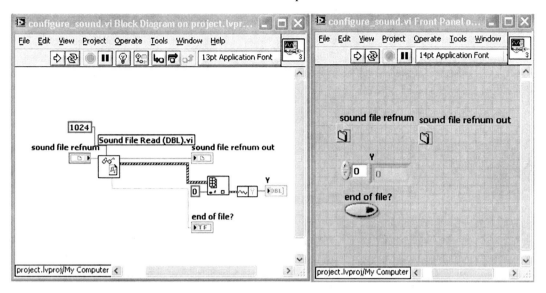

Figure L12.2: BD and FP of configure_sound.vi

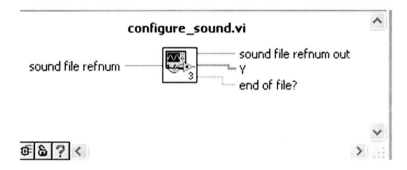

Figure L12.3: Terminals of configure_sound.vi

L12.4 Design of dwtd4conv.vi

The subVI **dwtd4conv.vi** (see Figures L12.6 and L12.7) computes the convolution of the convolution input memory and the highpass (HP) or lowpass (LP) coefficients, followed by decimation. Here, the HP and LP decomposition filters are implemented via convolution. This process is made clearer by following the m-file code that appears in Figure L12.5. This figure also shows a snapshot of the workspace variables. It is seen that the convolution of x=[1 3 2 6] with HP=[4 5] produces the outcome H=[4 17 23 34 30] with its length being equal to length of x + length of HP - 1. FIR filtering of the same input with the HP as a filter produces the outcome H_fltr=[4 17 23 34]. Thus, it is seen that the first length(x) elements of the convolution approach are equivalent to the filtering approach.

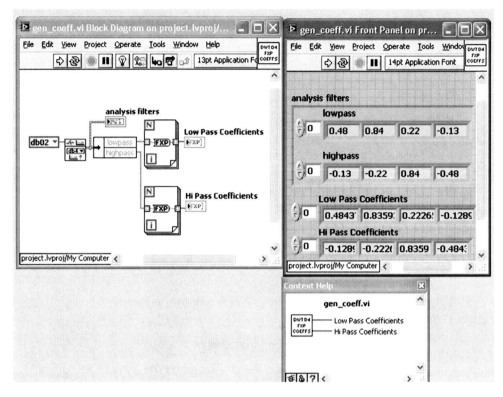

Figure L12.4: BD, FP and terminals of gen_coeff.vi

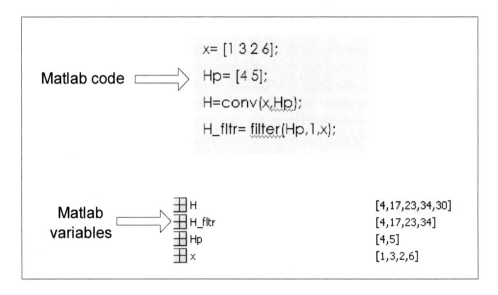

Figure L12.5: Filtering via convolution

Three memories (all LUT based) are used in this subVI, each having the data type <+/-12,4>.

Coeffs is a memory of size 8 for storing the highpass and lowpass filter coefficients, which for the dB2 wavelet type consist of four each. To capture the speech frame of size 512, the memory **X** with size 512 is used. Finally, the result of convolution and decimation for each of the highpass and lowpass coefficients is stored in the memory {**X**}, which is also of size 512. Furthermore, for a multi-level decomposition, **X** is used repeatedly to capture the result of the lowpass filtering for the next level. The pointer **X read pointer** is used to keep track of the input memory from which samples are read based on the decomposition level, while **Low Pass Write Pointer** and **High Pass Write Pointer** do the same for the coefficients memory.

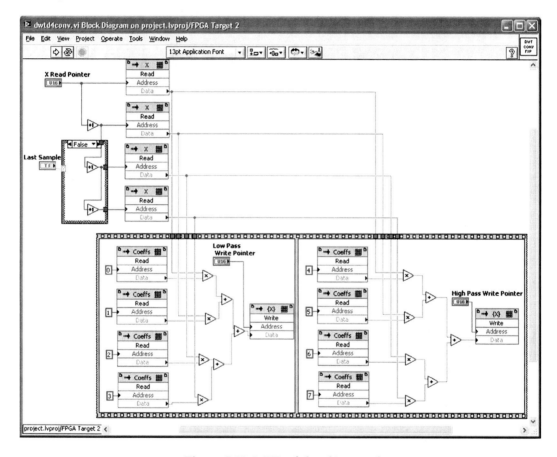

Figure L12.6: BD of dwtd4conv.vi

The convolution operation would require additional samples beyond the frame size to address overlapping frames. The samples from the beginning of the frame need to be wrapped back to the next frame to properly perform the convolution operation. Rather than implementing the convolutions using sequential loops, parallel multiplication and addition blocks are used here to gain efficiency. In addition, the convolution and decimation operations are combined into a single loop to reduce memory accesses. As part of the convolution operation, it is necessary to copy values between local memory banks. Since the number of required values to be copied decreases with each decomposition level, the execution time is optimized by only copying the necessary values.

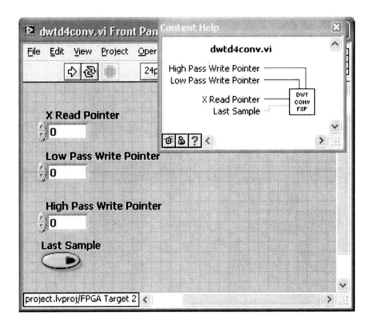

Figure L12.7: FP and terminals of dwtd4conv.vi

L12.5 Design of dwtd4fxp.vi

This is the top level FPGA VI that uses the subVI from the previous section. The BD of the FPGA VI contains a sequence diagram with the following three frames:

a. **Read in low pass coefficients:** As shown in Figure L12.8, this frame copies the low-pass coefficients from the host to target FIFO **Coeffs** (with the fixed-point <+/- 12,4> format and with 8 elements coerced to 21 elements) into the memory **Coeffs**. The coefficients are copied into the memory addresses 0, 1, 2, 3, respectively.

b. **Read in high pass coefficients:** As shown in Figure L12.8, this frame copies the highpass coefficients from the host to target FIFO **Coeffs** into the memory **Coeffs**. The coefficients are copied into the memory addresses 4, 5, 6, 7, respectively.

c. **Frame 3:** Figure L12.9 shows the BD of the third frame. It shows a while loop within the outermost sequence diagram as well as another sequence diagram with three frames within the while loop. Frame 1 within the while loop has a for loop to read data samples, which are passed from the host to the FPGA VI using the host to target FIFO **X** (with the fixed-point <+/- 12,4> format and with 512 elements coerced to 1029 elements) into the corresponding memory **X**.

Refer to figure L12.10. It displays Frame 2 within the while loop. This portion of the BD creates the data and coefficients pointers and passes them to the subVI **dwtd4conv**, as discussed in the previous section. The subVI performs the convolution, computes the wavelet coefficients for an 8-level decomposition, and writes these coefficients into memory {X}. These are then copied to the target to host FIFO {X} (with the fixed-point in <+/- 12,4> format and with 512 elements coerced to 2047 elements) as shown in Figure L12.11.

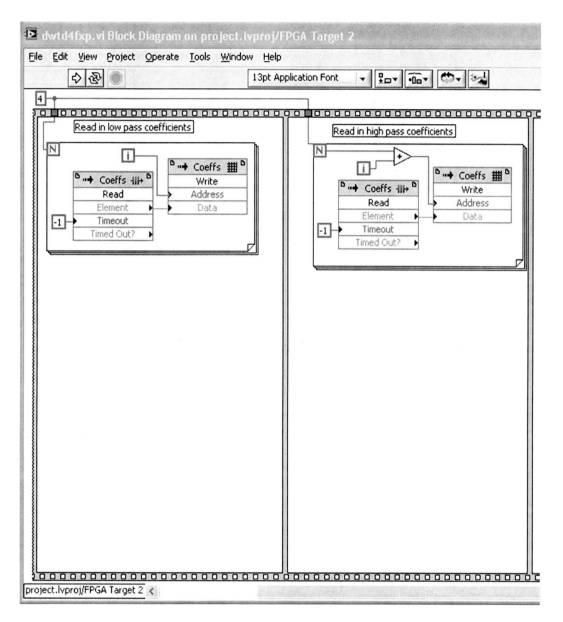

Figure L12.8: First and second frames of dwtd4fxp.vi

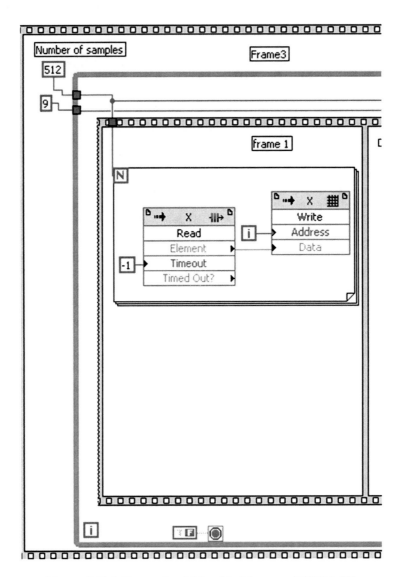

Figure L12.9: Frame 1 within third frame of FPGA VI

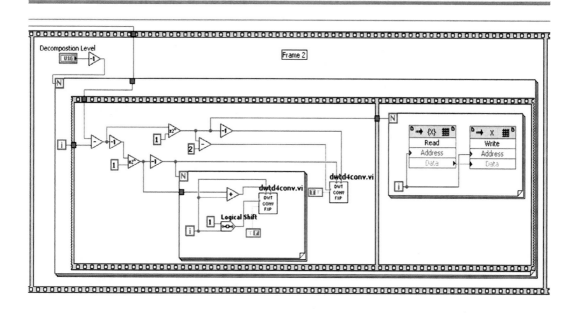

Figure L12.10: Frame 2 within third frame of FPGA VI

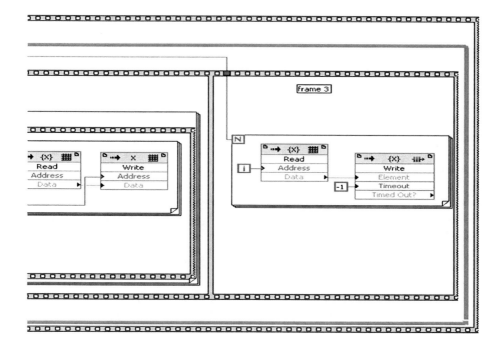

Figure L12.11: Frame 3 within third frame of FPGA VI

L12.6 Design of Host VI

Let us now go through the design of the top level host VI **host_soundfile.vi**. Refer to Figure L12.12.

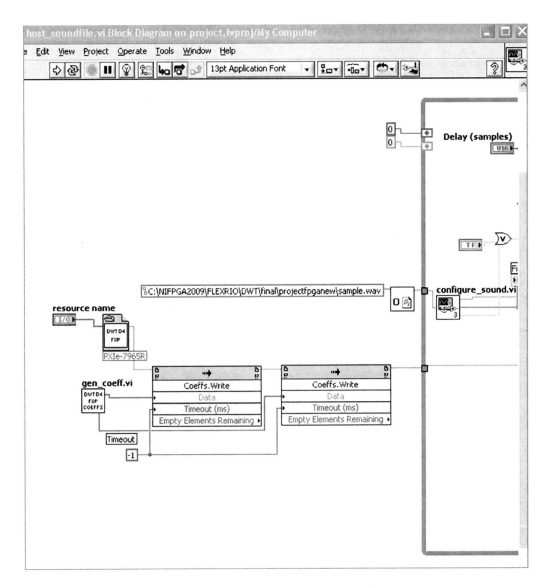

Figure L12.12: Host VI (part 1)

First, a reference to the top level FPGA VI is opened and the subVI gen_coeff.vi is called in order to generate the lowpass and highpass coefficients. These are then written to the host to target FIFO **Coeffs** using the **Invoke Method** function in write mode with the timeout of -1 (infinite timeout). Next, Figure L12.13 shows the portion of the VI that generates the sound data and passes the sound data to the FPGA VI. A while loop is created to continuously read sound data. **Sound File Read Open.vi** opens the wave file **sample.wav** for reading. The path for the file is provided as a constant. The reference from **Sound File Read Open** is passed to the subVI **configure_sound.vi** to configure the sound data as read buffers of size 1024. The frames considered are of size 512 out of each buffer of size 1024 depending upon the desired amount of overlap between the frames. The control **delay (samples)** provides the delay between the starting index of each frame. The indicators **Frames per buffer** and **Overlap (samples)** display the number of frames per buffer and the overlap between the frames in terms of number of samples, respectively. For example, with the buffer size of 1024 the frame size of 512, and a delay of 256, there will be three frames starting from 0 to 511, 255 to 767, and 511 to 1023, respectively, with an overlap of 256 or 50% between the frames.

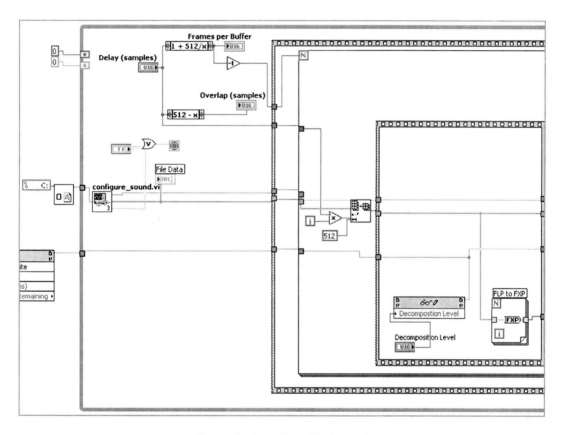

Figure L12.13: Host VI (part 2)

Next, a sequence diagram is created. A for loop (that runs for the number of times that there are frames per buffer) is placed within the sequence diagram. Another sequence diagram with four frames are placed inside the for loop. An array subset function generates the frames per buffer, and these are then converted to fixed-point format as shown in Figure L12.13. The number of decomposition levels for the input signal is passed on as a **Read write control**. Note that although this is a control, its value is fixed at 8 by editing the properties of the control and setting the **maximum** and **minimum** fields to 8.

Refer to Figure L12.14. In the second and third frames of the innermost sequence diagram, the data samples are passed to the host to target FIFO **X** and the analysis and detail coefficients are read back through the target to host FIFO {**X**}. In the last frame, the coefficients are converted back to floating-point format, and the function **Wavelet Transform Daubechies4 inverse.vi (Signal processing>>transforms>> Wavelet Transform Daubechies4 inverse.vi)** is used to obtain the reconstructed signal. The original input signal is also shown by passing the input frames through a combination of forward and inverse wavelet transforms via this function.

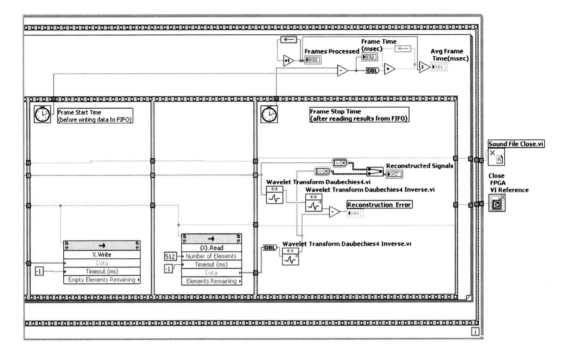

Figure L12.14: Host VI (part 3)

The wavelet transform Daubechies4.vi is followed by the wavelet transform Daubechies4 inverse.vi. Both the original and reconstructed signals are shown in the same chart indicator **reconstructed signals.** Similarly, the reconstruction error per frame is shown as a waveform in the chart indicator **Reconstruction error**. The frame time and the average frame time (averaged over the entire file) are also provided in msec.

Finally, the reference to the FPGA VI is closed which closes the sound file by using the functions **Close FPGA VI reference** and **Sound File Close.**

L12.7 DWT Results

Figure L12.15 shows the FP of the host VI. 512 sample frames are read from the sound buffer of size 1024 with 50% overlap. The wavelet type is considered to be dB2 with 8-level of decomposition.

The graph **Reconstructed signals** shows both the original and reconstructed signal overlaid on each other. It is seen that the reconstructed signal closely matches the original signal with a negligible error. The frame time, on average, is 1.9 ms, which is within the real-time limit.

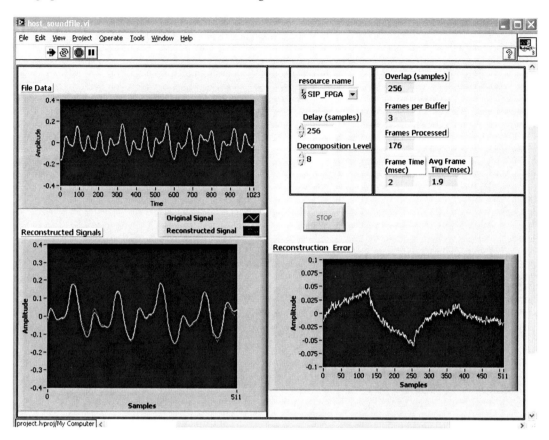

Figure L12.15: FP of host VI showing DWT results

CHAPTER 13

Application Project 2:
Software-Defined Radio

This chapter covers a software-defined radio system and its FPGA implementation using LabVIEW FPGA. A software-defined radio consists of a programmable communication system where functional changes can be made by merely updating software. For a detailed description of software-defined radio, the reader is referred to [1], [2].

4-QAM (Quadrature Amplitude Modulation) is chosen for the modulation scheme of the implemented software-defined radio system, noting that this modulation is widely used for data transmission applications over bandpass channels such as FAX modem, high speed cable, multi-tone wireless and satellite systems [2]. Here the communication channel is considered to be AWGN (Additive White Gaussian Noise).

13.1 QAM Transmitter

For transmission, pseudo noise (PN) sequences are generated to serve as the message signal. A PN sequence is generated with a five-stage linear feedback shift register structure, see Figure 13.1, whose connection polynomial is given by

$$h(D) = 1 + D^2 + D^5 \tag{13.1}$$

where D denotes delay and the summations represent modulo 2 additions.

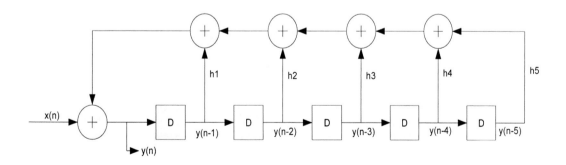

Figure 13.1: PN generation with linear feedback shift register

The sequence generated by the above equation has a period of 31 ($=2^5$-1). Two PN sequence generators are used in order to create the message sequences for both the in-phase and quadrature phase components. The constellation of 4-QAM is shown in Figure 13.2. For more details of PN sequence generation, refer to [2].

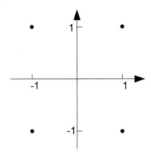

Figure 13.2: Constellation of 4-QAM

Note that frame marker bits are inserted in front of the generated PN sequences. This is done for frame synchronization purposes, discussed shortly in the receiver section. As illustrated in Figure 13.3, a total of ten frame bits are placed in front of each period of a PN sequence.

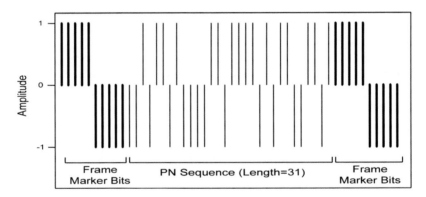

Figure 13.3: PN sequence generator

The generated message sequences are then passed through a raised-cosine FIR filter to create a band limited baseband signal. The frequency response of the raised cosine filter is given by

$$G(f) = \begin{bmatrix} 1 & for \ |f| \le (1-\alpha)f_c \\ \cos^2\left[\dfrac{\pi}{4\alpha f_c}\left(|f|-(1-\alpha)f_c\right)\right] & for \ (1-\alpha)f_c \le |f| \le (1+\alpha)f_c \\ 0 & elsewhere \end{bmatrix} \qquad (13.2)$$

where $\alpha \in [0\ 1]$ denotes a roll-off factor specifying the excess bandwidth beyond the Nyquist frequency f_c.

The output of the raised cosine filter is then used to generate a complex envelope, $\tilde{s}(t)$, of a QAM signal expressed by

$$\tilde{s}(t) = \sum_{k=-\infty}^{\infty} c_k g_T (t - kT) \tag{13.3}$$

where c_k indicates a complex message, made up of two real messages a_k and b_k, $c_k = a_k + jb_k$.

By modulating $\tilde{s}(t)$ with $e^{j\omega_c t}$, an analytical signal or pre-envelope, $s_+(t)$, is produced,

$$s_+(t) = \tilde{s}(t) e^{j\omega_c t} = \sum_{k=-\infty}^{\infty} c_k g_T (t - kT) e^{j\omega_c t} \tag{13.4}$$

The transmitted QAM signal, $s(t)$, is thus given by

$$\begin{aligned} s(t) &= \Re\left[s_+(t) \right] \\ &= a(t)\cos(\omega_c t) - b(t)\sin(\omega_c t) \end{aligned} \tag{13.5}$$

where $\Re[.]$ corresponds to the real part of the complex value inside the brackets. Figure 13.4 illustrates the block diagram of the QAM transmitter just discussed. Notice that the two data paths, indicated by a solid line and a dotted line, represent complex data. Again, the reader is referred to [2] for more theoretical details.

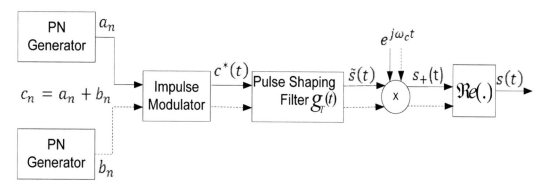

Figure 13.4: QAM transmitter as described in [2]

13.2 AWGN Channel

The signal from the transmitter is assumed to be corrupted by the addition of white Gaussian noise in the channel which is illustrated in Figure 13.5. Thus, the signal at the receiver end is expressed as

$$r(t) = s(t) + n(t) \tag{13.6}$$

where $n(t)$ denotes additive white Gaussian noise (AWGN) with the power spectral density of $\phi_{nn}(f) = \frac{N_0}{2}$. For more theoretical details, refer to [3].

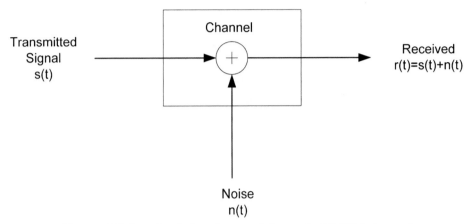

Figure 13.5: Received signal passed through AWGN channel

13.3 QAM Receiver

13.3.1 QAM Demodulation

Here, it is assumed that the exact phase and frequency information of the carrier is available. The received QAM signal is denoted by $r(t)$. To simplify the system, an ideal channel is assumed between the transmitter and the receiver, i.e. $r(t) = s(t)$.

If $r(nT)$ is considered to be the sampled received signal, the analytic signal $r_+(nT)$ is given by

$$r_+(nT) = r(nT) + j\hat{r}(nT) \tag{13.7}$$

where $\hat{r}(\cdot)$ indicates the Hilbert transform of $r(\cdot)$. Thus, the complex envelope of the received QAM signal $\tilde{r}(nT)$ can be expressed as

$$\begin{aligned}
\tilde{r}(nT) &= r_+(nT)e^{-j\omega_c nT} \\
&= a(nT) + jb(nT)
\end{aligned} \tag{13.8}$$

Such a QAM demodulation process is illustrated in Figure 13.6.

13.3.2 Frame Synchronization

Frame synchronization is required for properly grouping transmitted bits into an alphabet. To achieve this synchronization, a similarity measure consisting of cross-correlation is computed between the known marker bits and received samples. The cross-correlation of two complex values v and w is given by

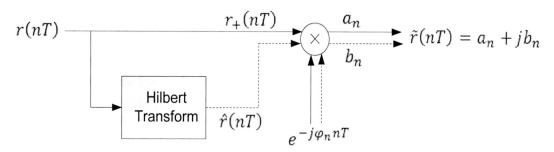

Figure 13.6: QAM Demodulation [2]

$$R_{wv}[j] = \sum_{n=-\infty}^{\infty} \overline{w}[n] \, v[n+j] \tag{13.9}$$

where the bar indicates complex conjugate.

An example of the cross-correlation outcome for frame synchronization is shown in Figure 13.7. The maximum value is found to be at the location of index 33. The subsequent message symbols are then framed from this index point.

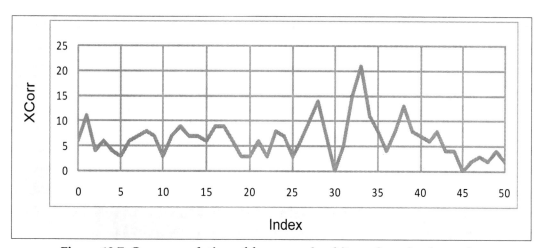

Figure 13.7: Cross-correlation of frame marker bits and received samples

13.3.3 Decision Based Carrier Tracking

Let us now consider the phase offset, denoted by θ, between the transmitter and the receiver. Based on this offset, the received signal can be written as

$$\tilde{r}(nT) = r_+(nT)e^{-j(\omega_c nT + \theta)}$$
$$= \hat{c}_n e^{-j\theta}$$

(13.10)

where $\hat{c}_n$ indicates the output of a slicer mapping a received sample to the nearest ideal reference in the signal constellation. As a result, the baseband error at the receiver is given by

$$\tilde{e}(nT) = \hat{c}_n - \tilde{r}(nT)$$

(13.11)

Next, the LMS update method is used to minimize a decision-directed cost function, $J_{DD}(\theta)$, consisting of the mean squared baseband error

$$J_{DD}(\theta) = avg\left[\left|\tilde{e}(nT)\right|^2\right]$$
$$= avg\left[\tilde{e}(nT)\overline{\tilde{e}(nT)}\right]$$

(13.12)

By differentiating $J_{DD}(\theta)$ with respect to θ, we get

$$\frac{dJ_{DD}(\theta)}{d\theta} = avg\left[\frac{d\left[\tilde{e}(nT)\overline{\tilde{e}(nT)}\right]}{d\theta}\right]$$
$$= 2avg\Re\left\{\overline{\tilde{e}(nT)}\frac{d\tilde{e}(nT)}{d\theta}\right\}$$

(13.13)

where

$$\frac{d\tilde{e}(nT)}{d\theta} = \frac{d}{d\theta}\left[\hat{c}_n - \tilde{r}(nT)\right] = -\frac{d\tilde{r}(nT)}{d\theta}$$

(13.14)

and

$$\frac{d\tilde{r}(nT)}{d\theta} = -j\hat{c}_n e^{-j\theta} = -j\tilde{r}(nT)$$

(13.15)

Equation (13.13) can thus be rewritten as

$$\frac{dJ_{DD}(\theta)}{d\theta} = 2avg\left[\Re e\left\{\overline{\tilde{e}(nT)}j\tilde{r}(nT)\right\}\right]$$

$$= -2avg\left[\Im m\left\{\overline{\tilde{e}(nT)}\tilde{r}(nT)\right\}\right] \qquad (13.16)$$

$$= -2avg\left[\Im m\left\{\overline{\tilde{c}_n}\tilde{r}(nT)\right\}\right]$$

where $\Im m[\cdot]$ corresponds to the imaginary part of the complex value inside the brackets. By writing the term $\Im m\left\{\overline{\tilde{c}_n}\tilde{r}(nT)\right\}$ in polar form, one gets

$$\Im m\left\{\overline{\tilde{c}_n}\tilde{r}(nT)\right\} = \Im m\left\{\overline{R_c e^{j\beta_c}} R_r e^{j\beta_r}\right\}$$

$$= R_c R_r \sin(\beta_r - \beta_c) \qquad (13.17)$$

Thus,

$$\sin(\beta_r - \beta_c) = \frac{\Im m\left\{\overline{\tilde{c}_n}\tilde{r}(nT)\right\}}{R_c R_r} \qquad (13.18)$$

Note that for small $\beta_r - \beta_c$,

$$\sin(\beta_r - \beta_c) \approx \beta_r - \beta_c$$

$$R_r \approx R_c = |c_n| \qquad (13.19)$$

As a result, the phase error $\Delta\theta(n)$ is given by

$$\Delta\theta(n) = \frac{\Im m\left\{\overline{\tilde{e}(nT)}\tilde{r}(nT)\right\}}{|c_n|^2} \qquad (13.20)$$

Figure 13.8 shows a block diagram representation of the above tracking equation.

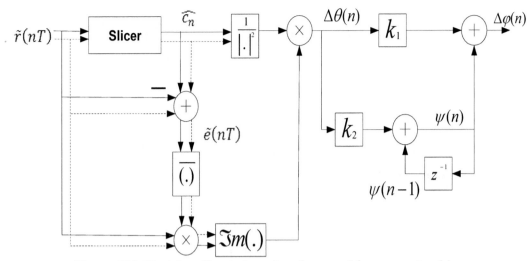

Figure 13.8: Decision directed carrier phase and frequency tracking

When both phase and frequency tracking are considered, the carrier phase of the receiver becomes

$$\varphi(n+1)=\varphi(n)+\Delta\varphi(n) \tag{13.21}$$

In this case, the phase update $\Delta\varphi(n)$ is given by

$$\Delta\varphi(n)=k_1\Delta\theta(n)+\psi(n) \tag{13.22}$$

where $\psi(n)$ denotes the contribution of frequency tracking, which is expressed as

$$\psi(n)=\psi(n-1)+k_2\Delta\theta(n) \tag{13.23}$$

The scale factors k_1 and k_2 are configured to be small here and usually $k_1/k_2 \geq 100$ is required for phase convergence [2].

13.4 Bibliography

[1] C. Johnson and W. Sethares, *Telecommunication Breakdown: Concepts of Communication Transmitted via Software-defined Radio*, Prentice-Hall, 2004.
[2] S.Tretter, *Communication System Design using DSP Algorithms*, Klumer Academic/Plenum Publishers, 2003.
[3] J. Proakis, *Digital Communications*, McGraw-Hill, 2001.

Lab 13:

FPGA Implementation of 4-QAM Receiver

In this lab, the discussed 4-QAM modem system is built using LabVIEW and the receiver part is implemented using the LabVIEW FPGA. As shown in Figure L13.1, this system consists of the following functional modules: message source, pulse shape filter, QAM modulator, AWGN channel, QAM demodulator, frame synchronizer and phase & frequency tracker. The system is divided into three parts: transmitter, AWGN channel and receiver. Sending the transmitted signal after passing through the AWGN channel to the FPGA board where the signal is demodulated. The building of each functional module is described in the sections that follow. Figure L13.1 illustrates the top level VI of the 4-QAM modem system.

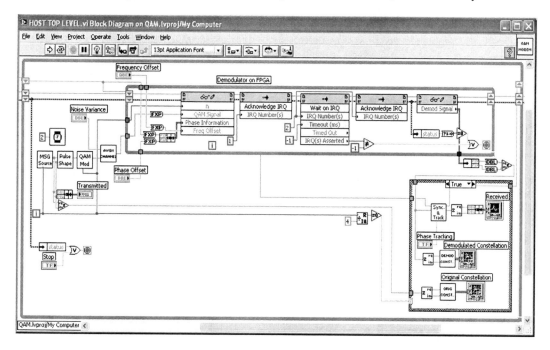

Figure L13.1: Top-level VI of 4-QAM modem system

L13.1 QAM Transmitter

Message source is the first component of the QAM modem. Here, PN sequences are used for generating the message source. In front of every sequence frame, marker bits are inserted to achieve frame synchronization. The BD of the **Message Source** VI is shown in Figure L13.2.

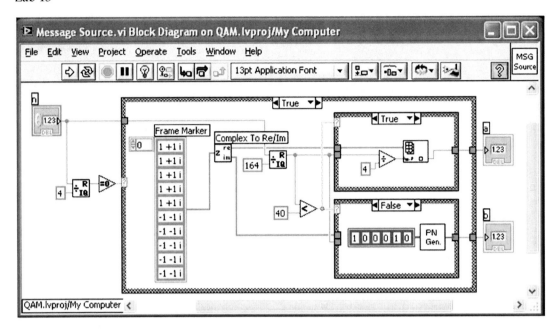

Figure L13.2: BD of Message Source.vi

The generated samples are oversampled 4 times. This is done by comparing with 0 the remainder of a global counter, indicated by n, divided by 4. Thus, out of four executions of this VI, one message sample (frame marker bit or PN sample) is generated. For the remaining three executions of the VI, zero samples are generated. The total length of the message for one period of a PN sequence and frame marker bits is 164, which is obtained by 4 (oversampling rate) x [10 (frame marker bits) + 31 (period of PN sequence)]. A constant array of ten complex numbers is used to specify the marker bits. Note that the real parts of the complex numbers are used as the frame marker bits of the in-phase samples and the imaginary parts as the frame marker bits of the quadrature-phase samples. In order to create complex constants, the representation of a numeric constant is changed by right-clicking on it and choosing **Representation>>Complex Double** (or **Complex Single**).

The BD of the **PN Generator** VI is shown in Figure L13.3. This subVI generates a pseudo noise sequence of length 31 by XORing the values of the second and fifth shift registers.

The **Shift Register**, **Rotate 1D Array**, **Index Array** and **Replace Array Subset** functions are used to compute a new PN sample and to rotate the shift register. A **For Loop** with one iteration and a **First Call?** function (**Functions>>Program- ming>>Synchronization>>First Call?**) are used in order to pass the shift register value of a current call to a next call of the subVI. The **First Call?** function checks whether a current call is occurring for the first time or not. If that is the case, the shift register values are initialized by their specified initial values. Otherwise, the old values of the shift registers are passed from the previous execution of the subVI. Alternatively, the built-in **Binary MLS** VI, (**Functions>>Signal Processing>>Signal Generation>>Binary MLS**) can be used for building this component. Next, the generated samples are passed to a pulse shape filter shown in Figure L13.4. A raised cosine filter is used to serve as the pulse shape filter. The **FIR Filter PtByPt** VI is utilized for this purpose.

The two outputs of the pulse shape filters are combined to construct the pulse shaped message signal by using the **Re/Im** to **Complex** function (**Functions>>Numeric>>Complex>Re/Im to Complex**).

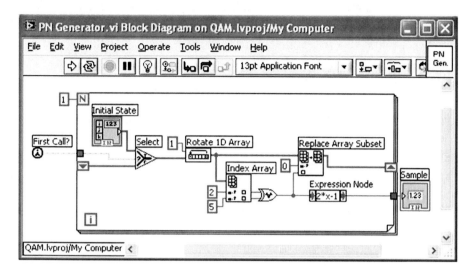

Figure L13.3: BD of PN Generator.vi

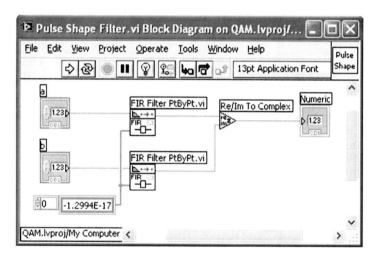

Figure L13.4: BD of Pulse Shape Filter.vi

Filter coefficients can be obtained by filter design tools such as LabVIEW DFD toolkit. These coefficients are then stored in an array of constants for a specific sampling rate, e.g. 4 in our case.

Next, the modulated signal is obtained by going through the pulse shape filter as shown in Figure L13.5. The QAM modulated signal s(t) is obtained by taking the real part of the pre-envelope signal s_+(t). This is achieved by performing a complex multiplication between the complex input and the complex carrier consisting of a cosine and a sine waveform. This completes the transmitter side.

Then, the transmitted signal is passed through a AWGN channel whose BD is shown in Figure L13.6. In this channel, the transmitted signal is added to the White Guassian noise. White Guassian noise is generated using the **Guassian White Noise PtByPt** VI (**Functions>>Signal Processing>>Point By Point>>Signal Generation PtByPt>>Guassian White Noise PtByPt**) whose standard deviation can be varied using the control **SD of Noise**. The output of this channel then becomes the input to the QAM receiver.

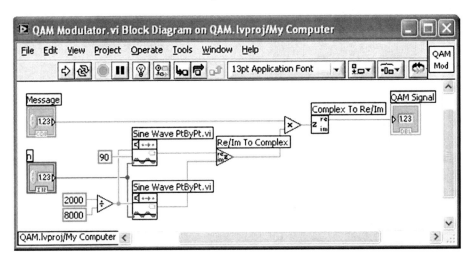

Figure L13.5: BD of QAM Modulator.vi

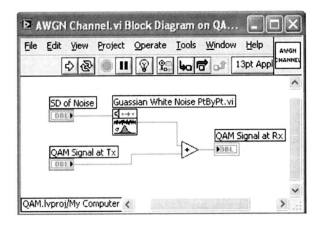

Figure L13.6: BD of AWGN Channel.vi

L13.2 QAM Receiver

On the receiver side, the received signal is demodulated on the FPGA target. The demodulated signal is sent to the host side for frame synchronization and phase/frequency tracking.

For the FPGA implementation of the demodulator, create a project. Add all the VIs previously created under **My Computer.**

The first module on the receiver side is the Hilbert transformer, which contains a FIR filter whose coefficients are computed previously on the host side. These coefficients are used on the FPGA side. The coefficients can be computed using the VI **DFD Remez Design (Functions>>Signal Processing>>Digital Filter Design>>Filter Design>>Advance FIR Filter Design>>DFD Remez Design)** of the DFD toolkit. In order to get an integer group delay, an even number, such as 32, is specified as the filter order. To analyze the group delay along with the magnitude and phase response, the **DFD Filter Analysis** tool is used as shown in Figure L13.7.

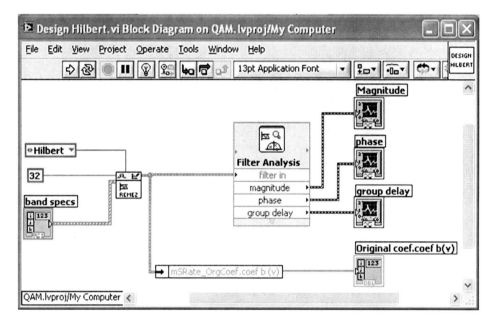

Figure L13.7: Building Hilbert transformer

The specifications of the Hilbert transformer are similar to a bandpass filter as indicated in Figure L13.8. Notice that only one element of the cluster array is needed to design the Hilbert transformer. However, when a control is created at the **band specs** terminal of the **DFD Remez Design** VI, there are two default cluster values. The second element, indexed at 1, should thus be deleted. To do this, select the element of the cluster array to be deleted, then right-click and choose **Data Operation>>Delete Element** from the shortcut menu.

By running the VI, the magnitude, phase response and group delay of the Hilbert transformer can be seen as shown in Figure L13.8.

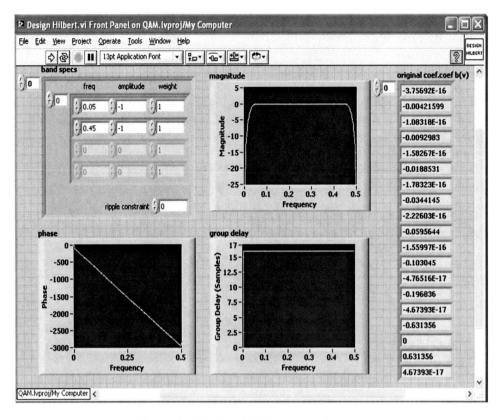

Figure L13.8: FP of Hilbert transformer

The array of indicators corresponding to the Hilbert transform coefficients is converted to an array of constants for utilization on the FPGA side. Note that the design and analysis of the Hilbert transformer are needed only in the designing phase not in the implementation phase.

L13.2.1 QAM Receiver FPGA Target Side

QAM FPGA TOP is the top level VI of the QAM demodulator on the FPGA side, see Figure L13.9. This VI contains a **While Loop** which runs continuously on the FPGA target. **Interrupts (Functions>>Programming>>Synchronization>>Interrupt)** are used for synchronizing the data transfer between the host and the target. To read correct values from the host side, first the **Interrupt** IRQ1 is used to make the VI to wait till the host VI writes data to controls and acknowledges IRQ1.

After acknowledging IRQ1, the host VI waits till the **QAM FPGA TOP** VI writes data to indicators after demodulating the signal and asserts the **Interrupt** IRQ2 as shown in Figure L13.9. Afterwards, the **HOST Top Level** VI acknowledges IRQ2 and reads data from the target. This process repeats for each demodulating sample.

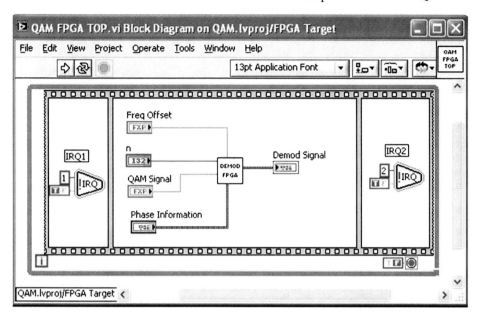

Figure L13.9: QAM Demodulator on FPGA Top level VI

More details of the **Demodulation FPGA** VI can be seen in Figure L13.10, containing the two components of Hilbert transformer and QAM demodulation.

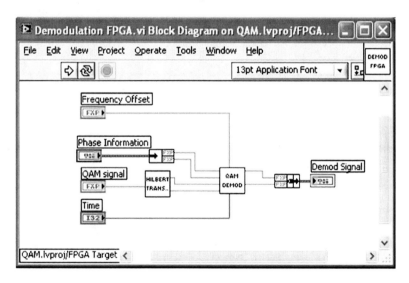

Figure L13.10: Demodulation on FPGA

In the Hilbert transformer, the coefficients previously computed are used. The BD of the **Hilbert Transform** VI is shown in Figure L13.11. This VI contains the subVIs **Data Queue** and **FIR filter_designed** and a sample number counter. For counting sample numbers, a **For Loop** with one iteration and a **First Call?** function (**Functions>>Programming>>Synchronization>>First Call?**) are used to pass the shift register

409

value of a current call to a next call of the subVI. The **First Call?** function checks whether a current call is occurring for the first time or not. If that is the case, the shift register values are initialized with zero. Otherwise, the old values of the shift registers are passed from the previous execution of the subVI. The function **Increment** (**Functions>>Programming>>Numeric>>Increment**) is used to increase the count by 1 each time.

A Data Queue VI as shown in Figure L13.12 is employed in order to synchronize the input and output of the Hilbert transformer. Here, the input samples are delayed until the corresponding output samples become available. This is needed due to the group delay associated with the filtering operation. For a FIR filter of 33 taps, the group delay is 16. Hence, the latest 16 elements are stored by creating a **FIFO** (right click on **FPGA Target** under the Project Explorer and select **(new>> FIFO)** with the properties as shown in Figure L13.13).

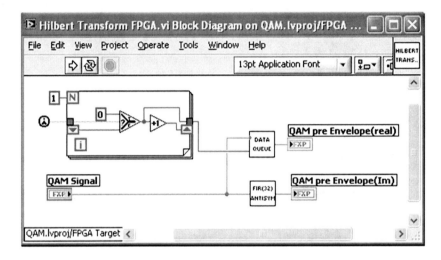

Figure L13.11: Hilbert Transform.vi

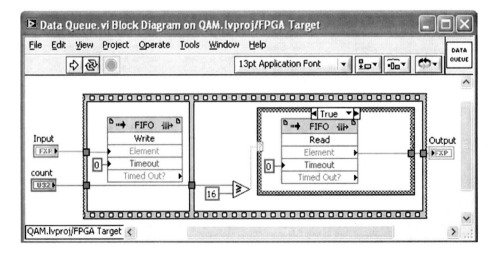

Figure L13.12: Data Queue.vi

The input of the **Data Queue** VI gets stored in the FIFO, and its output gets read from the FIFO if count equals or exceeds 16; till then output remains zero. This way, the output from the **Data Queue** VI becomes the QAM pre-envelope (real) part as indicated in Figure L13.11.

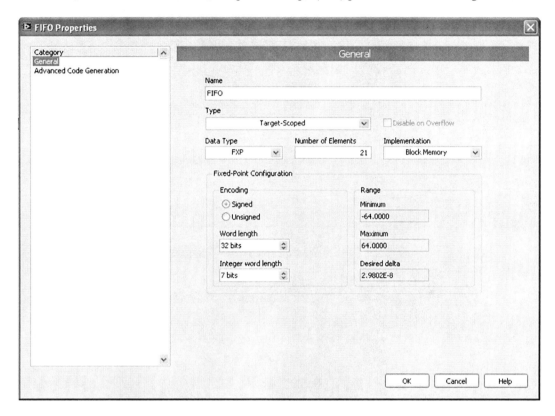

Figure L13.13: FIFO Properties

Next, the coefficients of the previously computed or designed FIR filter via **DFD Remez Design** are used. The direct form structure is used for the realization of the FIR filter. The number of delay elements is set to 32 as the number of coefficients is 33. As a result, the computational complexity involves 33 multiplications and 32 additions but since the coefficients are asymmetric, the number of multiplications is reduced to 16. A part of the **FIR filter_designed** VI is shown in Figure L13.14. Input samples are delayed by one sample using the **Delay One** VI which uses a **For Loop** with one iteration along with **Shift Registers** as shown in Figure L13.15.

The output from the FIR filter is the QAM pre-envelope (Imaginary) part as shown in Figure L13.11. The analytic signal produced from the Hilbert transformer is then demodulated by the QAM demodulator as shown in Figure L13.16. For demodulation, high throughput addition, multiplication, division, sine & cosine, rectangular to polar and polar to rectangular operations are used. The complex multiplication of the Sine & Cosine signals and the Inphase & Quadrature signals are realized using the functions **High Throughput Rectangular To Polar** and **High Throughput Polar To Rectangular** as shown in Figure L13.16.

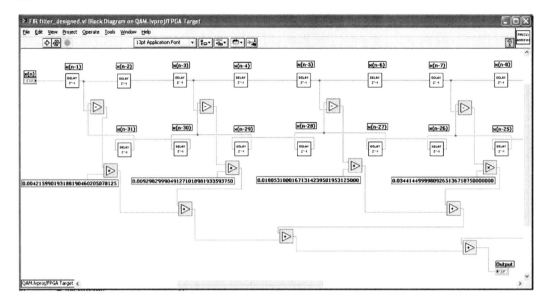

Figure L13.14: Part of FIR filter_designed.vi

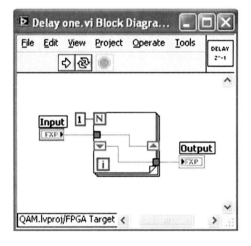

Figure L13.15: Delay one.vi

The demodulated signal is combined as a cluster into **Demod Signal** as indicated in Figure L13.10. This demodulation part completes the FPGA target side. The Demod Signal is then read into the **Host Top Level** VI, where the frame synchronization and phase/frequency tracking are done.

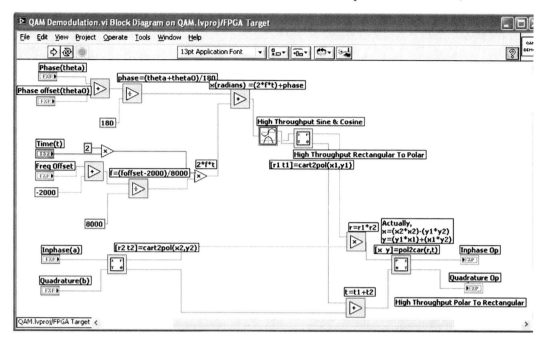

Figure L13.16: QAM Demodulation

Fixed-point data type is used for all the operations on the FPGA target whereas double data type is used on the host side. While transferring data from the host side to the target side, double data type is converted to fixed-point data type using **To Fixed Point(Functions>>Programming>>Numeric>>Conversion>>To Fixed Point)**. Also, for transferring data from the FPGA target, the host VI converts fixed-point data type to double using**To Double Precision Float(Functions>>Programming>>Numeric>>Conversion>>To Double Precision Float)**, see Figure L13.1.

L13.2.2 QAM Receiver Host Side

Next, on the host side, the QAM demodulated signal is decimated by 4 using a **Case Structure** which selects every fourth sample for further processing. This decimated signal is sent to the **Sync & Tracking** VI for the frame synchronization and phase/frequency tracking. The **Sync & Tracking** VI is an intermediate level subVI incorporating several sub-VIs/functions and operating in two different modes: frame synchronization and phase/frequency tracking. Figure L13.17 illustrates the BD of the Sync & Tracking VI.

The input samples are passed into the receiver queue, implemented via the **Complex Queue PtByPt** VI **(Functions>>Signal Processing>>Point by Point>>Other Functions PtByPt>>Complex Queue PtByPt)**, in order to obtain the beginning of frames by cross-correlating the frame marker bits and received samples in the queue. Filling the queue is continued until it is completely filled. Extra iterations are done to avoid including any transient samples due to delays associated with the filtering operations in the transmitter.

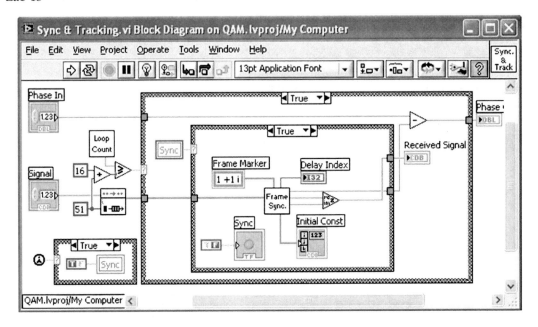

Figure L13.17: Sync & Tracking VI – frame synchronization mode

The length of the queue is configured to be 51 in order to include the entire marker bits in the queue. This length is decided based on this calculation: 31 (one period of PN sequence) + 2*10 (frame marker bits). Also, 16 extra samples are taken to flush out any possible transient output of the filter as mentioned previously. Bear in mind that the length of the queue or the number of extra reads varies based upon the specification of the transmitted signal such as the length of the frame marker bits and the number of taps of the phase shape filter. A counter, denoted by the **Loop Count** VI in Figure L13.18, is used to count the number of samples filling the queue. Once the queue is completely filled and extra reads are done, the frame synchronization module is initiated.

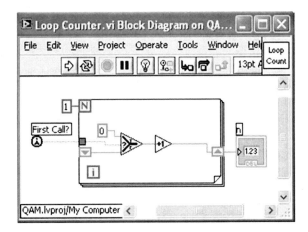

Figure L13.18: Loop Counter.vi

The subVI for frame synchronization is shown in Figure L13.19. In this subVI, the cross-correlation of the frame marker bits and the samples in the receiver queue are computed. The absolute value of the complex output is used to obtain the cross-correlation peak since the location of this peak coincides with the beginning of the frame. The **Array Max & Min** function is used to detect the index corresponding to the maximum cross-correlation value.

Figure L13.20 shows the **Complex CrossCorrelation** VI. This VI performs the complex cross-correlation operation by computing Equation (13.10). Once the index of the maximum cross-correlation value is obtained, all the data samples are taken at this location of the queue. Consequently, the data bits get synchronized.

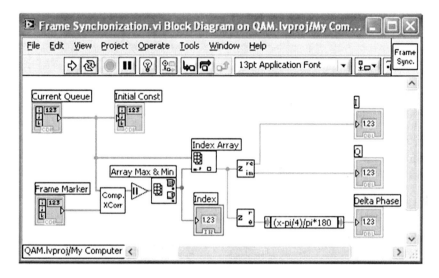

Figure L13.19: Frame Synchronization.vi

The initial phase estimation is achieved using the phase of the complex data at the beginning of the marker bits. Considering that the ideal reference is known for the first bit of the frame marker, $1 + i$ in our case, this allows us to obtain the phase difference between the ideal reference and the received frame marker bits. The real and imaginary parts of the data at the beginning of the marker bits are also passed to the **Phase** and **Frequency Tracking** VI to provide the initial constellation.

The subVI of the frame synchronization is now complete. Notice that three local variables are created in order to pass the indicator values to the other parts of the VI which cannot be wired. In the **Sync & Tracking** VI, a **Rounded LED** indicator labeled as **Sync** is placed on the FP. A local variable is created by right-clicking either on the terminal icon in the BD or on the **Rounded LED** indicator in the FP and choosing **Create>>Local Variable.** Next, a local variable icon is placed on the BD. More details on using local and global variables can be found in [1].

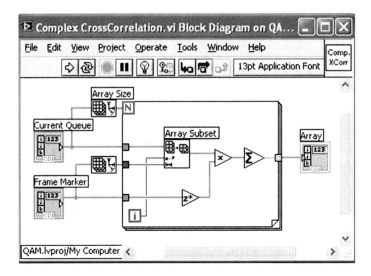

Figure L13.20: Complex CrossCorrelation.vi

The local variable **Sync** is used to control the flow of data for frame synchronization. The initial value of the local variable is set to **True** to execute the frame synchronization. Then, it is changed to **False** within the **Case Structure** so that it is not invoked again. The other two local variables, **Initial Const** and **Delay Index,** are used as the inputs of the phase and frequency tracking module, see Figure L13.21.

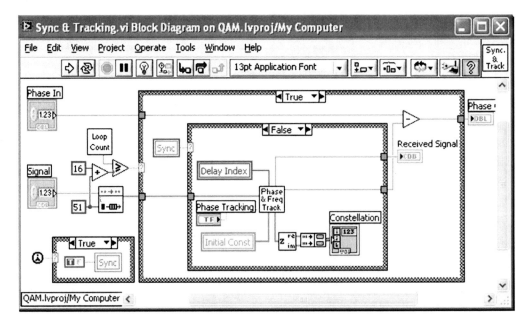

Figure L13.21: Sync & Tracking.vi – phase and frequency tracking mode

Now, let us describe the **Phase and Frequency Tracking** VI illustrated in Figure L13.22. A **Formula Node (Functions>>Programming>>Structures>>Formula Node)** is shown in the upper part of the BD, which acts as a slicer to determine the nearest ideal reference based on the quadrant on the I-Q plane. A **Formula Node** structure is capable of evaluating a script written in text-based C code. There are numerous built-in mathematical functions and variable which can be used in a **Formula Node**. For example, **pi** represents π in the formula node script shown in Figure L13.22. Further details on the formula node can be found in [1].

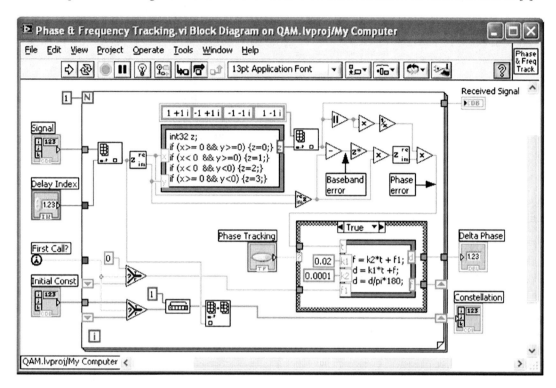

Figure L13.22: Phase & Frequency Tracking.vi

The phase error, see the BD in Figure L13.22, is computed from Equation (13.21). This error is multiplied by a small scale factor to determine the phase update $\Delta\varphi(n)$ in a second **Formula Node** implementing Equation (13.22).

The last two VIs of the modem are **Demod Constellation** and **Original Constellation**. The logic for these two VIs is the same as the BD of Demod Constellation shown in Figure L13.23. These VIs are used for storing the latest 50 original and demodulated signal samples in a **Complex Queue PtByPt** function **(Functions>>Signal Processing>>Point by Point>>Other Functions PtByPt>>Complex Queue PtByPt)** for drawing constellation diagrams. The output of these VIs can be directly connected to a **XY Graph (Controls>>Modern>>Graph>>XY Graph).**

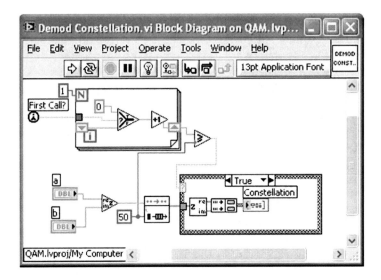

Figure L13.23: Demod Constellation.vi

Now, all the components of the modem system are in place. As the final step, **Waveform Charts** and **XY Graphs** are added to the Host Top Level VI as shown in Figure L13.1. The FP of the system can be seen in Figure L13.24 with zero noise. The received signal becomes nearly perfect reproduction of the transmitted signal except for the time delay. If there exist a phase and a frequency offset with no tracking and zero noise, the received signal appears as displayed in Figure L13.25 where the constellation of the received signal is rotated. By tuning on the **phase tracking** control, the constellation of the samples in the I-Q plane becomes that of the ideal reference as illustrated in Figure L13.26.

Up to now, noise variance is considered zero, i.e. zero channel noise. One can increase this noise by increasing the variance. The constellation of the received signal in the presence of this noise can be seen in Figure L13.27. As seen in this figure, signals appear scattered around the ideal reference points. The amount of scatter depends on the noise variance, that is to say increasing the noise variance increases the amount of scatter.

In summary, a 4-QAM transmitter is built in LabVIEW on the host side and its corresponding receiver is implemented on the FPGA side. By using the phase and frequency tracking module, the phase and/or frequency offset between the transmitter and receiver can be studied and compensated.

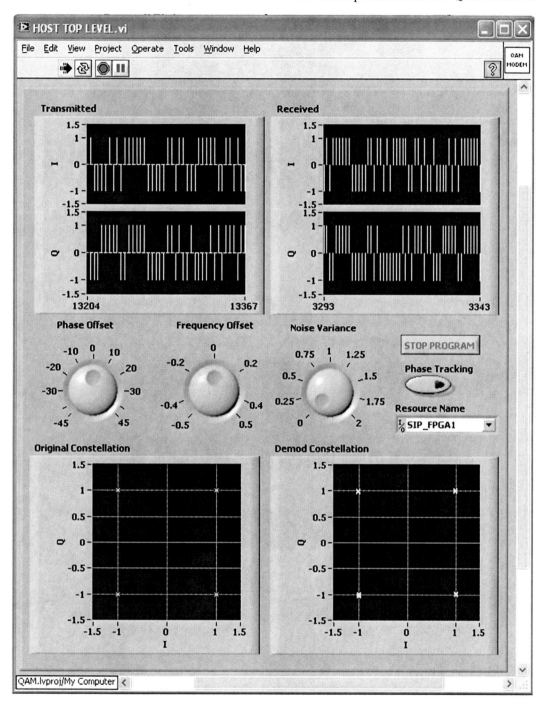

Figure L13.24: Initial phase estimation with no noise

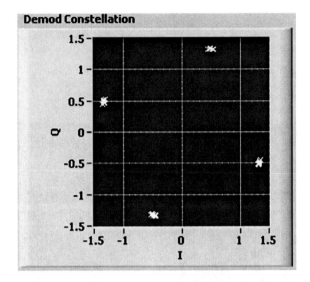

Figure L13.25: Received signal constellation with no phase & frequency tracking

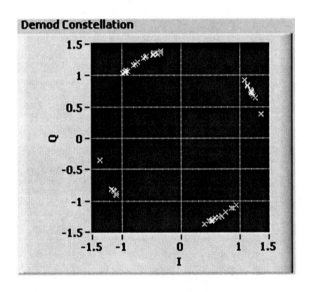

Figure L13.26: Phase and frequency tracking in IQ plane

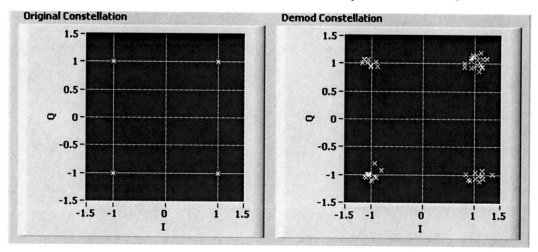

Figure L13.27: Received signal scatter constellation with noise

L13.3 Bibliography

[1] National Instruments, *LabVIEW User Manual,* Part Number 320999E-01, 2003.

CHAPTER 14

Application Project 3: MP3 Player

The International Organization of Standardization (ISO) standard MPEG-I layer-III, known as MP3, is one of the most widely used standards for digital compression and storage of audio data. The MP3 standard was developed by the Fraunhofer Institute to provide compression of audio files without any perceptible loss in audio quality [1]. This standard gives a compression ratio of 12:1 and yet preserves the CD quality audio. Besides MP3 players, there exist many software tools that are capable of playing MP3 files.

This chapter presents a LabVIEW implementation of single channel (mono) MP3 player using the LabVIEW FPGA module after providing an overview of its theory. The overview presented below is not meant to be a detailed description of MP3 decoding, rather to provide enough information for one to understand the building components or blocks associated with MP3 decoding. The interested reader is referred to [1-4] for theoretical details of MP3 decoding.

The application lab in this chapter covers the FPGA implementation of MP3 decoding using the LabVIEW FPGA. The functional blocks associated with an MP3 player are depicted in Figure 14.1. In what follows, the function of each block is briefly mentioned.

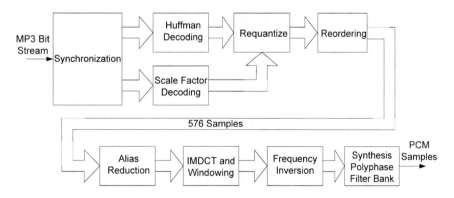

Figure 14.1: Functional blocks associated with MP3 player [2]

14.1 Synchronization Block

The first block is the Synchronization block. This block serves the purpose of receiving the incoming bitstream, extracting certain information from it and passing the extracted information to the succeeding blocks. This information consists of the Header information, the

Cyclic Redundancy Check (CRC) Information and the Side Information. The Header Information specifies the type of the MP3 file, the bitrate used for transmission, the sampling frequency and the nature of the audio. The CRC information provides details about the integrity of data while the Side Information provides the necessary parameters for decoding data as well as the reconstruction of scale factors.

An MP3 file is divided into smaller units called frames, as shown in Figure 14.2. Each frame is divided into five sections: Header, CRC, Side Information, Main Data, and Ancillary Data. Main Data is the coded audio, while Ancillary Data is optional and contains user-defined attributes such as song title, artist name or song genre. Main Data is further divided into two granules: Granule 0 and Granule 1.

Let us first mention some details on Header Information. Size of the Header Information is 4bytes. During decoding, the Header Information is identified by the occurrence of 12 consecutive '1s' [3]. Header Information is divided into various fields which are shown in Figure 14.3 along with the number of bits each field occupies. Version and Layer fields indicate information about MPEG version and layers. If Error Protection Bit is 'ON', then 16-bit CRC follows the Header. Bitrate used for transmission sampling frequency and the nature of the mode is decoded based on the remaining fields.

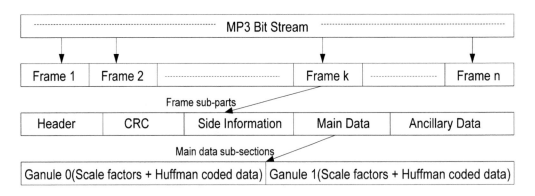

Figure 14.2: Anatomy of an MP3 file

4 Bytes	MP3 Sync Word (12)	Version (1)	Layer (2)	Error Prot.(1)	Bit Rate(4)	Freq. (2)	Pad. Bit(1)	Priv. Bit(1)	Mode (2)	Mode Ext.(2)	Other (4)

Figure 14.3: Header Information bit mapping (number of bits for each field)

Next, the Side Information is divided into four parts: Main Data Begin Pointer, Private Bits, SCFSI bits and Granule0/Granule1 information. The length of the Side Information varies depending upon the nature of the audio signal; 17 bytes of mono and 32 bytes of stereo, as indicated in Figure 14.4.

Main Data Begin Pointer indicates the beginning of Main Data in the bitstream. Noting that the MP3 encoding uses the 'bit reservoir' technique [1], Main Data for a current frame does not necessarily begin after the Side Information for that frame but reside in another frame, as illustrated in Figure 14.5. This is done to obtain more compression, as data from current frame may not completely fill that frame. Main Data Begin Pointer provides a negative offset in bytes form the header of a current frame to the location where Main Data for the current begins. This pointer is 9 bits long, so the bit reservoir can be at most $2^9 - 1$ bytes long.

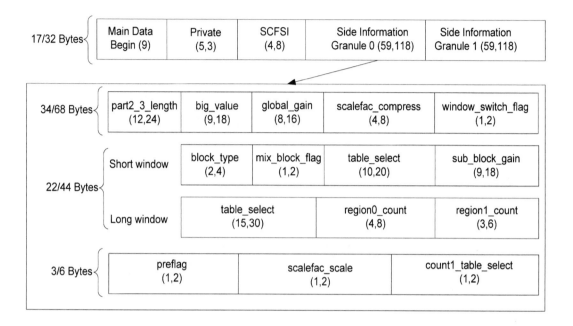

Figure 14.4: Side Information bit mapping (number of bits for mono/stereo) and its fields

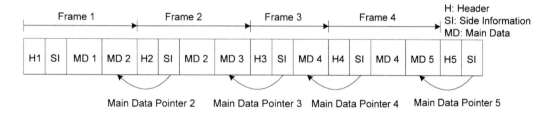

Figure 14.5: Bit reservoir technique [1]

Private bits are used for private use and not normally used for decoding. Scale Factor Selection Information (SCFSI) determines whether the same scale factor bands are transmitted for both granules considering that the scale factors for Granule 0 can be sometimes reused by Granule 1 in the Layer III standard. At encoding time, a total of 576 frequency lines of a granule are sorted into four groups of scale factor bands. Each group of scale factor bands corresponds to a bit of the 4-bit long SCFSI as shown in Table 14.1. If the bit corresponding to a group is set, this means the scale factors for that group are common to both granules and are transmitted only once. If it is zero, then they are transmitted separately for the granules.

The fields for the Granule 0/Granule 1 side information is illustrated in Figure 14.4. These fields are grouped into three categories based on their functions: Huffman decode (part2_3_length, big_value, table_select, region0_count, region1_count, count1table_select), window selection (window_switch_flag, block_type, mix_block_flag) and requantization (global_gain, scalefac_compress, subblock_gain, scalefac_scale, preflag). These fields are discussed further in the following sections.

SCFSI bits	Scale factor band
0	0,1,2,3,4,5
1	6,7,8,9,10
2	11,12,13,14,15

Table 14.1: Scale factor bands corresponding to SCFSI bits [1]

14.2 Scale Factor Decoding Block

The Scale Factor Decoding block, see Figure 14.1, decodes the scale factors to allow the reconstruction of the original audio signal. Scale factors are used to mask out the quantization noise during encoding by boosting the sound frequencies that are more perceptible to human hearing.

Scale factors decoded after the coded scale factors portion is separated from Main Data. The number of bits used for coded scale factors is specified by the part2_length field which is obtained from slen1 and slen2 values. These values are determined from the table of scale factors by using the scalefac_compress field, displayed in Figure 14.4, as an index to the table. Note that the method to calculate part2_length changes depending on the type of window used during encoding. The windowing is done during encoding to reduce the effect of aliasing and the type of window (short or long) is determined by the block type. In the following sections, the scale factor output is referred to as scalefac_l and scalefac_s for long windows and short windows, respectively.

14.3 Huffman Decoder

The Huffman decoding is the most critical block in the decoding process. This is due to the fact that the bitstream consists of contiguous variable length codewords which cannot be identified individually. Once the start of the first codeword is identified, the decoding proceeds sequentially by identifying the start of the next codeword at the end of the previous codeword. Consequently, any error in decoding propagates; in other words, the remaining codewords cannot be decoded correctly. To better understand how this block functions, some necessary information regarding the format of Huffman coded bits is mentioned next.

14.3.1 Format of Huffman Code Bits

In the MP3 standard, the frequency lines are partitioned into three regions called rzero, count1 and big_value. As Huffman coding is dependent on the relative occurrence of values, coding in each region is done with Huffman tables that correspond to the characteristics of that region.

A continuous run of all zero values is counted and grouped as one of the regions called rzero. This region is not coded because its size can be calculated from the size of the other two regions. The second region, count1, comprises a continuous run of -1, 0 or 1. The two Huffman tables for this region encode four values at a time, so the number of values in this region is a multiple of 4. Finally, the third region big_value covers all the remaining values. The 32 Huffman tables for this region encode the values in pairs. This region is further sub-divided into three sub-regions: region0, region1 and region2. The region boundaries are determined during encoding. Figure 14.6 depicts the output of the Huffman Decoder splitted into regions.

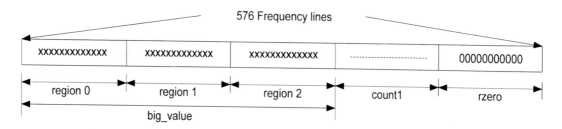

Figure 14.6 Regions of Huffman decoder output [4]

In the big_value region, a parameter called Escape is used in order to improve the coding efficiency. In this region, values exceeding 15 are represented by 15, and the difference is represented by the Escape value. Notice that the number of bits required to represent the Escape value is called Linbits, which is associated with the Huffman table used for encoding. A sign bit follows Linbits for nonzero values.

14.3.2 Huffman Decoding

The Huffman Decoding block consists of two components: Huffman Information Decoding and Huffman Decoding. Huffman Information Decoding uses the Side Information to set up the fields for Huffman Decoding. It acts as a controller and controls the decoding process by providing information on Huffman table selection, codeword region and how many frequency lines are decoded. This decoding is illustrated in Figure 14.7.

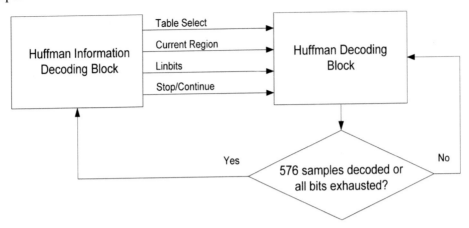

Figure 14.7: Huffman decoding block as a controller

Huffman Information Decoding starts by determining the part3_length value, which indicates the number of Huffman coded bits present in the current granule. This value is obtained by subtracting part2_length of the Scale Factor Decoder from part2_3_length of the Side Information.

The next step involves determining the codeword region considering that the selection of the Huffman tables is region specific. The decoding always starts with region0 of the big_value region. The start of region1 and region2 is determined using the region0_count and region1_count field of the Side Information. The start of the count1 region is not explicitly defined and begins after all the codewords in the big_value region have been decoded. The count1 region ends when the number of bits exceeds part3_length.

The third step consists of obtaining the table numbers for each region. The table_select field of the Side Information gives the table numbers for all the three regions of the big_value region. For the count1 region, the number 32 is added to the count1_table_select field of the Side Information to get the table number for this region. This block terminates the decoding process if 576 lines are decoded or the part3_length bits are used. If the decoding stops before 576 lines are decoded, zeros are padded at the end so that 576 lines are generated.

Huffman Decoding requires 34 Huffman tables for decoding Main Data. Since two of the 34 tables (table number 4 and 14) are not used, only 32 tables are required for decoding. Two tables out of these 32 tables are used for the count1 region while the rest are used for the big_value region. The process of Huffman Decoding can be understood better with the help of Figure 14.8, which shows the pattern of the Huffman coded bitstream.

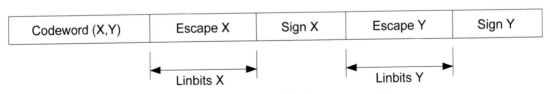

Figure 14.8: Bit format of Huffman coded bits

The decoder fetches one bit at a time and compares it with all the codewords in the table. The fetched bits represent the codeword(x,y) in the figure. If there is a match, then the corresponding value is repeated. If the returned value for the big_value region equals 15, the next bits are fetched and the number represented by them is added to the decoded value. As discussed earlier, the Linbits field is determined by the Huffman table in which a codeword match exists.

As the last step, the sign of the decoded codeword is determined. The same procedure is carried out for the codewords in the count1 region with the difference that Linbits are not used and one codeword from this region gives four decoded values.

14.4 Requantizer

The MP3 encoder incorporates a quantizer block that quantizes the frequency lines so that they can be Huffman coded. The output of the quantizer is multiplied by the scale factors to suppress the quantization noise. The function of the requantizer block is to combine the outputs of the Huffman Decoder and Scale Factor Decoder blocks, generating the original frequency lines. Figure 14.9 illustrates the function of the Requantizer block.

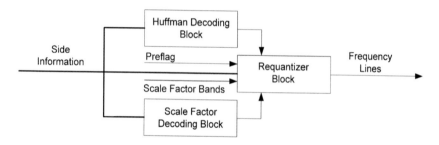

Figure 14.9: Block diagram of Requantizer block

In the Requantizer block every output sample from Huffman Decoder block needs to be raised to the power 4/3. This power is the inverse of the one used in the encoder quantization process, i.e. 0.75. The result is then multiplied by the sign of the Huffman decoded value and logarithmically quantized.

Requantization can be described by two equations, one for long and one for short windows, which are stated below [1].

For long window,

$$xr_i = sign(is_i) \cdot abs(is_i)^{\frac{4}{3}} \cdot \frac{2^{\frac{1}{4}(global\_gain[gr]-210)}}{2^{((scalefac\_mul \cdot scale\_l[gr][sfb])+(preflag[gr] \cdot pretab[sfb]))}} \tag{14.1}$$

For short window,

$$xr_i = sign(is_i) \cdot abs(is_i)^{\frac{4}{3}} \cdot \frac{2^{\frac{1}{4}(global\_gain[gr]-210-8 \cdot subblock\_gain[gr][sfb][window])}}{2^{(scalefac\_mul \cdot scale\_s[gr][sfb][window])}} \tag{14.2}$$

where xr_i's, is_i's and gr denote requantized arrays, Huffman decoded arrays and granule, respectively. The remaining terms are explained below. The preflag and global_gain parameters are obtained from the Side Information. The value 210 is a system constant and is defined in the ISO/IEC 11172-3 document [1]. The scalefac_mul parameter depends on the scalefac_scale field of the Side Information. If scalefac_scale is 0, then scalefac_mul is 0.5. If scalefac_scale is 1, then scalefac_mul is 1. Thus, in the above equations, the Huffman values are scaled by 2 or $\sqrt{2}$. The parameter pretab is used for further amplification of higher frequencies and is specified as follows:

$$pretab\,[21] = \{0,0,0,0,0,0,0,0,0,0,0,1,1,1,1,2,2,3,3,3,2\} \qquad (14.3)$$

The 21 elements in this array correspond to the 21 scale factor bands. The parameters scalefac_1 and scalefac_2 represent the decoded scale factors for long and short windows, respectively. The parameter **sfb** indicates the current scale factor band. The selection of the scale factor bands depends on the sampling frequency. One thing that must be ensured before requantization is that the scale factor bands should cover all the 576 frequency lines.

14.5 Reordering

Huffman coding gives better results if the inputs are ordered in an increasing order or have similar values. This is the reason the frequency lines are ordered in increasing order of frequency during encoding as values closer in frequency have similar values. Normally, the output of the Modified Discrete Cosine Transform (MDCT) in the encoder is ordered into subbands with increasing frequency values. However, for short blocks, the output samples are ordered into subbands first by increasing windows and then by frequency. In order to remove this dependency of the window type on output samples, the output of the MDCT is ordered first by subband then by frequency and lastly by window. Figure 14.10 illustrates the effect of the reordering block on the frequency lines.

Note that reordering is done only for subbands with short windows. Hence, the main task of the reordering block is to search for short windows to reorder the frequency lines. The output of the requantizer, for short windows, gives 18 samples in a subband. These samples are not dependent on the window used. The reordering block simply picks the samples and reorders them in groups of six for each window, thereby generating them as they were before reordering.

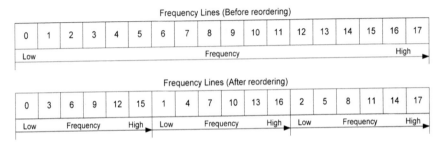

Figure 14.10: Reordering of frequency lines for short blocks [4]

14.6 Alias Reduction

During the encoding process, the pulse code modulated (PCM) samples are filtered into subbands using bandpass filters. However, due to the non-ideal nature of the bandpass filters, aliasing effects occur. To minimize aliasing artifacts, windowing is done after the MDCT block. That is why during the decoding process, an alias reduction block is used to generate the frequency lines similar to those generated by the MDCT in the encoder. This block adds the alias components which are specified in the ISO/IEC 11172-3 document [1]. Alias reduction is applied to all windows other than short windows.Basically, the Alias Reduction block consists of eight butterfly calculations, described in [2], per subband, similar to that in the FFT calculation. The alias components correspond to the scale factors with which the frequency lines are scaled. The formula to calculate the two scale factors ca_i and cs_i is stated below

$$cs_i = \frac{1}{\sqrt{1+c_i^2}}, \ ca_i = cs_i \cdot c_i, \quad 0 \le i \le 7 \tag{14.4}$$

where ca_i and cs_i denote the aliased components and c_i's are defined in [1].

14.7 IMDCT and Windowing

The Inverse MDCT (IMDCT) block is responsible for generating samples which serve as the input to the Polyphase filter. The IMDCT takes in 18 input values and generates 36 output values per subband in each granule. The reason for generating twice as many output values is that the IMDCT contains a 50% overlap. This means that only 18 out of 36 values are unique, and the remaining 18 values are generated by a data copying operation. For a fast implementation of the IMDCT block, its symmetry property is used, which is described by the following equation:

$$x[i] = \begin{cases} -x\left[\dfrac{n}{2}-i-1\right], & i=0,\ldots,\dfrac{n}{4}-1 \\[2ex] -x\left[\dfrac{3n}{2}-i-1\right], & i=0,\ldots,\dfrac{3n}{4}-1 \end{cases} \tag{14.5}$$

After performing IMDCT, windowing is done so as to generate time samples that are similar to those obtained after the filterbank in the encoder. This is because windowing is carried out on the output of the filterbank which provides the input to the MDCT block during encoding. The type of window used depends on the block_type field in the Side Information: long or short. The window functions that are used for different blocks are given by [1]

for block type = 0,

$$z[i] = x[i] \cdot \sin\left(\frac{\pi}{36}\left(i+\frac{1}{2}\right)\right), \quad 0 \le i \le 35 \tag{14.6}$$

for block type = 1,

$$
z[i] = \begin{cases} x[i] \cdot \sin\left(\frac{\pi}{36}\left(i+\frac{1}{2}\right)\right) & 0 \le i \le 17 \\ x[i], & 18 \le i \le 23 \\ x[i] \cdot \sin\left(\frac{\pi}{12}\left(i-18+\frac{1}{2}\right)\right), & 24 \le i \le 29 \\ 0, & 30 \le i \le 35 \end{cases} \tag{14.7}
$$

for block type = 2,

$$
z[win][i] = x[win][i] \cdot \sin\left(\frac{\pi}{12}\left(i-18+\frac{1}{2}\right)\right), \quad 0 \le i \le 11, \, 0 \le win \le 2 \tag{14.8}
$$

for block type = 3,

$$
z[i] = \begin{cases} 0, & 0 \le i \le 5 \\ x[i] \cdot \sin\left(\frac{\pi}{12}\left(i-6+\frac{1}{2}\right)\right), & 6 \le i \le 11 \\ x[i], & 12 \le i \le 17 \\ x[i] \cdot \sin\left(\frac{\pi}{36}\left(i+\frac{1}{2}\right)\right), & 18 \le i \le 35 \end{cases} \tag{14.9}
$$

After the output of the IMDCT is multiplied with a window function, the 36 output values per subband are overlapped and added to produce 18 output values for every subband of a granule. The upper 18 values of the previous subband are stored and added to the lower 18 values of the current subband. Figure 14.11 illustrates the flow diagram for this overlap and add operation.

14.8 Polyphase Filter Bank

Finally, a Polyphase filter bank is used to transform the 32 subbands, each with 18 time samples from every granule, to 18 bands of 32 PCM samples. PCM is a standard format of storing digital data in uncompressed format, CD audio being the prime example. PCM samples are defined depending on the sampling frequency and bitrate. A higher sampling frequency implies that higher frequencies are present and a higher bitrate produces a better resolution. Generally, CD audio uses 16 bits at 44.1 kHz. Here, after a brief overview of the PCM format, the generation of PCM samples is briefly explained with the help of the flow diagram shown in Figure 14.12.

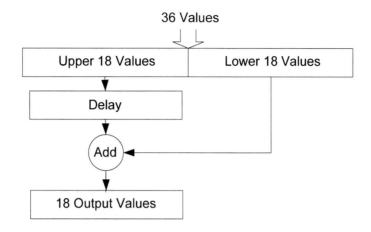

Figure 14.11: Overlap and Add operation

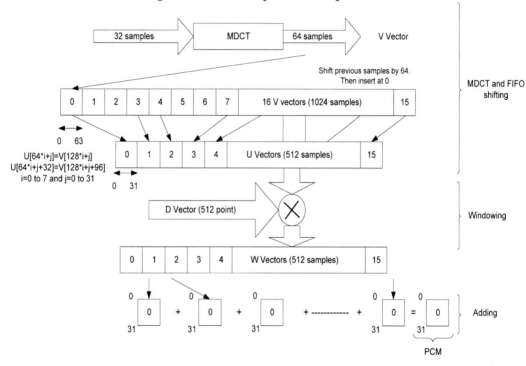

Figure 14.12: Steps in generation of PCM samples [4]

14.8.1 MDCT

The MDCT block transforms the 32 input samples, one from each subband, to a 64-point V vector given by the following equation:

$$V[i] = \sum_{k=0}^{31} \cos\left[\pi(16+i)(2k+1)/2n\right]S[k], \qquad 0 \le i \le 63 \qquad (14.10)$$

where $S[k]$ denotes 32 input samples and $n = 64$.

For a fast implementation of this block, as mentioned earlier, the symmetry property of MDCT is used which requires only the computation of 50% of the values. An alternative way for a fast implementation of MDCT is via Lee's method [5], which uses a FFT type approach.

14.8.2 FIFO Shifting

The 64 output values (V vector) from the MDCT block are then fed to a 1024 sample first-in-first-out (FIFO) shift register. The MDCT operation is repeated 18 times per granule. However, due to the size of the shift register, only the last 16 V vectors are saved. Each time a V vector is generated, the shift register is shifted by 64 places to accommodate for a new V vector. The shift register is reset only at the beginning of the decoding process and not during the decoding of each frame. The last part of the IMDCT block involves the generation of a 512-sample U vector. This vector is generated by selecting alternate subbands from the shift register.

14.8.3 Windowing and Adding

As the last part towards generating PCM samples, the windowing shown in Figure 14.12 is done. This consists of multiplying the U vector by a 512-point window function. The resulting 512 point vector is transformed into 16 vectors consisting of 32 values each. These 16 vectors are then added sample-wise to generate one subband of PCM samples. The sum is then represented in 16-bit format. This operation is repeated 18 times to generate the 18 subbands of the 32 PCM samples.

14.9 Bibliography

[1] ISO/IEC 11172-3, *Information Technology-Coding of Moving Pictures and Associated Audio for Digital Storage media at 1.5 Mbits/s-Part 3: Audio*, First Edition, August 1993.

[2] K. Lagerström, *Design and Implementation of an MPEG-1 layer III Audio Decoder*, Masters Thesis, Chalmers Institute of Technology, Sweden, 2001.

[3] http://www.id3.org/mp3frame.html

[4] S. Gadd and T. Lenart, *A Hardware Accelerated MP3 Decoder with Bluetooth Streaming*

[5] B. Lee, "A New Algorithm to Compute Discrete Cosine Transform," *IEEE Trans. on Accoust., Speech, and Signal Processing*, vol. 32, pp. 1243-1245, Dec 1984.

Lab 14:

FPGA Implementation of MP3 Player

In this application lab, the MP3 decoder system briefly discussed in this chapter is implemented using the LabVIEW FPGA.

L14.1 Top-Level VI

Figure L14.1 illustrates the top-level BD of the implemented MP3 decoder. The subVIs located in the While Loop are repeated as many as number of frames.

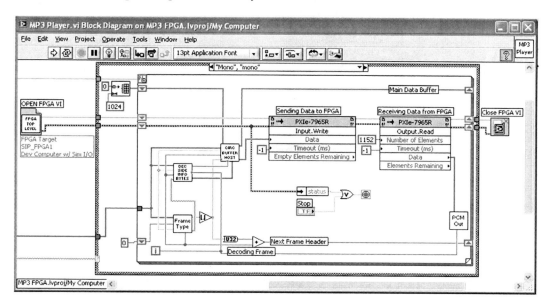

Figure L14.1: System-level Block Diagram of MP3 decoder

The VI **Frame Type** finds the frame header from the bit stream and extracts the decoding bitrate and sampling frequency. The location of the frame header is passed to the VI **Dec Side Info Bytes** where data related to the Side Information and sampling frequency is bundled.

A **Shift Register** is placed to serve as a buffer for Main Data as part of the bit reservoir technique. The VI **Circ Buffer Host** is used to fill the buffer with Main Data, to calculate a pointer marking the start of Main Data of a current frame, and to bundle Main Data and Side Info Bytes as Frame Data Bytes for sending to an FPGA target.

DMA FIFO's are used for transferring data between the host and FPGA. Decoding of frame data to PCM samples is carried out on the FPGA target. PCM samples are saved to a file using the VI **PCM Out**. The decoded PCM file can be played by any audio application software.

L14.2 Host Side Implementation

L14.2.1 MP3 Read

As the first step of the MP3 decoding process, an MP3 file is opened and read by the VI **MP3 Read**, see Figure L14.2. This VI reads an MP3 file as specified by the **File Path** control via the VI **Read from Text File (Functions>>Programming>>File I/O>>Read from Text File)**. The data stream is converted to unsigned integer byte using the **String to Byte Array** function (**Functions>>Programming>>String>> String/Array/Path Conversion>>String to Byte Array**).

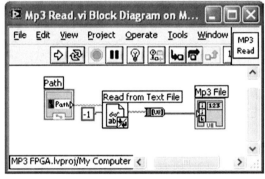

Figure L14.2: BD of MP3 Read VI

L14.2.2 MP3 File Info

The VI **MP3 File Info** extracts and displays the information about an MP3 file that is present in the file header, see Figure L14.3. It consists of two subVIs: **Frame Type** and **MP3 Info Display.** The former finds the header and extracts the decoding information, while the latter displays this information in an ordered fashion.

The VI **Frame Type**, see Figure L14.4, uses the subVI **Finder Header** to find the new location of the header by searching for twelve consecutive ones. Once the header is found, three bytes of the header which contain the frame information are passed to the VI **Find Type**. Then, this VI performs the actual decoding of the header information using table look-ups.

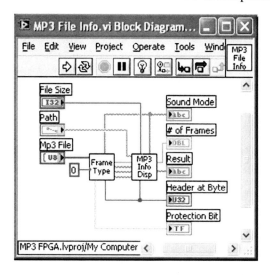

Figure L14.3: BD of MP3 File Info.vi

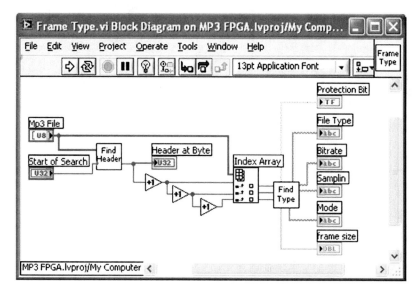

Figure L14.4: BD of Frame Type.vi

The VI **MP3 Info Display** generates a string array based on the output of the VI **Frame Type**. The output string array of this VI is displayed on the FP.

L14.2.3 Decode Side Info Bytes

The VI **Dec Side Info Bytes** extracts the 17 bytes of the Side Information from the bit stream. Figure L14.5 illustrates the extraction of the 17 bytes of the Side Information using the function **Array Subset (Functions>>Programming>> Array>>Array Subset).** In case an optional CRC is present, indicated by the Protection bit, the starting location of the Side Information is moved by two bytes after the header.

437

At the end of 17 bytes, two more bytes corresponding to First Frame Indicator and Sampling Frequency are inserted using the function **Insert into Array (Functions>> Programming>>Array>>Insert into Array)**, where the First Frame Indicator byte will be one if the current processing frame is first; otherwise zero. The sampling frequency is encoded into the Sampling Frequency byte. Table 14.2 shows the encoded byte for possible sampling frequencies of a mono MP3 file. The Main Data begin pointer is extracted from the first byte of the Side Info using the subVI **Get Bits Host**.

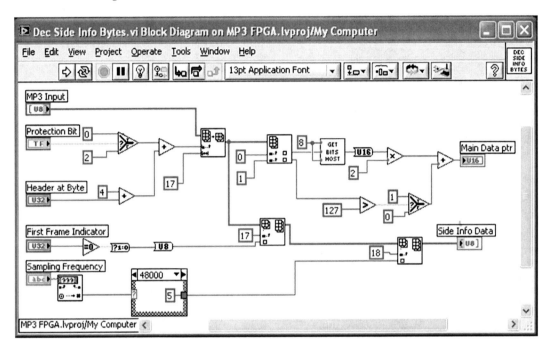

Figure L14.5: A BD portion of Dec Side Info Bytes.vi

Sampling frequency	Sampling Frequency Byte
0	0
22050	1
24000	2
16000	3
44100	4
48000	5
32000	6

Table 14.2: Encoded Sampling Frequency Byte

The VI **Get Bits Host** which is shown in Figure L14.6 performs 'logical AND' to extract bits from the input byte. This VI provides a mask that contains one at the positions to be extracted from the input byte. The length and location of these ones in the mask are specified by the input parameters to the VI.

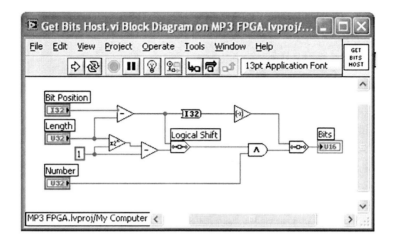

Figure L14.6: BD of Get Bits Host VI

L14.2.4 Circular Buffer Host

Main Data of a frame is obtained from previous frames as well as a current frame depending on the value of the Main Data Begin pointer. The VI **Circ Buffer Host**, see Figure L14.7, extracts the Main Data section of the current frame and fills a 1024 point buffer. Thus, a total of 1024 Main Data samples from the previous and current frame are stored in the buffer. This buffer is rotated in such a way that a new Main Data section is inserted to the end of the buffer. In order to fetch Main Data for decoding of a current frame, the index of the start of Main Data is provided by subtracting the sum consisting of the size of the new Main Data section and **Main Data Ptr** from the length of the buffer. 19 bytes of the Side Info bytes which are extracted using the VI **Dec Side Info Bytes** are inserted at the end of the 1024-byte Main Data to make 1043 bytes of Frame Data which is then send to the FPGA using a DMA FIFO as shown in Figure L14.1.

After sending Frame Data Bytes, decoding of these bytes to 1152 integer PCM samples is carried out on the FPGA target. Decoded PCM samples are sent back to the host side using the DMA FIFO.

L14.2.5 PCM Out

After receiving PCM samples from the FPGA target, the VI **PCM Out** writes these samples to a file. This VI uses (i) the **Build Path** function (**Functions>> Programming>>File I/O>>Build Path**) to build the path of the saving file, (ii) the **Open/Create/Replace File** function (**Functions>>Programming>>File I/O>>Open/Create/Replace File**) to replace a file if the file already exists; otherwise it will create a new file, (iii) the **Set File Position** function (**Functions>> Programming>>File I/O>>Advanced File Functions>> Set File Position**) to set the position to start for first frame; otherwise to the end of the file, (iv) the **Write to Binary File** function (**Functions>>Programming>>File I/O>> Write to Binary File**) to write PCM samples to the file, and (v) the **Close File** function (**Functions>> Program-**

439

ming>>File I/O>> Close File) to close the opened file. The controls are connected to the appropriate functions as illustrated in Figure L14.8.

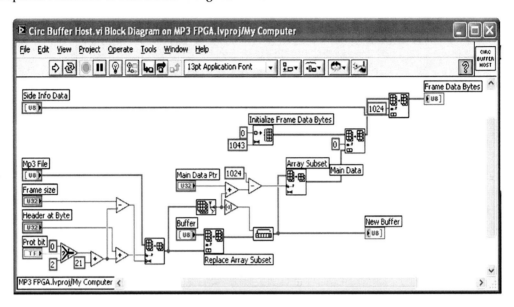

Figure L14.7: BD of Circular Buffer.vi

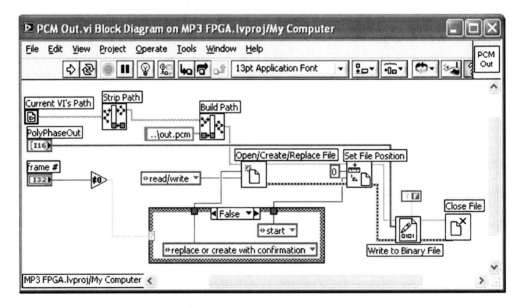

Figure L14.8: BD of PCM Out.vi

L14.3 Target Side Implementation

The VI **FPGA_Top** is the top level VI on the FPGA target side, see Figure L14.9. This VI contains a **While Loop** which runs continuously on the FPGA target. DMA FIFO's are used for transferring data between the host and the target. The **Input** FIFO waits till data are made available from the host, once data are ready; they get transferred to the **Input** memory unit. Using this memory, the subVI **FPGA Process** decodes the frame data to PCM samples which are saved to the **Out_Samples** memory unit. From this memory, data get transferred to the **Output** FIFO and then to the host. The function **Flat Sequence Structure (Functions >>Programming>> Flat Sequence Structure)** is used to ensure that the VI **FPGA Process** is executed only after receiving data from the host. This process is repeated for each frame.

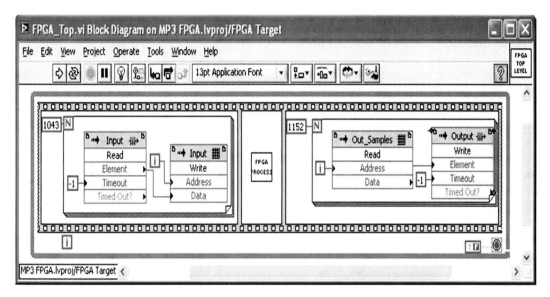

Figure L14.9: BD of FPGA Top Level VI corresponding to MP3 decoder

L14.3.1 FPGA Process

Figure L14.10 illustrates the BD of the VI **FPGA Process**. As each frame consists of two granules, all the subVIs located in the **For Loop** are repeated twice. The sampling frequency is decoded from the last byte of the **Input** memory unit. As discussed earlier, the sampling frequency is encoded using Table 14.2 on the host side, the same table is used in the form of a **Case Structure** for decoding purposes.

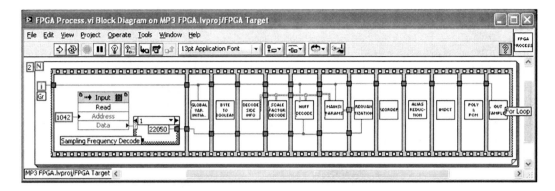

Figure L14.10: BD of FPGA Process.vi

Flat Sequence Structure is used to ensure that each subVI is executed only after the execution of the preceding subVI. The VI **Global Dec** initializes all global variables needed for decoding. The VI **Byte To Boolean Array** converts the frame data bytes into bits. The VI **Decode Side Info** decodes the Side Information and bundles the decoded parameters into a cluster for easy access by other VIs.

The **Scale Factor Decode** VI decodes the scale factors that are used to suppress quantization and any noise. The **Huffman Decode** VI incorporates a number of subVIs to determine the required information about Huffman decoding, to calculate the length of Huffman coded bits and to decode different regions of Huffman coded data. The **Requantization** VI combines the output of the **Scale Factor Decode** and **Huffman Decode** VIs. This VI implements the requantization equations.

The **Reorder** VI arranges the frequency lines of short blocks in the same order as in the IMDCT block of the encoder. The **Alias Reduction** VI computes the anti-alias coefficients and weighs the frequency lines accordingly. The **IMDCT** VI computes IMDCT, does windowing on the IMDCT output and performs overlap/add on the windowed output to generate the polyphase filter input.

The **Poly & PCM** VI carries out three operations. It multiplies every odd sample of each odd subband by '-1' (denoted by Frequency Inversion), implements the polyphase filter and generates PCM samples. The **Out samples** VI combines decoded PCM samples of the two granules.

L14.3.2 Global Variable Initialization

Three global variables are used for MP3 decoding. These variables are initialized only for the first granule of the first frame. The global variable **bufferOffset** is initialized to 64. **Vvector_G** is a 1024-element memory unit and **PrevBlck_G** is a 576-element memory unit, which are initialized to zero. Figure L14.11 shows the BD of this **Global Dec** VI.

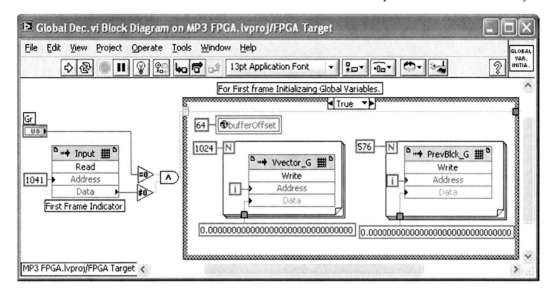

Figure L14.11: BD of Global Dec.vi

L14.3.3 Byte to Boolean Array

The first 1024 elements of the frame data bytes correspond to the Main Data bytes. These 1024 bytes are converted to 8192 bits using the function **Number to Boolean Array (Functions>> Programming>>Boolean>>Number to Boolean Array)** and stored in the **Input_bits** memory unit. The BD of this VI is shown in Figure L14.12.

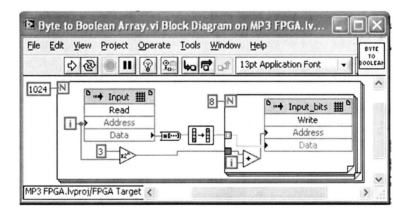

Figure L14.12: BD of Byte to Boolean Array VI

L14.3.4 Decode Side Info

As shown in Figure L14.13, 17 bytes of the Side Information are read from the **Input** memory unit to form an array. Each byte of the Side Information is wired to **Timed Loop (Functions>> Programming>>Timed Structures>>Timed Loop)** in order to extract the individual parameters.

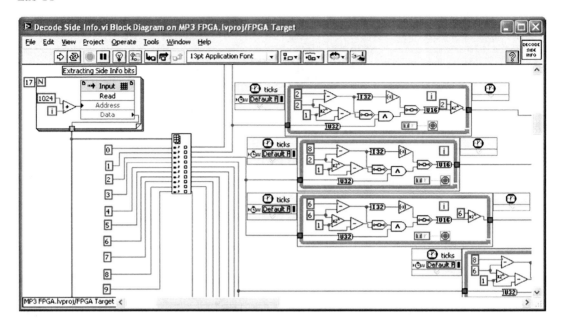

Figure L14.13: A BD portion of Decode Side Info.vi

The logic inside the **Timed Loop** is similar to the **Gets Bits Host** VI which was explained earlier except that a constant is connected to the **Scale by Power of 2** function (**Functions>>Programming>>Numeric>>Scale by Power of 2**) instead of a variable for reducing the FPGA resource usage [1]. All the parameters of the Side Information extracted from the bit stream are bundled to form a cluster, indicated by **Side Info.**

L14.3.5 Scale Factor Decode

The **Scale Factor Decode** VI consists of two VIs: **Scale Factor Decode0** and **Scale Factor Decode1**, which are used for decoding Granule0 and Granule1, respectively. Considering that the basic components of these VIs are the same, here only a general description of the **Scale Factor Decode0** VI is mentioned. A BD section of the VI is illustrated in Figure L14.14. First, the block type of the current granule is obtained (long, short or mix) via the Side Information. Based on this block, scale factors are decoded. Figure L14.14 shows the Scale Factor Decode in case of long block. **slen_1_2 Con** is a memory unit with constant values, where these values are provided in the **slen_1_2 Initia** VI as part of the project folder. Memory is initialized with these constant values using the **Initialization** VI as discussed in Lab 4. Decoded Scale Factors for Granule0 are stored in the **Sfac_gr0** memory unit, similarly scale factors for Granule1 are stored in the **Sfac_gr1** memory unit. Some VIs require scale factor values of Granule1 while decoding Granule0, so each time the scale factors for both granules are decoded.

L14.3.6 Huffman Decode

The **Huffman Decode** VI decodes coded data by performing a table look-up using thirty-four standard Huffman tables. This VI consists of the following subVIs: **Part2 length0, Part2** length1, Huffman Info, Increment and Replace Bit, Big Value Search, Big value Sign, Count1 search and Count1 Sign. These subVIs are explained below.

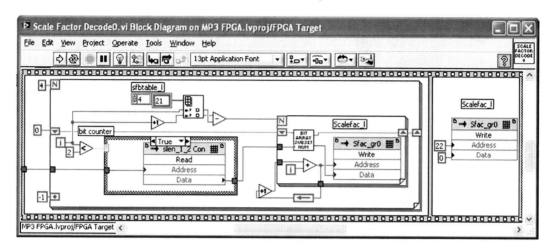

Figure L14.14: A BD portion of Scale Factor Decode0.vi

The VIs **Part2 length0** and **Part2 length1** are used to calculate the part2 length of Granule0 and Granule1, respectively, using the unbundled Side Information.

The VI **Huffman Info**, see Figure L14.15, extracts the individual table_select for each of the three sub-regions of the big_value region from the table_select part of the Side Information.

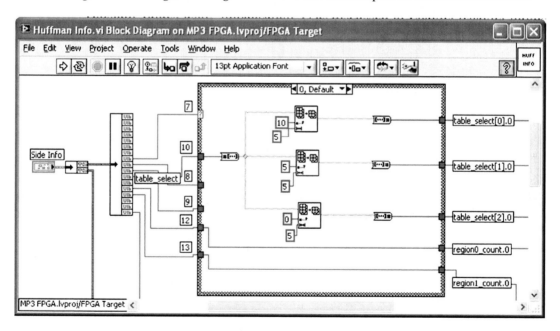

Figure L14.15: A BD portion of Huffman Info.vi

Note that the two parameters region0_count and region1_count are used to determine the boundaries of the big_vlaue region. Once table_select and the boundaries are determined in the **Huffman Info** VI, the decoding of the Big Value and Count1 regions begins.

The big_value stage of the **Huffman Decode** VI consists of the **search loop** section which is located in the **While Loop** shown in Figure L14.16. This section searches for a codeword match in the big_value region. Before this search takes place in the **Search Loop,** the current region to be decoded must be determined, which is achieved by using the **Case Structure** as shown in Figure L14.16. This is important as the table_select value is region dependent. This loop extracts one bit at a time from the **Input_bits** memory unit, appends it to the previously extracted bit using the **Increment and Replace Bit** subVI, and passes it to the **Big Value Search** VI. The **Search Loop** terminates when a match is found.

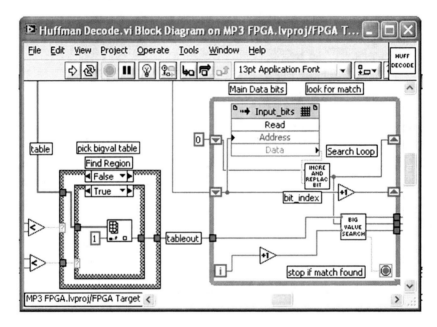

Figure L14.16: Search loop part of big_value stage.vi

The VI **Increment and Replace Bit** left shifts the previous number and replaces the LSB with the input bit from the Input_bits stream and finds the new number. As shown in Figure L14.17, this VI uses the functions **Scale by Power by 2, Number to Boolean Array, Replace Array Subset and Boolean Array to Number** inside a **Timed Loop**.

The VI **Big Value Search** performs the search for the number generated by the **Increment and Replace Bit** VI. Figure L14.18 illustrates the BD of the **Big Value Search** VI. Three **Case Structures** are used for decoding the number. The outer **Case Structure** is used for selecting the table number, the middle **Case Structure** is used for Number of bits which is equivalent to the iteration number of **Search Loop**, and the inner **Case Structure** is used to extract the decoded values, X and Y; if there is no Case corresponding to the input number then the value will be zero, i.e. no match found. The Boolean indicator, denoted by **Match**, is used to terminate the **Search Loop** of the **Huffman Decode** VI.

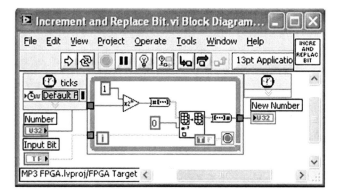

Figure L14.17: BD of Increment and Replace Bit.vi

Next, the **Big Value Sign** VI, see Figure L14.19, is used to determine the Escape and the Sign values for X and Y. As mentioned earlier, the Escape value is only evaluated when the decoded value is equal to fifteen. This value is determined by fetching Linbits, obtained from **Big Value Search**. Then, the value represented by the fetched bits is added to the decoded outputs as shown in **Escape Value** case structure in the BD. The decoded outputs are updated with Escape and Sign values to form the final output. Decoded Big values are stored in the 'is' memory unit.

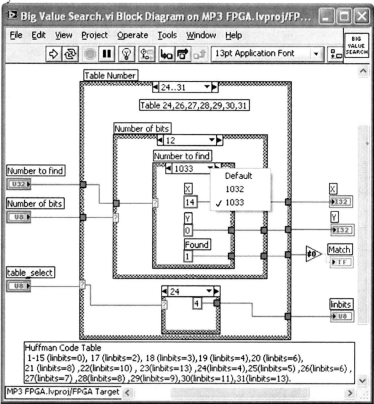

Figure L14.18: BD of Big Value Search.vi

447

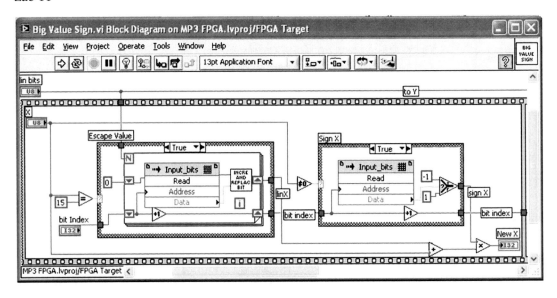

Figure L14.19: Escape and Sign bit Decoding.vi

After the big_value stage, the decoding of the count1 stage follows. This stage is similar to the big_value decoding, except that this time four decoded outputs V, W, X and Y are obtained. The process of determining the sign bit is similar to the big_value sign process. Figure L14.20 illustrates the count1 stage of the Huffman decoding. Decoded values are stored in the 'is' memory unit.

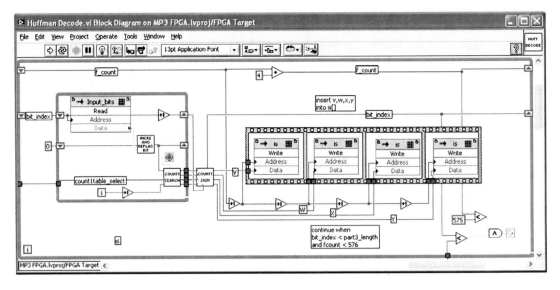

Figure L14.20: Count1 stage of Huffman Decoding.vi

L14.3.7 Main Side Information Parameters

This VI is used to extract the frequently used fields of the Side Information including **glob-al_gain, window_switch_flag, block_type, mix_block_flag, scalefac_scale** and **preflag.** The parameters for both the granules are extracted once. As explained earlier, the scale factors are decoded for both the granules at once. Hence, by using **Case Structure,** the MainSI parameters and the scale factors of the present granule are selected as shown in Figure L14.21.

L14.3.8 Requantization

The **Requantization** VI combines the outputs of the **Scale Factor Decode** and **Huffman Decode** VIs to implement the Requantization equations given by Equations (14.1) and (14.2). Figure L14.22 shows one section of the **Requantization** VI, which uses the **IP Block** node **(Functions>>Addons>>IP Integration Node).** The VHDL code used in the **IP Block** is provided in the project folder, where **requantize_long.vhd** is the top level VHDL file using **requan_types.vhd** and **mul.vhd.** As explained in Lab 4, **IP Block** is compiled using these VHDL files. Similarly, in the other section for short blocks, **requantize_short.vhd, requan_types.vhd** and **mul.vhd** are used for compiling the IP block node.

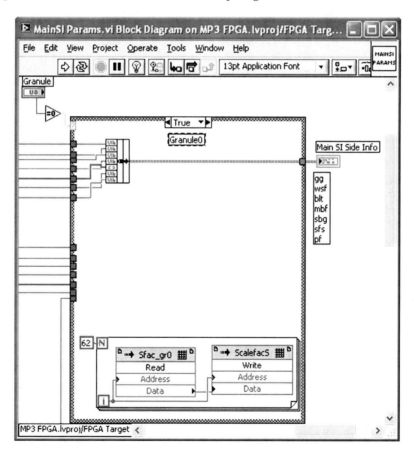

Figure L14.21: A BD portion of MainSI Params.vi

In this VI, the memory constants **Sfac_l Con**, **Sfac_s Con** and **Pretab_Con** are used whose values are given in the VIs **Scalefac_l Const Initia**, **Scalefac_s Const Initia** and **pretab_con Initia,** respectively. Requantized samples are stored in the **XRO** and **Xr** memory units. Only three MainSI parameters are needed from here on, so those are stored at the end of requantized samples in the **XRO** memory unit. Note that requantized values from the **IP Block** node are converted to 32-bit fixed-point data type with 30 bits of fractional bits using the **Boolean Array to Number** function as shown in Figure L14.22, considering that floating-point data type is not supported by FPGA.

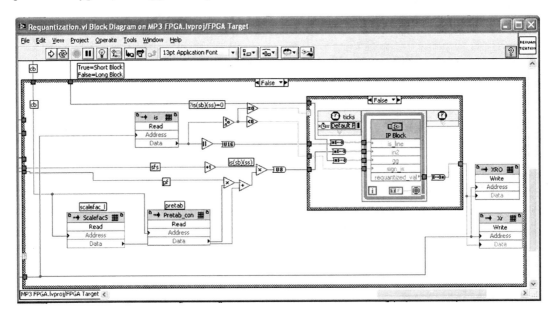

Figure L14.22: A BD portion of Requantization.vi

L14.3.9 Reordering

The **Reorder** VI changes the order of frequency lines or the output of the Requantization block. Since reordering is required for only short and mix blocks, the **Reorder** VI first identifies the block type before performing reordering. The BD of the **Reorder** VI for short blocks is illustrated in Figure L14.23. This VI simply replaces the elements in the **XRO** memory unit with the elements taken from a different index of the **Xr** memory unit. Note that previously both of these memory units contained the same elements.

For mix blocks, the reordering is not carried out for the first two subbands as they consist of long blocks. In this VI, the constant memory unit **sfBand_Const** whose constant values are given in the **sfBand_Const Initia** VI in the project folder.

L14.3.10 Alias Reduction

The **Alias Reduction** VI scales the reordered frequency lines with the alias coefficients. Since alias reduction is performed only for long blocks, this VI identifies such blocks and performs alias reduction on them. The BD shown in Figure L14.24 carries out the butterfly calculation.

Note that the alias coefficients are defined in the **ca and cs const** memory unit whose constant values are given in the **ca and cs_const Initia** VI.

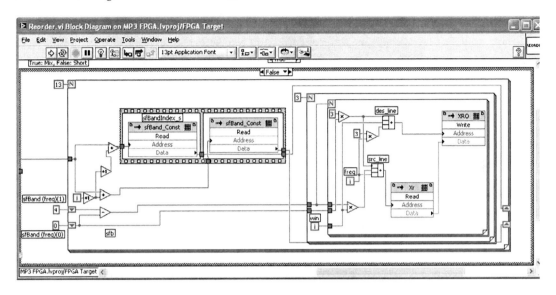

Figure L14.23: A BD portion of Reorder VI

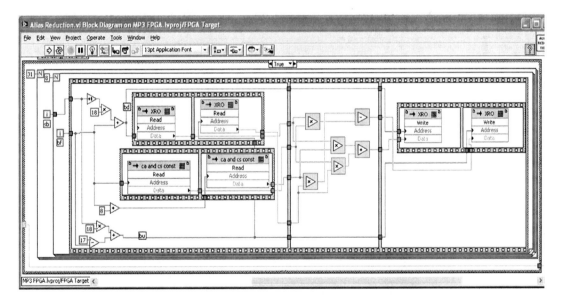

Figure L14.24: A BD portion of Alias Reduction.vi

L14.3.11 IMDCT

The **IMDCT** VI converts the frequency domain samples to time domain samples, thereby providing the input samples to the Polyphase filter. This VI also performs the windowing and overlap/add operations on the output samples as shown in Figure L14.25. The global memory unit **PrevBlck_G** is used to pass the output values used for the overlap/add opera-

451

tion from one frame to another. As mentioned earlier, the values of this memory unit are initialized only once at the beginning of decoding.

The **IMDCT Calc** subVI performs the actual IMDCT computations. Since the formula used for the calculation of IMDCT is different for long and short blocks, the implementation of each case is done separately.

The BD section of calculating IMDCT for long blocks is illustrated in Figure L14.26. Input values are convolved with the IMDCT coefficients, which are stored in the **Cos_l_const** memory unit whose values are provided in the **Cos_l Initia** VI. Windowing is performed on the convolution output. This is done by multiplying the IMDCT output with predefined window coefficients stored in the **Win_Const** memory unit whose values are provided in the **win_Const_initia** VI.

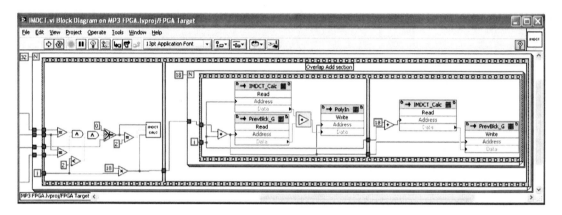

Figure L14.25: A BD portion of IMDCT.vi

Similar to long blocks, for short blocks inputs are convolved with the IMDCT coefficients which are stored in the **Cos_s_const** memory unit whose values are provided in the **Cos_s Initia** VI. Here, the convolution is done in the **IMDCT Calc Mul** subVI as shown in Figure 14.27.

After performing windowing, 36 output values of IMDCT are stored in the **IMDCT_calc** memory unit. These values are overlapped and added for the final IMDCT array, which are stored in the **PolyIn** memory unit as shown in Figure 14.25.

L14.3.12 Poly & PCM

The **Poly & PCM** VI generates the final output of the MP3 decoding process, i.e. **PCM** samples. This VI transforms the thirty-two subbands of eighteen samples each to eighteen subbands of thirty-two PCM samples. As shown in Figure L14.28, **Poly & PCM** consists of two subVIs (**Freqinv** and **MDCT & WVec_ADD**) and a global variable **bufferOffset** for transferring data from a current frame to a next frame. The **FreqInv** VI performs frequency inversion on the input samples by negating every odd sample of every odd subband. This inversion is done to compensate for the negation of values during the MDCT stage.

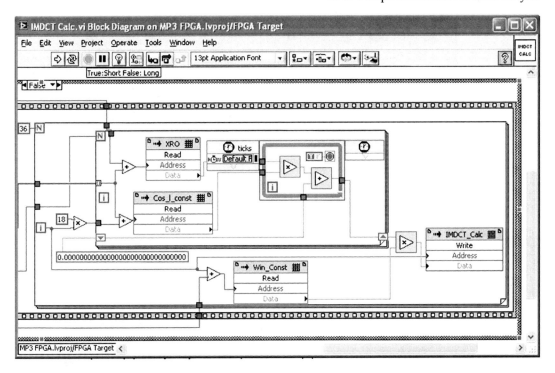

Figure L14.26: A BD portion of IMDCT Calc.vi (long block case)

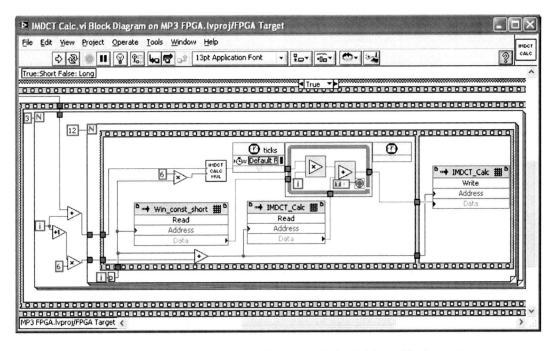

Figure L14.27: A BD portion of IMDCT Calc.vi (short block case)

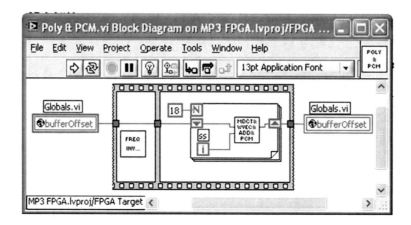

Figure L14.28: BD of Poly & PCM.vi

The **MDCT & Wvec_ADD** VI, see Figure L14.29, computes the MDCT array and adds the generated windowed vector. This VI uses the two subVIs **Matrix_mul** and **Window_mul**.

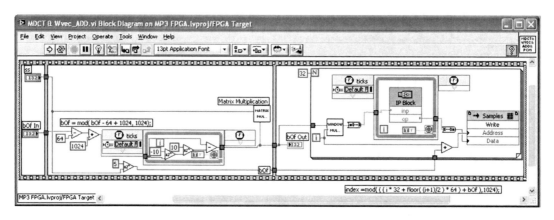

Figure L14.29: BD of MDCT & Wvec_ADD.vi

The **Matrix_mul** VI implements Equation (14.10). Here, the symmetry property of MDCT is exploited by using **Case Structure** as shown in Figure L14.30. Filter coefficients are stored in the **Filter_Const** memory unit, where their values are specified in the **Filter_Const Initia** VI. The global memory unit **Vvector_G** plays a circular buffer role where the elements are shifted circularly and a new element is inserted at the index specified by **bOf**.

After updating the **Vvector_G** memory unit, the values are windowed by multiplying with the window function specified by the **Window_mul** VI. In the **Window_mul** VI, 16 samples from the **Vvector_G** and **Window** memory unit are multiplied and added together to get a windowed sample. Each windowed sample is sent to the **IP Block** node, which is compiled using the file **Sample.vhd**, whose output is converted to I16 to form a PCM sample. This is repeated for 32 times to get 32 PCM samples as shown in Figure L14.29. The **MDCT & Wvec_ADD** VI is repeated for 18 times to form the final 18 X 32 output samples which are stored in the **Samples** memory unit.

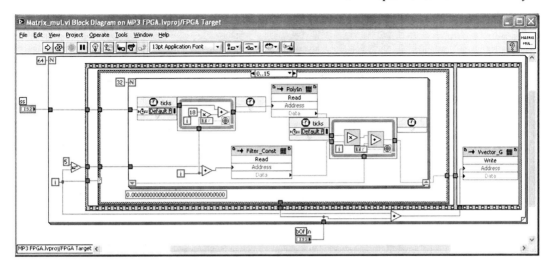

Figure L14.30: BD of Matrix_mul.vi

L14.3.13 Output Samples

Decoded 576 samples of the two granules are then stored in the **Out_Samples** memory unit using the **Out samples** VI. A **Case Structure** is used in order to store Granule1 samples at the end of Granule0 samples, as shown in Figure L14.31.

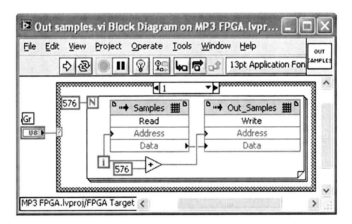

Figure L14.31: BD of Out samples.vi

At this point, all the necessary VIs on the FPGA side are complete, which need to be placed under the FPGA target in the project explorer as shown in Figure L14.32. Also, all the memory elements used on the FPGA target for decoding are added under the FPGA target in the project explorer. Table L14.3 shows all the memory elements used including memory constants. Note that all the VIs are constructed in an optimized way to minimize resource usage and time consumption.

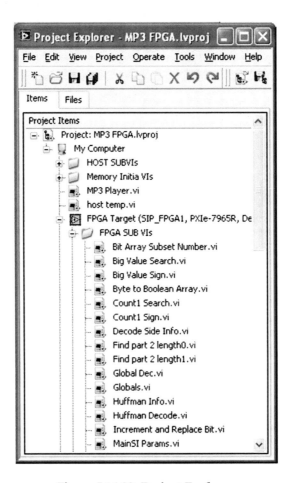

Figure L14.32: Project Explorer

14.4 MP3 Player

The **MP3 Player** VI is the top-level VI of the MP3 decoder system. It integrates both the host and FPGA side implementations. The **MP3 Read** VI is placed in its BD. The file path of the MP3 file is obtained from the path terminal of the **File Dialog** function (**Functions>>Programming>>File I/O>>Advanced File Functions>>File Dialog**). The output of the **MP3 Read** VI is wired to an **Array Size** function and the **MP3 Frame Info** VI. The output of the **Array Size** function is wired to the **MP3 Frame Info** VI. Indicators for all the outputs **MP3 Frame Info** VI are created to display the MP3 file information on the FP. A 1024-point array is initialized with zero.

Finally, a **For Loop** is created and the **Frame #** output of the **MP3 Frame Info** is wired to the loop count N. The **Open FPGA VI Reference** function (**Functions>> FPGA Interface>>Open FPGA VI Reference**) is placed on the BD and is made to point to the **FPGA_Top** VI by configuring as done in earlier labs. All the VIs and the **Invoke Method** of the DMA FIFO's are connected to the **Open FPGA VI Reference** function as shown in Figure L14.1. At the end, the **Close FPGA VI** function (**Functions>>FPGA Interface>>Close FPGA VI**) is placed and wires are connected to it from the **Invoke Method**.

Memory	Size	Data Type	Remarks
ca and cs const	16	FXP (32,2)	Constant
Cos_l_const	648	FXP (32,2)	Constant
Cos_s Const	72	FXP (32,2)	Constant
Filter_Const	2048	FXP (32,2)	Constant
IMDCT_Calc	36	FXP (32,2)	
Input	1043	U8	
Input_bits	8192	Boolean	
is	576	I32	
Out_Samples	1152	I16	
PolyIn	576	FXP (32,2)	
pretab_con	21	U8	Constant
PrevBlck_G	576	FXP (32,2)	Global
Samples	576	I16	
ScalefacS	62	I32	
Sfac_gr0	62	I32	
Sfac_gr1	62	I32	
sfac_l Con	69	U16	Constant
sfac_s Con	42	U8	Constant
sfBand_Const	42	U8	Constant
slen_1_2 Con	32	U8	Constant
Vvector_G	1024	FXP (32,2)	Global
Win_Const	144	FXP (32,2)	Constant
Win_const_short	12	FXP (32,2)	Constant
Window	512	FXP (32,2)	Constant
Wvector	576	FXP (32,2)	
Xr	576	FXP (32,2)	
XRO	581	FXP (32,2)	

Table L14.3: Memory elements details

Running the VI brings up a file dialog to choose an MP3 file to play. The decoded PCM file can be played by any audio application software. Figure L14.33 displays the FP of the MP3 player VI.

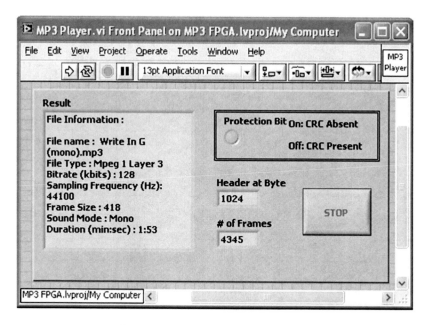

Figure L14.33: FP of MP3 Player.vi

L14.5 Modifications Made for Successful Compilation

The FPGA VI can run on an actual FPGA board or in emulation mode to test the behavior. The FPGA top level VI should be compiled before running it on an actual board. When compiling the **FPGA_Top** VI, that is the FPGA top level VI for the MP3 decoder, for the NI PXIe 7965R board, one may get an out of memory error. The reason for this type of error is due to over usage of the resources on the FPGA board or the utilization of only 2GB memory for 32-bit machines on which the program is running [1]. To compile it successfully, some blocks implemented on the FPGA target can be moved to the host side. The relative processing or consumption times of the FPGA VIs running in the emulation mode are provided in Table L14.4. As the VIs **Huffman Decode** and **Requantization** constitute less than 5% of total processing time, these VIs can be moved to the host side. By moving these VIs, still about 95% of the processing is done on the FPGA target.

To do so, requantized samples along with the required Side Information are sent to the FPGA board for decoding. The **Huffman Decode** and **Requantization** VIs need to be modified before placing them on the host side. These modifications are explained below.

L14.5.1 Huffman Decode

In the FPGA execution version of this VI, memory elements are used to store all array elements and **Case Structures** are used in the search algorithm. Here, all memory elements are replaced with arrays and the search algorithm is modified. All the other connections are kept the same. Mainly, changes are made in the **Search Loop**, the **Big Value Search** VI and the **Count1 Search** VI.

In the **Search Loop,** the search is done on multiple bits of Huffman coded data. The length of the bits to be extracted is determined by the predefined codeword lengths in the selected table. A **Case Structure** is created where its contents are arrays corresponding to the codeword lengths of the tables. The array for the table is selected using the **tableout** parameter. Figure L14.34 illustrates the implementation of the **Search Loop.**

The **Big Value Search** VI creates the 2D Huffman tables consisting of integers. The Huffman tables are divided such that codewords with the same length occur together. A 2D array is constructed with two columns and rows given by the number of codewords of the selected length. The elements are specified by the decoded values X and Y. Another 1D array is constructed with the integer values of the codewords. For example, the codeword '111001' is represented in the array as 57. The two arrays are then combined into a cluster and a 2D array of all such clusters is formed. The clusters in the 2D array are indexed by **table_select** and **length.** As a result, the search is only done on the table specified by **table_select** and **length.** Also, a 1D array whose elements are **Linbits** is created for each table.

SubVI	Relative Time Consumption (%)
Huffman Decode VI	1.20
Requantization VI	1.08
Reorder VI	0.01
Alias Reduction VI	1.09
IMDCT VI	2.20
Poly & PCM VI	92.27

Table L14.4: Processing time percentages of FPGA VIs

After indexing the correct table, codeword searching is carried out using the **Search 1D Array (Functions>>Programming>>Array>>Search 1D Array)** VI. This VI returns '-1' for no match. Figure L14.35 shows the BD of the **Big Value Search** VI.

Next, the three controls of **table_select, length** and **Bits to Search** are created. The three functions of **Index Array, Unbundle** and **Search 1-D Array** are then placed. The **Big Value Search** VI is implemented using these functions as illustrated in Figure L14.35. The **Big Value Search** VI is linked to the **Big Value Sign** VI. The same changes are made in the **Search Loop** for the **Count1 Search** VI and the above steps are repeated to construct a cluster of LUTs. The difference is that the 2D array in the cluster has four columns. The four columns correspond to the decoded values of V, W, X and Y. The bits are searched as before and elements from the 2D array are indexed out at the row index given by the output of the **Search 1-D Array** function and columns 0, 1, 2, and 3. All the indicators are placed and the **Huffman Decode** VI is completed by linking this VI to the **Count1 Sign** VI.

Decoded values are combined to form an array which is passed to the **Requantization** VI. The scale factors are passed to the **Huffman Decode** VI in the form of arrays instead of storing them in memory units.

L14.5.2 Requantization

In the **Requantization** VI, all the memory elements are also replaced with arrays, and the calculation of 4/3 power is done using the **Power of X** function (**Functions>>Mathematics>> Elementary & Special Functions>>Exponential Functions>>Power of X**). Then, requantized values are stored as double data type in arrays. The BD section of the **Requantization** VI is shown in Figure L14.36. 576 samples of requantized values for each granule are sent to the FPGA target for further decoding. All the remaining VIs on the FPGA target remains the same.

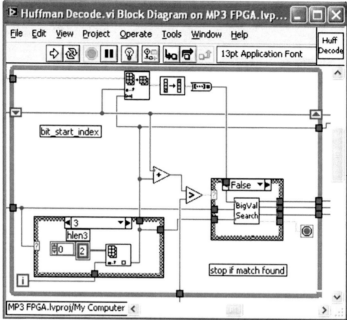

Figure L14.34: Search loop of Huffman Decode.vi

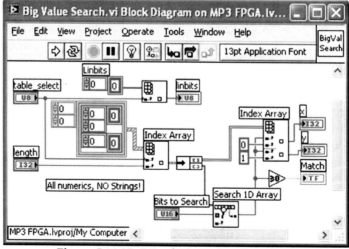

Figure L14.35: BD of Big Value Search.vi

Now, after moving the **Huffman Decode** and **Requantization** VI to the host, the **FPGA_Top** VI is compiled, the compilation is concluded successfully, that is a bit file is successfully generated. **Open FPGA VI Reference** can be configured to point to the generated bit file. The bit file can be run on the actual FPGA board by changing the debugging option of the FPGA target as discussed in Chapter 11. Figure L14.37 shows the BD of the modified **MP3 player** VI.

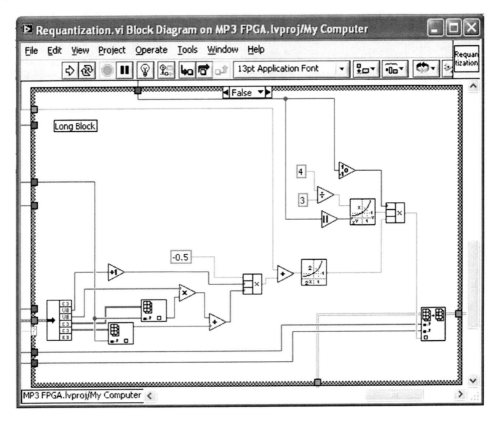

Figure L14.36: A BD portion of Requantization.vi

The discussed changes are necessary to make the program run successfully on the FPGA target in real-time, i.e. an MP3 file getting decoded within the playing time of the file. As the MP3 file gets decoded in real-time, sound VIs can be incorporated to play MP3 files while getting decoded. A Waveform Chart of the decoded samples along with the other indicators are shown in Figure L14.38.

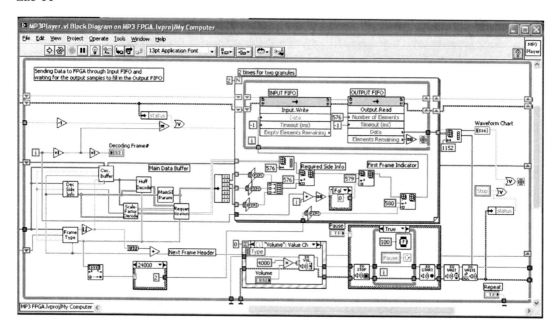

Figure L14.37: BD of modified MP3Player.vi

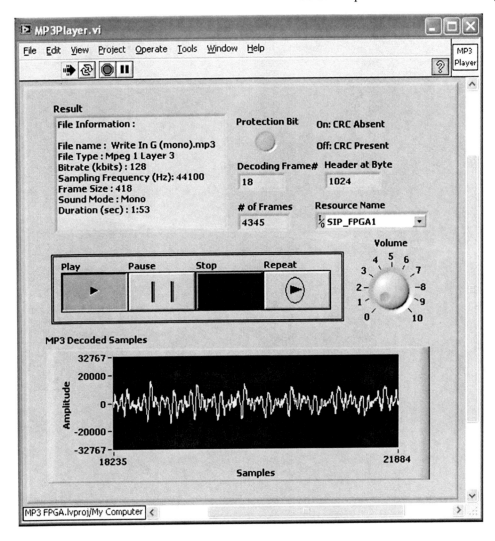

Figure L14.38: FP of modified MP3 Player

L14.6 Bibliography

[1] National Instruments, *LabVIEW User Manual*, Part Number 320999E-01, 2003.

Index

(Indices appearing in bold pertain to LabVIEW functions)

bit, 140

..sion, 142